Fundamentals of Soil Science

Dr. A. Rathinasamy
Professor (SS & AC),
Agricultural College & Research Institute,
Thanjavur, TNAU

Dr. B. Bakiyathu Saliha
Assistant Professor of Soil Science
Department of Soils and Environment
Agricultural College and Research Institute
Madurai

Published by:

SCIENTIFIC PUBLISHERS (INDIA)
5A, New Pali Road, P.O. Box 91
Jodhpur 342 001, India
E-mail: info@scientificpub.com
Website: www.scientificpubonline.com

Print: 2023

ISBN: 978-93-86652-17-1
eISBN: 978-93-86237-34-7

Printed in India

PREFACE

Soil Science is not a new science, but one that has gained much attention and interest in the past decade. Dealing with soil at any age? Believe it or not, soil (to some known as "dirt") is part of all of our lives on a daily basis. As environmental issues such as water quality, waste management, ecological biodiversity, land resource carrying capacity, and alternative land uses continue to gain more attention from the public.

The importance of soil and soil management is being increasingly recognized in our country. Which have a high pressure of population on the available land and find no alternative in meeting the demand for food and other agricultural raw materials except through increase in agricultural production per unit land.

Soil Science is a unique discipline concerning a complex material that is part of many natural and utilitarian systems. As such, the teaching of Soil Science requires principles that reflect the nature of soil and the practices of soil scientists. The goal of this book is to introduce students and teachers the basic principles of soil science as an integral part of the curriculum for soil science, environmental sciences, earth science, and water quality.

The content is structured around basic soil concepts, beginning with the significance of soil in our everyday lives and progressing through soil formation, the physical and chemical properties of soils, and the role, soil and the earth play in environmental management today and in the future. It is necessary to develop better understanding among the students on the relationship between soil and water quality, crop and vegetation management, and environmental science.

In view of the concern of humanity regarding pollution and environment, I felt it is necessary to add a chapter on soil pollution and soil quality. A special chapter on Organic farming is also provided.

One can learn about the soil in their own back yard, who to contact to get local "experts". How to get our students more interested in soil and environmental studies.

There is a great need for the book on Fundamentals of Soil Science which explain the basic principles and concepts and draw upon Indian data so as to make it useful to Indian students, teachers in agriculture. It is hoped that the book "fundamentals of Soil Science" will meet the need of all students of soil science and other disciplines related to agriculture, horticulture and forestry.

AUTHORS

CONTENTS

Sl. No	Contents	Page No
1	Introduction	1-5
2	Rocks and Minerals	6-16
3	Weathering and Soil Formation	17-28
4	Soil Forming Factors	29-35
5	Soil Forming Processes	36-46
6	Soil Classification	47-61
7	Soil Survey and Mapping	62-82
8	Soils of India	83-90
9	Phases of Soils	91-95
10	Physical Properties of Soil	96
11	Soil Texture	97-107
12	Soil Structure	108-116
13	Soil Consistency	117-124
14	Densities of Soil	125-129
15	Soil Colour	130-134
16	Soil Temperature	135-141
17	Soil Air	142-146
18	Soil Moisture	147-161
19	Soil Colloids	162-175
20	Ion Exchange	176-187
21	Problem Soils	188-197
22	Soil Physical Constraints	198-202
23	Quality of Irrigation Water	203-206-

24	Essential Nutrients	207-211
25	Soil Fertility and Its Evaluation	212-226
26	Fertilizers and Bio-fertilizers	227-247
27	Soil Organic Matter	248-264
28	Soil Organisms	265-269
29	Organic Farming	270-280
30	Soil Pollution	281-289
31	Soil Erosion and Conservation	290-302
32	Soil Quality	303-308
	Appendix	309-314
	Glossary	315-331
	Model Questions	332-351
	References	352-354

1 INTRODUCTION

The term soil is derived from the latin world 'solum' which means floor. **For a layman** - soil is the dirt or debris on the surface of the earth. To a **mining engineer** the soil is the debris covering the rocks (or) minerals that he must quarry. To a **highway engineer** the soil may be the material on which a roadbed is to be placed. To an **agriculturist** soil is a natural medium for plant growth. The **farmer** makes a living from the soil and he is forced to pay more alteration to its characteristics. To the former the soil is more than useful it is indispensable.

The Concept of Soil

The **concept of soil** has been changing with the accumulation of more data and knowledge. **Whitney (1892)** developed the concept of soil as a nutrient bin, ie, they hold all the nutrients required for the plant growth. **Geologists** considered soil to be the product of rock weathering. **Ruffin (1968)** considered soil as a mixture of earth's uppermost mantle of weathered rock and organic matter. According to **Hilgard (1992),** "Soil is more or less a loose and friable material in which plants by means of their roots finds nourishment as well as other condition of growth.

SOIL IS A NATURAL BODY

The concept of soil as a natural body originated from its ability to produce and sustain crop. **Justus Von Liebig** (1840) in his famous publication **"Chemistry applied to Agriculture and physiology"** put forward the theory that plants assimilate mineral nutrients from soils. He is called as **father of agricultural chemistry.**

The fundamental principles of soil science were first laid down by **V.V. Dokuchaev (father of soil science)** and his students in USSR in 1879. According to them, soils are the surface mineral and organic formations, always more or less coloured by humans. According to them soils are formed by weathering of rocks and minerals.

Soil is a Three Dimensional Body

Soil should be recognized as three dimensional body having length, breadth and depth. For a thorough study of soils, the study of surface soil will not be sufficient. We have to dig a **pit (soil profile)** and study the physical, chemical and biological characteristics.

Scope of Soil Science

Soil Science has six well defined and developed disciplines. Scope of soil Science is reflected through these disciplines.

Soil Science : The science dealing with soil as a natural resource on the surface of the earth, including Pedology (soil genesis, classification and mapping) and the physical, chemical and biological and fertility properties of soil and these properties in relation to their management for crop production .

1. **Soil fertility:** Nutrient supplying properties of soil
2. **Soil chemistry:** Chemical constituents, chemical properties and the chemical reactions
3. **Soil physics:** Involves the study of physical properties
4. **Soil microbiology:** Deals with micro organisms, its population, classification, its role in transformations
5. **Soil conservation:** Dealing with protection of soil against physical loss by erosion or against chemical deterioration i.e excessive loss of nutrients either natural or artificial means.
6. **Pedology:** Dealing with the genesis, survey and classification

SOIL – DEFINITION

Joffe (1936): Soil is a natural body of mineral and organic constituents differentiated into horizons usually unconsolidated, of variable depth which differs among themselves as well as from the underlying parent material in morphology, physical makeup, chemical properties and composition and biological characteristics.

SSSA (1970): (i) The unconsolidated mineral matter on the surface of the earth that has been subjected to and influenced by genetic and environmental factors of parent material, climate (including moisture and temperature effects), macro and microorganisms and topography, all affecting over a period of time and producing a product, that is "SOIL" that differs from the material from which it is derived in many, physical, chemical, biological and morphological properties and characteristics.

(ii) The unconsolidated mineral material on the immediate surface of the earth that serves as a natural medium for the growth of land plants.

V.V. Dokuchaev (1900): "Soils are applied solely to those superficial or nearly superficial horizons of rocks, that have been more or less modified naturally by the interaction of water, air and various kinds of organisms, either living or dead; this being reflected in a certain manner in the composition, structure and colour of such formations. Where this conditions are absent, there are no natural soils, but either artificial mixtures or rock"

Buckman and Brady: Soil may be defined as "A dynamic natural body on the surface of the earth in which plants grow, composed of mineral and organic materials and living forms".

APPROACHES OF SOIL STUDY

Two Concepts: One treats soil as a natural body, weathered and synthesized product in nature while other treats soil as a medium for plant growth.

Pedological Approach: The origin of the soil, its classification and its description are examined in Pedology. (from Greek word *pedon*, means soil or earth). Pedology is the study of soil as a natural body and does not focus on the soil's immediate practical use. A pedologist studies, examines and classifies soil as they occur in their natural environment.

Edaphological Approach: Edophology (from Greek word *edaphos*, means soil or ground) is the study of soil from the stand point of higher plants. Edaphologists consider the various properties of soil in relation to plant production. They are practical and have the production of food and fibre as their ultimate goal. They must determine the reasons for variation in the productivity of soils and find means for improvement.

ORIGIN OF EARTH

Earth is one of the 9 planets (8 excluding Pluto) orbiting the SUN in the Solar System. The Universe is composed of several Galaxies. Our solar system is part of Milky Way galaxy which is disk shaped with about 1, 00,000 million stars of varying sizes. Our solar system consists of 9 planets and 31 satellites, a belt of asteroids

THEORIES

- ***Nebular Hypothesis***
- ***Planetesimal Hypothesis***

Nebular Hypothesis

Solar system formed through the condensation of a nebula which once encircled the Sun. The outer planets formed first, followed by Mars, the Earth, Venus, and Mercury. This hypothesis suggests a sequential origin from outermost planet to innermost. As per this hypothesis, Mars evolved earlier than the Earth. This hypothesis is widely accepted to explain origin of different planets.

Planetesimal Hypothesis

A catastrophic hypothesis, proposed by Thomas Chamberlin and Forest Moulton (1905), in which the planets of the Solar System are seen to arise from an encounter between the Sun and another star. In this scenario, the gravity of the passing star tears a succession of bolts from the solar surface. Bolts coming from the side nearer the star are thrown out to distances comparable with those of the giant planets, while those from the far side of the Sun are ejected less violently to the distances of the terrestrial planets.

Table 1. Radius, distance from sun and revolution period

S.No	Planet	Radius (km)	Distance from sun(m.k)	Revolution period (day)
1	Mercury	2439	57.9	88
2	Venus	6052	108.2	245
3	Earth	6378	149.4	365.25
4	Mars	3397	227.9	687
5	Jupiter	71398	778.3	4333
6	Saturn	60000	1427	10743
7	Uranus	23620	2870	30700
8	Neptune	24300	4496.5	60280
9	Pluto	1150	5970	90130

Table 2. Geological Time Scale of Earth – Development of Life

Era	Period	Age (m.Yr)	Organism
Archean	Archean	> 3500	Lifeless
Precambrian	Precambrian	-do-	Soft bodied plant & animals
Palaeozoic	Cambrian	500	Algae and shell bearing molluscus
	Ordovician	400	Molluscus & sea weeds
	Silurian	360	Land plants & breathing land animals
	Devonian	320	True fishes and ampibians
	Carboniferous	280	Coal forming materials , non flowering plants
	Permian	246	Reptiles
Mesozoic	Triassic	235	Marine life & reptiles
	Jurassic	185	Dinosours
	Cretaceous	139	Many plants & fishes

Cenozoic	Tertiary		
	Eocine	20	Mammals
	Myocine	29	Mammals
	Pliocine	10	Mammals
	Quartanary		
	Plistocine	01	Mammals
	Recent time		Mammals

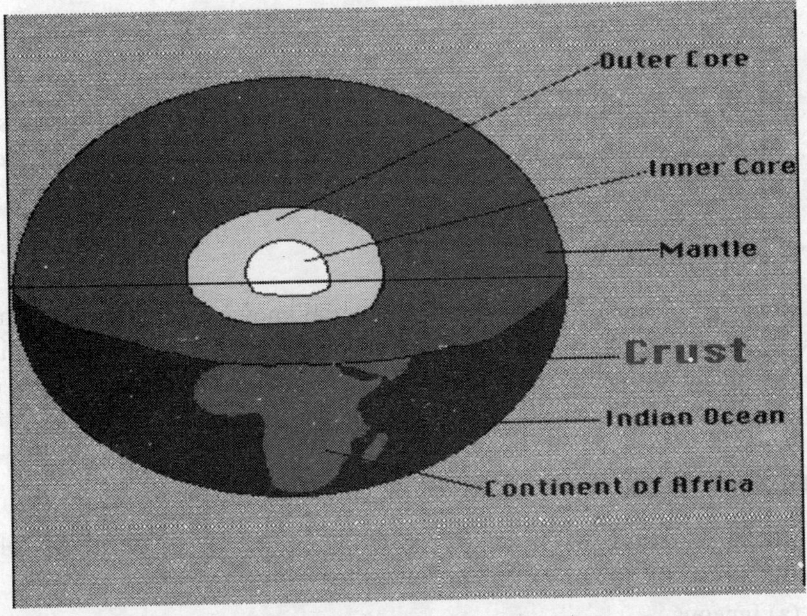

Fig. 1. Earth Crust with cores

Table 3. Average composition of Earth's crust (% by weight)

Non - metallic	Oxygen	O^{2-}	46.60%	74.32% (3/4th)
	Silica	Si^{4+}	27.72%	
Metallic	Aluminium	Al^{3+}	8.13%	1/4th of the total
	Iron	Fe^{2+}	5.00%	
	Calcium	Ca^{2+}	3.63%	
	Sodium	Na^{+}	2.83%	
	Potassium	K^{+}	2.59%	
	Magnesium	Mg^{+2}	2.09%	
	Others	-	1.41%	

Eight elements are abundant – 98.6%

2 ROCKS AND MINERALS

Rocks are the materials that form the essential part of the Earth's solid crust. "*Rocks are hard mass of mineral matter comprising one or more rock forming minerals*".

ROCK FORMATION

Rocks are formed from the molten material known as **magma**. The study of rocks is called Petrology (in Greek, petra means rock, logos means science). Petrology deals with the description of rocks; petrogenesis is the study of the origin of rocks

Soil Forming Rocks

Rocks are solid mass, an aggregation of minerals on the basis of their formation rocks are generally grouped into three classes

(a) Igneous

(b) Sedimentary

(c) Metamorphic

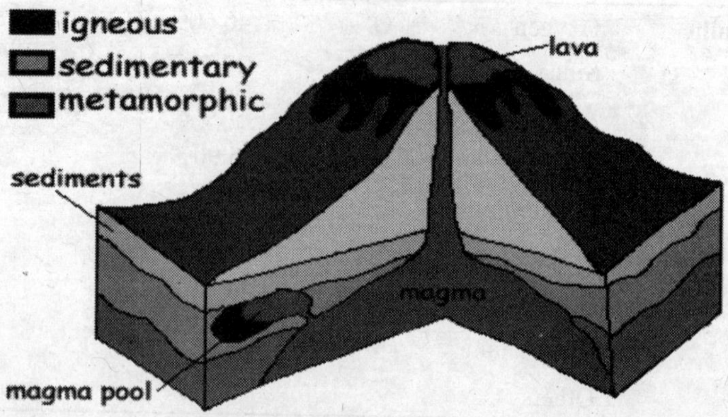

Fig: 2. Soil Forming Rocks

1. IGNEOUS ROCKS (primary or massive rocks) (Latin ignis - fire)

These are first formed in the earth crust due to the solidification of molten magma. Based on the mode of formation, they are further classified as extrusive and intrusive rocks

Extrusive rocks (or volcanic rocks)

These rocks are formed due to the consolidation of magma on the surface of the earth. The magma, when it flows on the Earth surface is called LAVA. E.g. Basalt

Intrusive rocks (or plutonic rocks)

These rocks are produced due to solidification of magma below the surface of the earth.

Plutonic – intrusive rocks solidifies at greater depth

Hypabassal rocks -- solidifies at shallow depth from the surface.

E.g. Granite, syenite, diorite, Gabbro *etc.*

Rocks formed in vertical cracks are called **dykes** and in horizontal cracks are called **sills. Vesicular rocks:** Molten magma cools on the surface. Steam of water is entrapped into rocks and forms vesicles.

Based on the silica content, rocks are also classified as

1. Acid rocks : >66% SiO_2 (Granite, Rhyolite)

2. Intermediate : 56 to 65% SiO_2

(Sub acid rocks 60 to 66% SiO_2 (Syenite and Trachyte))

(Sub basic rocks 56 to 60% SiO_2 (Diorite and Andesite))

3. Basic rocks : 40 to 55% (Gabbro, basalt)

Table 4. Igneous rocks and their Charecteristics

S.N	Rocks	Origin	Essential minerals	Common minerals	Average specific gravity	Remarks
i.	Granite	Plutonic holocry-stalline	Quartz (20 to 30%)	Hornblende, magnetite, mica	2.64	Light coloured white or reddish
ii.	Syenite	Plutonic Holocry-stalline	Quartz, orthoclase	Hornblende, magnetite, biotite	2.80	Light coloured white or reddish
iii.	Diorite	Plutonic Holocry-stalline	Quartz	Hornblende, magnetite, biotite	2.85	Darker

iv	Gabbro	Plutonic Holocry-stalline	Labradorite, augite, olivine	Hornblende, ilmenite	3.0	Blakish
v.	Dolerite	Hypabasal	Labradorite, augite, olivine	Hornblende, ilmenite	3.0	Blakish
vi.	Basalt	Volcanic cry-stalline with glassy mass	Labradorite, augite, olivine	Hornblende, ilmenite	3.0	Dark coloured

Brief description of important igneous rocks

Basalt

This is the most abundantly formed rock from molten material. It is a fine grained and dark coloured rock which contains 50 per cent feldspars and 50 per cent ferromagnesian minerals, including pyroxene and olivine. The coarse-grained rock with the comparable composition is gabbro.

Pumice

It is a very light weight rock which has a lower specific gravity than water. As a result, it floats on water. In composition, it is comparable to granite / rhyolite and in texture, it is like a sponge.

Granite

A coarse-textured and light coloured rock that contains 60 to 70 per cent feldspars, of which orthoclase constitutes 40 to 45 per cent, plagioclase 20 to 25 per cent, ferromagnesian minerals 3 to 10 per cent and quartz 20 to 30 per cent. The fine grained rock with the above composition is rhyolite.

2. SEDIMENTARY ROCKS (Latin sedimentum - settling)

These rocks are formed from the consolidation of sediments accumulated through wind or water action at the surface of the earth. Many are deposited in layer or formed through chemical reactions as precipitates from aqueous solutions. Sediments may contain various size particles cemented together by substances like SiO_2, Fe_2O_3 or lime. These rocks are also called as clastic rocks.

Based on the origin, the sedimentary rocks are classified as

1. Residual : Laterite

2. Transported

a. Deposited as solids in suspension: Sandstone, shale

b. Deposited by chemical precipitation: Limestone, ironstone

c. Deposited through agency of organic matter : Peat, Phosphatic deposits

Based on the grain size, sedimentary rocks are classified as

1. Rocks with boulder pebbles sized minerals (Rudaceous) : : Conglomerate
2. Rocks with sand size particles (Arenaceous) : Sandstone
2. Rocks with silt size particles (silt rocks) : Siltstone
4. Rocks with clay size particles (Argillaceous): Shale

Table 5. Sedimentary rocks and their Characteristics

S.No	Rock	Mineral composition	Colour and structure
1.	Sandstone	Mainly quartz with some $CaCO_2$, iron oxides and clay	Light to red, granular
2.	Shale	Clay minerals, quartz and some organic matter	Light to dark thinly laminated
3.	Limestone	Mainly calcite with some dolomite, iron oxides, clay, phosphate and organic matter	Light grey to yellow, fine grained and compact

The sedimentary rocks are formed from sediments, derived from the breaking down of pre-existing rocks. The sediments are transported to new places and deposited in new arrangements and cemented to form secondary rocks. Stratification is the most common feature of these rocks and as such these are also termed as stratified rocks.

Formation

In the formation of sedimentary rocks, four stages are recognized. They are

i. Weathering

The igneous and other rocks disintegrate owing to physical, chemical and biological weathering. The percolating water having carbonic and humic acids decompose the rocks and minerals giving rise to quartz and other secondary minerals and soluble substances like calcium, magnesium, iron and salts.

$$2KAlSi_3\,O_8 + H_2CO_3 + H_2O \rightarrow 4l_2Si_2O_5\,(OH)_4 + K_2CO_3 + 4SiO_2$$

The products of these weathering agencies provide the basic materials (gravels, sand, silt, mud) formation of sedimentary rocks.

ii. Transportation

The disintegrated material is transported by the agencies such as water, wind, glaciers, run off and gravity.

Water	:	Alluvium
Wind	:	Loess or Aeolian

Ice	:	Moraine
Gravity	:	Colluvial
Sea	:	Marine
Lake	:	Lacustrine

iii. Deposition or sedimentation

The detrital materials, comprising minerals and rock fragments are deposited when the carrying agent has no longer energy enough to move it further. For instance, a stream flowing at a certain velocity possesses enough energy to move particles of a certain size. With the decreasing velocity, the carrying capacity also decreases. This is how the coarser particles settle first and the finer particles later. This kind of deposition is called graded bedding.

iv. Diagnosis

It refers to the transformation of unconsolidated sediments to hard rock. It involves compaction and cementation.

Compaction: The weight of the upper incumbent layers – thousands of metre thick – causes compression of the lower layers or deposits. The interstitial water and air are removed by the pressure of overlying sediments. The fine grained deposits under such environments are transformed to clays, shales etc.

Cementation: The most common materials that serve as cementing agents are lime, silica and iron oxide. The sedimentary rocks are divided into six groups. They are

1. **Arenaceous** rocks are formed by the deposits of coarse grained particles. (eg.) Sandstone, grit, Conglomerate and Breccia.
2. **Argillaceous** rocks consist of small size particles usually known as clay (eg.) clay, mudstone, and shale.
3. **Calcareous** rocks consist of carbonate of lime or carbonate of lime and magnesia (eg.) limestone, chalk, ferruginous limestone.
4. **Carbonaceous** rocks are formed from the decomposing vegetation under anaerobic conditions (eg.) peat, lignite, coal, anthracite etc.
5. **Siliceous** rocks are also of organic origin having been formed from parts of minute plants and animals like diatoms and deposited either from salt or fresh water (eg.) Flint, Chert
6. **Precipitated salts** consists mainly of deposits formed as rock masses either by cooling, evaporation or by direct chemical precipitation

(Eg) Oxides	:	Haematite, limonite, bauxite, quartz
Carbonates	:	Limestone, Magnesite
Sulphates	:	Gypsum
Chlorides	:	Rock salt

Brief description of important sedimentary rocks

Conglomerate

It is a detrital rock made up of more or less rounded fragments. If the fragments are more angular than rounded, the rock is called *breccia.*

Sandstone

It is an intermediate between fine grained shale or mudstone and coarse grained conglomerate. The predominant mineral present is quartz; it is called *arkose* if, quartz and feldspars are predominantly observed.

Shale: Fine grained detrital rock composed of clay and silt sized particles.

Limestone

A rock, chiefly composed of calcite mineral, that has been deposited by an inorganic process. Some may be formed by an organic chemical process.

3. METAMORPHIC ROCKS

These are formed from igneous and sedimentary rocks under the influence of heat, pressure, chemically active liquids and gases. Change may occur in mineral composition or texture or both. The changes due to water is called hydrometamorphosis and due to pressure is called dynamometamorphosis

Sand stone : Quartizite
Shale : Slate/mica, schist
Lime stone : Marble
Granite : Granite gneiss
Dolerite : Hornblende gneiss

Table 6. Metamorphic rocks and their Characteristics

S.No.	Rock	Mineral composition	Colour and structure
1.	Gneiss	Formed from granite	Alternating light and dark colours, banded and foliated
2.	Schist	Formed from basalt or shale	As original rock, foliated
3.	Quartzite	Formed from sandstone	Light or brown, compact and uniform texture, foliated structure
4.	Slate	Formed from shale	Grey to black, compact and uniform texture, foliated structure
5.	Marble	Formed from lime stone	Light red, green, black, compact fine to coarse texture, foliated structure

Table 7. Pre-existing and their equivalent metamorphosed rocks

Pre-exiting /original rock	**Metamorphosed rock**
Conglomerate	Gneiss
Sandstone	Quartzite – quartz – schist
Shale	Slate – phyllite – mica schist
Limestone	Marble
Dolomite	Dolomite marble – soapstone or serpentine
Iron ores	Haematite – schist
Basalt	Schists
Gabbro	Gneiss – chlorite schist
Granite and Syenite	Gneiss – mica schist
Coal	Graphite

Brief description of important metamorphic rocks

Gneiss: A crystalline rock, with a banded appearance. The light coloured minerals, such as feldspars, quartz and mica roughly alternate with bands of dark coloured ferro - magnesian minerals.

Schist: A finely foliated or laminated rock, composed dominantly of micas and chlorite, together with some quartz and ferro-magnesian minerals. It can be derived from igneous and sedimentary rocks.

Marble: A non-foliated crystalline rock, composed of calcite and dolomite.

Slate: A very finely foliated rock with mineral particles too small to be seen. It splits into thin, smooth sheets. Mica, quartz and chlorite are the chief minerals in it.

Table 8. Classification of metamorphic rocks (based on texture and structure)

Texture	**Rock name**	**Mineral composition**	**Parent rock**
Foliated (Parallel structure): Latin foliatus, means leaved or leafy: The rocks that contain micas and ferro- magnesian minerals and show foliation as the minerals are flattened and arranged in parallel layers.			
Coarse – grained	Gneiss	Feldspars, quartz, micas, amphiboles	Granite
Coarse-grained	Schist	Micas and other elongated silicates with minor amounts of quartz (and feldspars)	Gabbro, basalt
Medium-grained	Phyllite	Micaceous (mica and chlorite) with larger grains representing a transition from schist to slate	Shale
Very fine grained	Slate	Micaceous minerals (mica, chloirte) with quartz	Shale

Unfoliated (massive structure): The rocks containing quartz and feldspars do not show foliation even under pressure because of large-sized crystals.

Fine-grained	Anthracite	Carbon (92-93%)	Peat, lignite, coal
	Talc – schist	-	Talc
	Amphibolite	-	Hornblende

Granular: The rocks consisting of mostly equidimensional grains.

Medium to coarse – grained	Quartzite	Quartz and quartz cement	Sandstone
Fine – grained	Marble	Calcite, dolomite	Dolomite, Calcite

MINERALS

Minerals are naturally occurring solids with a definite chemical composition and crystal structure. "Solid substances composed of atoms having an orderly and regular arrangement"

Table 9. Minerals and their constituents

Minerals (arranged in the order of their crystallization)	Important constituents	Percent distribution
Primary minerals		
Ferro magnesium minerals		
Ortho-ino silicates		16.8
Olivine	Fe, Mg	
Pyroxenes	Ca, Na, Fe, Mg	
Amphiboles	Ca, Na, Fe, Mg, Al, OH	
Phyllo Silicates		3.6
Biotite	K, Fe, Mg, Al, OH	
Muscovite	K, Al, OH	
Non-Ferro Magnesium		
Tecto Silicates		
Feldspars		61.0
Anorthite	Ca, Al	
Albite	Na, Al	
Orthoclase	K, Al	
Quartz		
Secondary clay minerals		
Minerals	Na, K, Ca	11.6
Others	Mg, Fe, Al, OH	6.0

When molten magma solidifies, different elements present in them freely arrange in accordance with the attractive forces and geometric form. Silica tetrahedron is the fundamental building blocks for the formation of different minerals. (SiO_2). Different silicate minerals are ortho silicates, ino-silicates, phyllosilicates and tecosilicates. There are non-silicate minerals also. These are different oxides, carbonates, sulphates, phosphates *etc*.

Minerals that are original components of rocks are called **primary minerals** (feldspar, mica, *etc*.). Minerals that are formed from changes in primary minerals and rocks are called **secondary minerals** (clay minerals). Those minerals that are chief constituents of rocks are called as **essential minerals** (Feldspars, pyroxenes micas *etc*) and those which are present in small quantities, whose presence or absence will not alter the properties of rocks are called **accessory minerals** (tourmaline, magnetite *etc*) of **the >2000 known minerals, only few occur in abundance in the Earth crust.**

SECONDARY CLAY MINERALS

Clay minerals in soils are formed from primary minerals due to weathering processes. These clay minerals are of size <0.002 mm and are considered to be the most reactive part of soil. Important soil properties like nutrient and water holding capacity are controlled by clay minerals. These minerals are layered silicates consisting of silica tetrahedron and aluminium octahedron.

(1) 1:1 clay mineral (1 silicon tetrahedron + 1 aluminium octahedron) - Kaolinite

(2) 2: 1 clay mineral

(i) Non-expanding clay mineral

i. Black mica (Biotite)

ii. White mica (Muscovite)

iii. Weathered mica (illite)

(ii) Expanding clay mineral

i. Partially expanding (Vermiculite)

ii. Fully expanding (Montmorillonite)

(3) 2 : 2 clay mineral:chlorite

Non-silicate minerals

Oxides: Haematite (Fe_2O_3)

Limonite (Fe_2O_3, $3H_2O$)

Goethite (FeO (OH) H_2O)

Gibbsite ($Al_2O_3H_2O$)

The red, yellow or brown colours in soils are due to the presence of goethite and haematite, which occur as coatings on the surface of soil particles.

Carbonates: Calcite ($CaCO_3$)

Dolamite ($CaMgCO_3$)

Sulphates: Gypsum ($CaSO_4.2H_2O$)

Phosphates: Apatite (Rock phosphate $Ca_3(PO_4)_2$ - primary source of phosphorus

PHYSICAL PROPERTIES OF MINERALS

1. Color
2. Streak
3. Fracture/ cleavage
4. Hardness
5. Luster
6. Crystal form
7. Taste
8. Specific gravity
9. Magnetism
10. Effervescence (fizz)
11. Birefringence
12. Fluorescence

1. Color

2. Streak (Hematite)

3. Fracture/ cleavage

4. Hardness

Table 2.2 Mohs Scale of Hardness

Mineral	Scale Number	Common Objects
Talc	1	
Gypsum	2	——— Fingernail
Calcite	3	——— Copper coin
Fluorite	4	
Apatite	5	——— Knife blade
Orthoclase	6	——— Window glass
Quartz	7	——— Steel file
Topaz	8	
Corundum	9	
Diamond	10	

5. Luster

Table 2.3 Mineral Luster

Metallic	Strong reflections produced by opaque substances
Vitreous	Bright, as in glass
Resinous	Characteristic of resins, such as amber
Greasy	The appearance of being coated with an oily substance
Pearly	The whitish iridescence of such materials as pearl
Silky	The sheen of fibrous materials such as silk
Adamantine	The brilliant luster of diamond and similar minerals

6. Crystal form

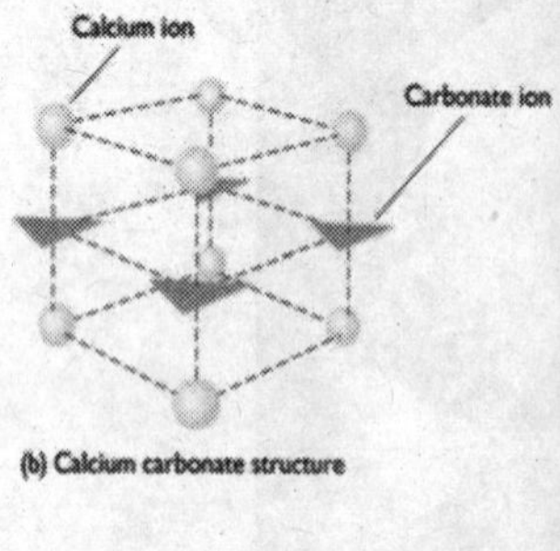

(b) Calcium carbonate structure

7. Effervescence (fizz)

3 WEATHERING AND SOIL FORMATION

Rocks and minerals are formed under a very high temperature and pressure, exposed to atmospheric conditions of low pressure and low temperature and they become unstable and weather. Soils are formed from rocks through the intermediate stage of formation of Regolith which is the resultant of weathering.

WEATHERING

The sequence of processes in the formation of soils is:

Weathering of rocks and minerals → formation of regolith or parent material → formation of true soil from regolith.

Rock → Weathering → Regolith → soil forming factors and processes → True soil (otherwise)

Two processes involved in the formation of soil are

(i) Formation of regolith by breaking down (weathering) of the bed rock.

(ii) The addition of organic matter through the decomposition of plant and animal tissues, and reorganization of these components by soil forming processes to form soil.

Weathering: A process of disintegration and decomposition of rocks and minerals which are brought about by physical agents and chemical processes, leading to the formation of Regolith (unconsolidated residues of the weathering rock on the earth's surface or above the solid rocks).

Parent material: It is the regolith or at least its upper portion. May be defined as the unconsolidated and more or less chemically weathered mineral material from which soil is developed.

Weathering involves three processes

Physical (or) mechanical - disintegration

Chemical - decomposition
Biological

Table 10. Different agents of weathering

Physical/ Mechanical (disintegration)	Chemical (decomposition)	Biological (disintegration + decomposition)
1.Physical condition of rock	1.Hydration	1.Man & animals
2.Change in temperature	2.Hydrolysis	2. Higher plants & their roots
3.Action of H_2O	3.Solution	3.Micro organisms
- fragment & transport	4.Carbonation	
- action of freezing	5.Oxidation	
- alternate wetting & drying	6.Reduction	
- action of glaciers		
4. Action of wind		
5. Atmospheric electric phenomenon		

I. PHYSICAL WEATHERING

The rocks are disintegrated and are broken down to comparatively smaller pieces, with out producing any new substances

1. Physical condition of rocks

- The permeability of rocks is the most important single factor. Coarse textured (porous) sand stone weather more readily than a fine textured (almost solid) basalt.
- Unconsolidated volcanic ash weather quickly as compared to unconsolidated coarse deposits such as gravels.

2. Action of Temperature:

- The variations in temperature exert great influence on the disintegration of rocks. During day time, the rocks get heated up by the sun and expand.
- At night, the temperature falls and the rocks get cooled and contract. This alternate expansion and contraction weakens the surface of the rock and crumbles it because the rocks do not conduct heat easily.
- The minerals with in the rock also vary in their rate of expansion and contraction. The cubical expansion of quartz is twice as feldspar.
- Dark coloured rocks are subjected to fast changes in temperature as compared to light coloured rocks.

- The differential expansion of minerals in a rock surface generates stress between the heated surface and cooled un-expanded parts resulting in fragmentation of rocks.
- This process causes the surface layer to peel off from the parent mass and the rock ultimately disintegrates. This process is called Exfoliation

Fig: 3. Exfoliation of rocks

3. Action of Water: Water acts as a disintegrating, transporting and depositing agent.

(i) Fragmentation and transport: Water beats over the surface of the rock when the rain occurs and starts flowing towards the ocean

- Moving water has the great cutting and carrying force.
- It forms gullies and ravines and carries with the suspended soil material of variable sizes.
- Transporting power of water varies. It is estimated that the transporting power of stream varies as the sixth power of its velocity i.e the greater the speed of water, more is the transporting power and carrying capacity

Speed/Sec	Carrying capacity
15 cm	fine sand
30 cm	gravel
1.2 m	stones (1kg)
9.0 m	boulders (several tons)

The disintegration is greater near the source of river than its mouth

Fig. 4. Weathering of rocks along the River Trail

(ii) Action of freezing

Frost is much more effective than heat in producing physical weathering

- In cold regions, the water in the cracks and crevices freezes into ice and the volume increases to one tenth
- As the freezing starts from the top there is no possibility of its upward expansion. Hence, the increase in volume creates enormous out ward pressure which breaks apart the rocks

Fig. 5. Weathering of rocks due to freezing

(iii) Alternate wetting and Drying

Some natural substances increase considerably in volume on wetting and shrink on drying. (e.g.) smectite, montmorillonite

- During dry summer/ dry weather - these clays shrink considerably forming deep cracks or wide cracks.
- On subsequent wetting, it swells.
- This alternate swelling and shrinking / wetting or drying of clay enriched rocks make them loose and eventually breaks

(iv)Action of glaciers:

- In cold regions, when snow falls, it accumulates and changes into an ice sheet.
- These big glaciers start moving owing to the change in temperature and/or gradient.
- On moving, these exert tremendous pressure over the rock on which they pass and carry the loose materials
- These materials get deposited on reaching the warmer regions, where its movement stops with the melting of ice

Fig. 6. Weathering of rocks due to glaciers

4. Action of wind

- Wind has an erosive and transporting effect. Often when the wind is laden with fine material *viz.*, fine sand, silt or clay particles, it has a serious abrasive effect and the sand laden winds itch the rocks and ultimately breaks down under its force
- The dust storm may transport tons of material from one place to another. The shifting of soil causes serious wind erosion problem and may render cultivated land as degraded

  (e.g) Rajasthan deserts

Fig. 7. Weathering of rocks due to wind

5. Atmospheric electrical phenomenon

It is an important factor causing break down during rainy season and lightning breaks up rocks and/or widens cracks.

II. CHEMICAL WEATHERING

Decomposition of rocks and minerals by various chemical processes is called chemical weathering. It is the most important process for soil formation. Chemical weathering takes place mainly at the surface of rocks and minerals with disappearance of certain minerals and the formation of secondary products (new materials). This is called chemical transformation.

Feldspar + water → clay mineral + soluble cations and anions

Table 11. Average mineralogical composition (%)

Composition	Granite	Basalt	Shale	Sand Stone	Lime Stone
Feldspar	52.4	46.2	30.0	11.5	-
Quartz	31.3	-	2.3	66.8	-
Pyroxines- amphiboles	-	44.5	-	-	-
FeO minerals	2.0	9.3	10.5	2.0	-
Clay minerals	14.3	-	25.0	6.6	24.0
Carbonates	-	-	5.7	11.1	76.0

Chemical weathering becomes more effective as the surface area of the rock increases. Since the chemical reactions occur largely on the surface of the rocks, therefore, the smaller the fragments, the greater the surface area per unit volume available for reaction. The effectiveness of chemical weathering is closely related to the mineral composition of rocks.

(e.g) quartz responds far slowly to the chemical attack than olivine or pyroxene.

Chemical Processes of weathering

1. Hydration

It is a Chemical combination of water molecules with a particular substance or mineral leading to a change in structure. Soil forming minerals in rocks do not contain any water and undergo hydration when exposed to humid conditions. Upon hydration there is swelling and increase in volume of minerals. The minerals loose their luster and become soft. It is one of the most common processes in nature and works with secondary minerals, such as aluminium oxide and iron oxide minerals and gypsum. (e.g)

a) $2Fe_2O_3 + 3HOH \rightarrow 2Fe_2O_3 .3H_2O$
(Haematite) (red) (Limonite) (yellow)

b) $Al_2O_3 + 3HOH \rightarrow Al_2O_3 .3H_2O$
(Bauxite) (Hyd. aluminium Oxide)

c) $CaSO_4 + 2H_2O \rightarrow CaSO_4 .2H_2O$
(Anhydrite) (Gypsum)

d) $3(MgO.FeO.SiO_2) + 2H_2O \rightarrow 3MgO.2SiO_2.2H_2O + SiO_2 + 3H_2O$
(Olivine) (Serpentine)

2. Hydrolysis

It is most important process in chemical weathering. It is due to the dissociation of H_2O into H^+ and OH^- ions which chemically combine with minerals and bring about changes, such as exchange, decomposition of crystalline structure and formation of new compounds. Water acts as a weak acid on silicate minerals.

$KA1Si_3O_8 + H_2O \rightarrow HA1Si_3O_8 + KOH$
(Orthoclase) (Acid silti clay)

$HA1Si_3O_8 + 8\,HOH \rightarrow Al_2O_3.3H_2O + 6H_2SiO_3$
(recombination) (Hyd. Alum. oxide) (Silicid acid)

This reaction is important because of two reasons.

(i) clay, bases and silicic acid - the substances formed in these reactions - are available to plants

(ii) Water often containing CO_2 (absorbed from atmosphere), reacts with the minerals directly to produce insoluble clay minerals, positively charged metal ions (Ca^{++}, Mg^{++}, Na^+, K^+) and negatively charged ions (OH^- HCO_3^-) and some soluble silica – all these ions are made available for plant growth.

3. Solution

Some substances present in the rocks are directly soluble in water. The soluble substances are removed by the continuous action of water and the rock no longer remains solid and form holes, rills or rough surface and ultimately falls into pieces or decomposes. The action is considerably increased when the water is acidified by the dissolution of organic and inorganic acids. (e.g) halides, NaCl

$NaCl + H_2O \rightarrow Na^+, Cl^- H_2O$ (dissolved ions with water)

4. Carbonation: Carbon di oxide when dissolved in water it forms carbonic acid.

$$2H_2O + CO_2 \rightarrow H_2CO_3$$

This carbonic acid attacks many rocks and minerals and brings them into solution. The carbonated water has an etching effect up on some rocks, especially lime stone. The removal of cement that holds sand particles together leads to their disintegration.

$CaCO_3 + H_2CO_3 \rightarrow Ca(HCO_3)_2$
(calcite) slightly soluble → (Ca bi carbonate) readily soluble

5. Oxidation

It is the process of addition and combination of oxygen to minerals. The absorption is usually from O_2 dissolved in soil water and that present in atmosphere. The oxidation is more active in the presence of moisture and results in hydrated oxides.(e.g) minerals containing Fe and Mg.

$4FeO$ (Ferrous oxide) + $O_2 \rightarrow 2Fe_2O_3$ (Ferric oxide)

$4Fe_3O_4$ (Magnetite) + $O_2 \rightarrow 6Fe_2O_3$ (Haematite)

$2Fe_2O_3$ (Haematite) + $3H_2O \rightarrow 2Fe_2O_3 .3H_2O$ (Limonite)

6. Reduction

The process of removal of oxygen and is the reverse of oxidation and is equally important in changing soil colour to grey, blue or green as ferric iron is converted to ferrous iron compounds. Under the conditions of excess water or water logged condition (less or no oxygen), reduction takes place.

$2Fe_2O_3$ (Haematite) - $O_2 \rightarrow 4FeO$ (Ferrous oxide) -reduced form

In conclusion, during chemical weathering igneous and metamorphic rocks can be regarded as involving destruction of primary minerals and the production of secondary minerals. In sedimentary rocks, which is made up of primary and secondary minerals, weathering acts initially to destroy any relatively weak bonding agents (FeO) and the particles are freed and can be individually subjected to weathering.

III BIOLOGICAL WEATHERING

Unlike physical and chemical weathering, the biological or living agents are responsible for both decomposition and disintegration of rocks and minerals. The biological life is mainly controlled largely by the prevailing environment.

1. Man and Animals

- The action of man in disintegration of rocks is well known as he cuts rocks to build dams, channels and construct roads and buildings. All these activities result in increasing the surface area of the rocks for attack of chemical agents and accelerate the process of rock decomposition.
- A large number of animals, birds, insects and worms, by their activities they make holes in them and thus aids for weathering.
- In tropical and sub tropical regions, ants and termites build galleries and passages and carry materials from lower to upper surface and excrete acids. The oxygen and water with many dissolved substances, reach every part of the rock through the cracks, holes and galleries, and thus brings about speedy disintegration.
- Rabbits, by burrowing in to the ground, destroy soft rocks. Moles, ants and bodies of the dead animals, provides substances which react with minerals and aid in decaying process.
- The earthworms pass the soil through the alimentary canal and thus brings about physical and chemical changes in soil material.

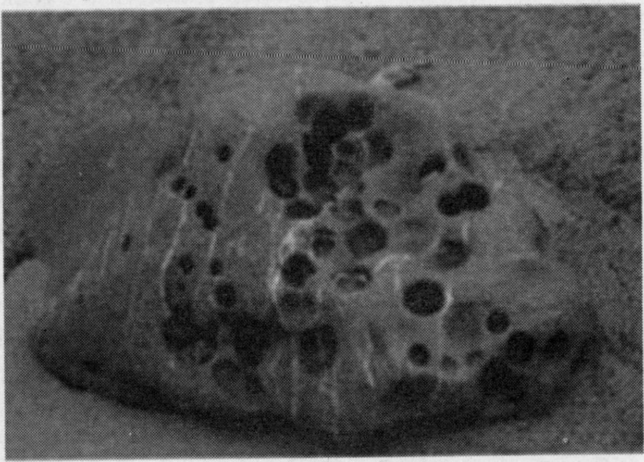

Fig. 8. Wearthering of rocks by animals, such as these Piddock shells,

2. Higher Plants and Roots

The roots of trees and other plants penetrate into the joints and crevices of the rocks. As they grew, they exert a great disruptive force and the hard rock may break apart.

(e.g.) pipal tree growing on walls/ rocks

The grass roots form a sponge like mass, prevent erosion and conserve moisture and thus allowing moisture and air to enter in to the rock for further action. Some roots penetrate deep into the soil and may open some sort of drainage channel. The roots running in crevices in lime stone and marble produces acids. These acids have a solvent action on carbonates. The dead roots and plant residues decompose and produce carbon dioxide which is of great importance in weathering.

Fig. 9. Weathering due to roots

Trees put down roots through joints or cracks in the rock in order to find moisture. As the tree grows, the roots gradually prize the rock apart.

3. Micro- organisms

In early stages of mineral decomposition and soil formation, the lower forms of plants and animals like, mosses, bacteria and fungi and actinomycetes play an important role. They extract nutrients from the rock and N from air and live with a small quantity of water. In due course of time, the soil develops under the cluster of these micro-organisms. These organisms are closely associated with the decay of plant and animal remains and thus liberate nutrients for the next generation plants and also produce CO_2 and organic compounds which aid in mineral decomposition.

Fig. 10. Weathering of rocks due to tiniest bacteria, algae and lichens

WEATHERING OF MINERALS

There are many factors which influence the weathering of minerals.

- Climatic conditions
- Physical characteristics
- Chemical and structural characteristics

1. Climatic Conditions

The climatic condition, more than any other factor- tends to control the kind and rate of weathering. Under conditions of low rainfall, there is a dominance of physical weathering which reduces the size and increases the surface area with little change in volume.The increase in moisture content encourages chemical as well as mechanical changes and new minerals and soluble products are formed.The weathering rates are generally fastest in humid tropical regions as there is sufficient moisture and warmth to encourage chemical decomposition. The easily weatherable minerals disappear on account of intense chemical weathering and more resistant products (hydrous oxides of Fe and Al) tend to accumulate

Climate controls the dominant type of vegetation which in turn controls the bio chemical reactions in soils and mineral weathering.

2. Physical Characteristics

(i) Differential composition

(ii) Particle size

(iii) Hardness and degree of cementation

3. Chemical and structural characteristics

- Chemical: For minerals of given particle size, chemical and crystalline characteristics determine the ease of decomposition. (e.g.) gypsum – sparingly soluble in water, is dissolved and removed in solution form under high rainfall.
- Ferro magnesium minerals are more susceptible to chemical weathering than feldspar and quartz
- Tightness of packing of ions in crystals: Less tightly packed minerals like olivine and biotite are easily weathered as compared to tightly packed zircon and muscovite (resistant)
- Chemical weathering of Silicates: The most important silicates are quartz, feldspar and certain Ferro magnesium minerals

Table 12. Weathering products of common silicate minerals

Minerals	**Composition**	**Decomposition products**	
		Minerals	**Others**
Olivine	$(Fe,Mg)_2$ SiO_2	serpentine,limonite haematite, quartz	Some Si in solution, carbonates of Fe and Mg
Pyroxenes	Fe, Mg	Clay, calcite, limonite	Some Si in solution, carbonates of Ca and Mg
Amphibole	Ca- silicates	Haematite,quartz	-do -
Biotite	Al		-do -
Plagioclase	Calcic	Clay, quartz,calcite	Some Si in solution, Na and Ca carbonates
	Sodic		-do-
Orthoclase	Potassic	Clay, quartz	Some Si in solution, potassium carbonate
Quartz		Quartz grains	Some Si in solution

Weathering sequence

The minerals remain in the soil and regolith have led to the following generalized order of weathering.

Quartz (most resistant) > muscovite > K feldspar > Na and Ca feldspar > biotite, hornblende, augite > olivine > dolomite and calcite > gypsum (least resistant).

4 SOIL FORMING FACTORS

The soil formation is the process of two consecutive stages.

1. The weathering of rock (R) into Regolith
2. The formation of true soil from Regolith

The evolution of true soil from regolith takes place by the combined action of soil forming factors and processes.

- The first step is accomplished by weathering (disintegration & decomposition).
- The second step is associated with the action of Soil Forming Factors.

Weathering

Factors

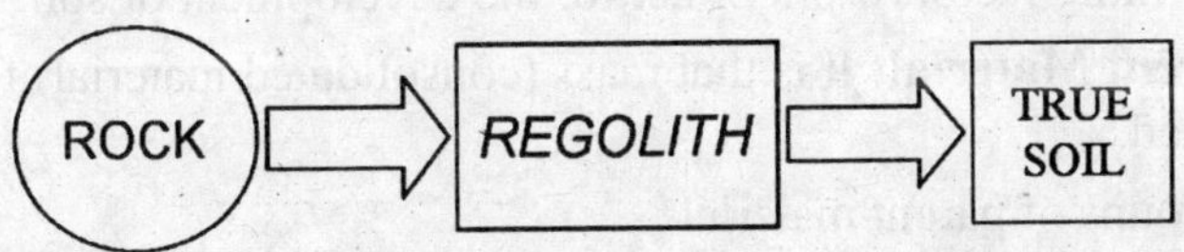

Dokuchaiev (1889) established that the soils develop as a result of the action of soil forming factors

$$S = f(P, Cl, O)$$

Further, Jenny (1941) formulated the following equation

$$S = f(Cl, O, R, P, T, \ldots)$$

Where,

Cl – Environmental climate

ơ – Organisms and vegetation (biosphere)

r – Relief or topography

p – Parent material

t- Time

... - additional unspecified factors

The five soil forming factors, acting simultaneously at any point on the surface of the earth, to produce soil.

Two groups – Passive: (i) Parent material, (ii) Relief, (iii) Time

Active: (iv) Climate, (v) Vegetation & organism

A. PASSIVE SOIL FORMING FACTORS

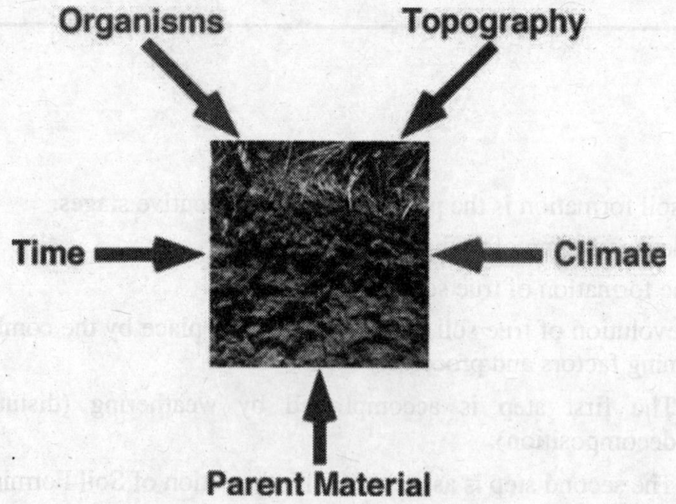

Fig. 11. Passive Soil forming factors

The passive soil forming factors are those which represent the source of soil forming mass and conditions affecting it. These provide a base on which the active soil forming factors work or act for the development of soil.

(1) Parent Material: It is that mass (consolidated material) from which the soil has formed.

Two groups of parent material

(i) Sedentary: Formed in original place. It is the residual parent material. The parent material differ as widely as the rocks

(ii) Transported: The parent material transported from their place of origin. They are named according to the main force responsible for the transport and redeposition.

(a) by gravity - Colluvial

(b) by water -Alluvial (river), Marine (ocean) Locustrine(lakes)

(c) by ice - Glacial

(d) by wind - Eolian

Colluvium: It is the poorly sorted materials near the base of strong slopes transported by the action of gravity.

Alluvium: The material transported and deposited by water is, found along major stream courses at the bottom of slopes of mountains and along small streams flowing out of drainage basins.

Lacustrine: Consists of materials that have settled out of the quiet water of lakes.

Moraine: Consists of all the materials picked up, mixed, disintegrated, transported and deposited through the action of glacial ice or of water resulting primarily from melting of glaciers.

Loess or Aeolian: These are the wind blown materials.

When the texture is silty - loess; when it is sand - Eolian.

The soils developed on such transported parent materials bear the name of the parent material; viz. Alluvial soils from alluvium, colluvial soils from colluvium etc. In the initial stages, however, the soil properties are mainly determined by the kind of parent material.

Endodynamomorphic soils: With advanced development and excessive leaching, the influence of parent material on soil characteristics gradually diminishes. There are soils wherein the composition of parent material subdues the effects of climate and vegetation. These soils are temporary and persist only until the chemical decomposition becomes active under the influence of climate and vegetation.

Ectodynamomorphic soils: Development of normal profile under the influence of climate and vegetation.

Soil properties as influenced by parent material: Different parent materials affect profile development and produce different soils, especially in the initial stages.

- Acid igneous rocks (like granite, rhyolite) produce light-textured soils (Alfisols).
- Basic igneous rocks (basalt), alluvium or colluvium derived from limestone or basalt, produce fine-textured cracking-clay soils (Vertisols).
- Basic alluvium or aeolian materials produce fine to coarse-textured soils (Entisols or Inceptisols).
- The nature of the elements released during the decaying of rocks has a specific role in soil formation. (e.g.) Si and Al forms the skeleton for the production of secondary clay minerals.
- Iron and manganese are important for imparting red colour to soils and for oxidation and reduction phenomena.
- Sodium and potassium are important dispersing agents for clay and humus colloids.
- Calcium and magnesium have a flocculating effect and result in favorable and stable soil structure for plant growth.

2. Relief or Topography: The relief and topography sometimes are used as synonymous terms. They denote the configuration of the land surface. The topography refers to the differences in elevation of the land surface on a broad scale.

The prominent types of topography designations, as given in FAO Guidelines (1990) are:

	Land surface	with slopes of
1	Flat to Almost flat	0 – 2%
2	Gently undulating	2 - 5%
3	Undulating	5 – 10%
4	Rolling	10 – 15%
5	Hilly	15 –3 0%
6	Steeply dissect	> 30% with moderate range of elevation (<300 m)
7	Mountainous	> 30% with great range of elevation (>300 m)

Soil formation on flat to almost flat position

On level topographic positions, almost the entire water received through rain percolates through the soil. Under such conditions, the soils formed may be considered as representative of the regional climate. They have normal solum with distinct horizons. But vast and monotonous level land with little gradient often has impaired drainage conditions.

Soil formation on undulating topography

The soils on steep slopes are generally shallow, stony and have weakly-developed profiles with less distinct horizonation. It is due to accelerated erosion, which removes surface material before it has the time to develop. Reduced percolation of water through soil is because of surface runoff, and lack of water for the growth of plants, which are responsible for checking of erosion and promote soil formation.

Soil formation in depression

The depression areas in semi-arid and sub humid regions reflect more moist conditions than actually observed on level topographic positions due to the additional water received as runoff. Such conditions (as in the Tarai region of the Uttar Pradesh) favour more vegetative growth and slower rate of decay of organic remains. This results in the formation of comparatively dark- coloured soils rich in organic matter (Mollisols).

Soil formation and Exposure/ Aspect

Topography affects soil formation by affecting temperature and vegetative growth through slope exposures (aspect). The southern exposures (facing the sun) are warmer and subject to marked fluctuations in temperature and moisture. The northern exposures, on the other hand are cooler and more humid. The eastern and western exposures occupy intermediate position in this respect.

3. Time

Soil formation is a very slow process requiring thousands of years to develop a mature pedon. The period taken by a given soil from the stage of weathered rock (i.e. regolith) up to the stage of maturity is considered as time. The matured soils mean the soils with fully developed horizons (A, B, C). It takes hundreds of years to develop an inch of soil. The time that nature devotes to the formation of soils is termed as Pedologic Time.

It has been observed that rocks and minerals disintegrate and/or decompose at different rates; the coarse particles of limestone are more resistant to disintegration than those of sandstone. However, in general, limestone decomposes more readily than sandstone (by chemical weathering).

- The soil properties also change with time, for instance nitrogen and organic matter contents increase with time provided the soil temperature is not high.
- $CaCO_3$ content may decrease or even lost with time provided the climatic conditions are not arid
- In humid regions, the H^+ concentration increases with time because of chemical weathering.

Table 13. Weathering stages in soil formation

Stages	Characteristic
1 Initial	Un weathered parent material
2 Juvenile	Weathering started but much of the original material still un weathered
3 Virile	Easily weatherable minerals fairly decomposed; clay content increased, slowly weatherable minerals still appreciable
4 Senile	Decomposition reaches at a final stage; only most resistant minerals survive
5 Final	Soil development completed under prevailing environments

B. ACTIVE SOIL FORMING FACTORS

The active soil forming factors are those which supply energy that acts on the mass for the purpose of soil formation. These factors are climate and vegetation (biosphere).

1. Climate

Climate is the most significant factor controlling the type and rate of soil formation. The dominant climates recognized are:

- **Arid climate:** The precipitation here is far less than the water-need. Hence the soils remain dry for most of the time in a year.
- **Humid climate**: The precipitation here is much more than the water need. The excess water results in leaching of salt and bases followed by translocation of clay colloids.
- **Oceanic climate**: Moderate seasonal variation of rainfall and temperature.
- **Mediterranean climate:** The moderate precipitation. Winters and summers are dry and hot.
- **Continental climate:** Warm summers and extremely cool or cold winters.
- **Temperate climate:** Cold humid conditions with warm summers.
- **Tropical and subtropical climate:** Warm to hot humid with isothermal conditions in the tropical zone.

Climate affects the soil formation directly and indirectly

Directly, climate affects the soil formation by supplying water and heat to react with parent material. Indirectly, it determines the fauna and flora activities which furnish a source of energy in the form of organic matter. This energy acts on the rocks and minerals in the form of acids, and salts are released. The indirect effects of climate on soil formation are most clearly seen in the relationship of soils to vegetation. Precipitation and temperature are the two major climatic elements which contribute most to soil formation.

Precipitation: Precipitation is the most important among the climatic factors. As it percolates and moves from one part of the parent material to another. It carries with it substances in solution as well as in suspension. The substances so carried are re deposited in another part or completely removed from the material through percolation when the soil moisture at the surface evaporates causing an upward movement of water. The soluble substances move with it and are translocated to the upper layer. Thus rainfall brings about a redistribution of substances both soluble as well as in suspension in soil body.

Temperature

- Temperature is another climatic agent influencing the process of soil formation.
- High temperature hinders the process of leaching and causes an upward movement of soluble salts.

- High temperature favors rapid decomposition of organic matter and increase microbial activities in soil while low temperatures induce leaching by reducing evaporation and there by favour the accumulation of organic matter by slowing down the process of decomposition. Temperature thus controls the rate of chemical and biological reactions taking place in the parent material.

Jenney (1941) computed that in the tropical regions the rate of weathering proceeds three times faster than in temperate regions and nine times faster than in arctic.

2. Organism & Vegetation

Organism

- The active components of soil ecosystem are plants, animals, microorganisms and man.
- The role of microorganisms in soil formation is related to the humification and mineralization of vegetation
- The action of animals especially burrowing animals to dig and mix-up the soil mass and thus disturb the parent material
- Man influences the soil formation through his manipulation of natural vegetation, agricultural practices etc.
- Compaction by traffic of man and animals decrease the rate of water infiltration into the soil and thereby increase the rate of runoff and erosion.

Vegetation

- The roots of the plants penetrate into the parent material and act both mechanically and chemically.
- They facilitate percolation and drainage and bring about greater dissolution of minerals through the action of CO_2 and acidic substances secreted by them.
- The decomposition and humification of the materials further adds to the solubilization of minerals
- Forests – reduces temperature, increases humidity, reduce evaporation and increases precipitation.
- Grasses reduce runoff and result greater penetration of water in to the parent material.

5 SOIL FORMING PROCESSES

The pedogenic processes, although slow in terms of human life, yet work faster than the geological processes in changing lifeless parent material into true soil full of life.

- The pedogenic processes are extremely complex and dynamic involving many chemical and biological reactions, and usually operate simultaneously in a given area.
- One process may counteract another, or two different processes may work simultaneously to achieve the same result.
- Different processes or combination of processes operate under varying natural environment.

The collective interaction of various soil–forming factors under different environmental conditions set a course to certain recognized soil–forming processes.

The basic process involved in soil formation (Simonson, 1959) includes the following:

- Gains or additions of water, mostly as rainfall, and organic and mineral matter to the soil
- Losses of the above materials from the soil
- Transformation of mineral and organic substances within the soil
- Translocation or the movement of soil materials from one point to another within the soil. It is usually divided into:
  - o Movement of solution (leaching)
  - o Movement in suspension (eluviation) of clay, organic matter and hydrous oxides

5.1. Soil–forming processes

5.1.1. Fundamental Soil–forming Processes

1. Humification

Humification is the process of transformation of raw organic matter into humus. It is an extremely complex process involving various organisms. First, simple compounds such as sugars and starches are attacked followed by proteins and cellulose, and finally very resistant compounds such as tannins, are decomposed, and the dark–coloured substance, known as humus, is formed.

Fig. 12. Humification process

2. Eluviation

It is the mobilization and translocation of certain constituents viz., clay, Fe_2O_3, Al_2O_3, SiO_2, humus, $CaCO_3$, other salts etc. from one point of soil body to another. Eluviation means washing out. It is the process of removal of constituents in suspension or solution by the percolating water from the upper to lower layers. The eluviation encompasses mobilization and translocation of mobile constituents resulting in textural differences. The horizon formed by the process of eluviation is termed as eluvial horizon (A_2 or E horizon).

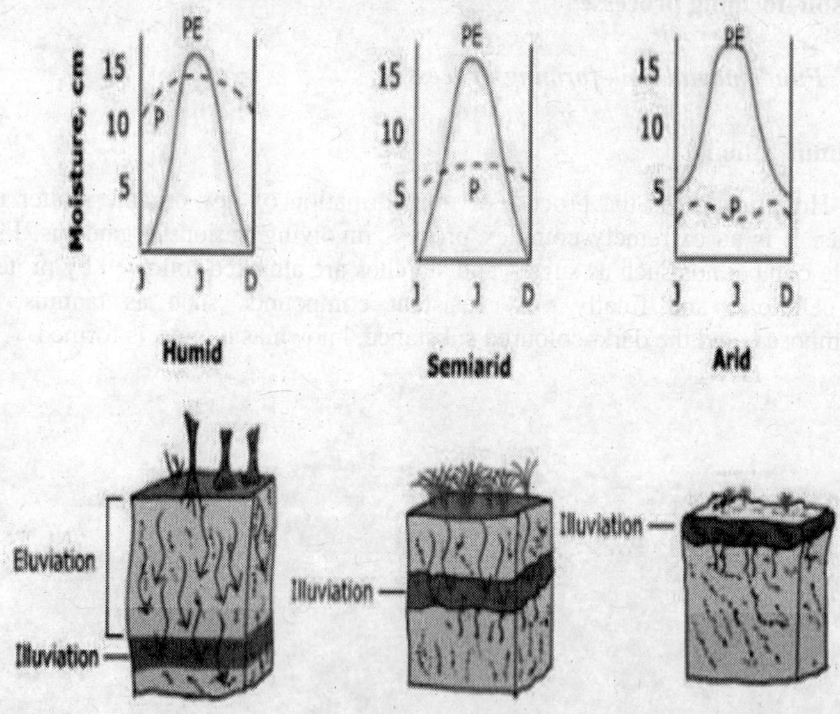

Fig. 13. Translocation depends upon relative mobility of elements and depth of percolation.

3. Illuviation

The process of deposition of soil materials (removed from the eluvial horizon) in the lower layer (or horizon of gains having the property of stabilizing translocated clay materials) is termed as Illuviation. The horizons formed by this process are termed as illuvial horizons (especially lower horizons) The process leads to textural contrast between E and Bt horizons, and higher fine: total clay ratio in the Bt horizon.

4. Horizonation

It is the process of differentiation of soil in different horizons along the depth of the soil body. The differentiation is due to the fundamental processes, humification, eluviation and illuviation.

5.1.2. Specific Soil–forming Processes

The basic pedologic processes provide a framework for later operations of more specific processes.

1. Calcification

It is the process of precipitation and accumulation of calcium carbonate ($CaCO_3$) in some parts of the profile. The accumulation of $CaCO_3$ may result in the development of a calcic horizon. Calcium is readily soluble in acid soil water and/or when CO_2 concentration is high in root zone as:

$$CO_2 + H_2O \rightarrow H_2CO_3$$

$$H_2CO_3 + Ca \rightarrow Ca\,(HCO_3)_2 \text{ (soluble)}$$

$$Ca\,(HCO_3)_2 \xrightarrow[CO_2]{\text{Temp.}} CaCO_3 + H_2O \text{ (precipitates)}$$

The process of precipitation after mobilization under these conditions is called calcification and the resulting illuviated horizon of carbonates is designated as Bk horizon (Bca).

Fig. 14. Calcification in lower horizons

2. Decalcification

It is the reverse of calcification i.e., the process of removal of $CaCO_3$ or calcium ions from the soil by leaching.

$$\underset{\text{(insoluble)}}{CaCO_3 + CO_2 + H_2O} \xrightarrow[CO_2]{\text{Temp.}} Ca\,(HCO_3)_2 \text{ (soluble)}$$

3. Podzolization

It is a process of soil formation resulting in the formation of Podzols and Podzolic soils. In many respects, podzolization is opposite of calcification. The calcification process tends to concentrate calcium in the lower part of the B horizon, whereas podzolization leaches the entire solum of calcium carbonates.

Apart from calcium, the other bases are also removed and the whole soil becomes distinctly acidic. In fact, the process is essentially one of acid leaching.

The process operates under favourable combination of the following environments:

(i) **Climate** - A cold and humid climate is most favourable for podzolization.

(ii) **Parent material** - Siliceous (Sandy) material, having poor reserves of weatherable minerals, favours the operation of podzolization as it helps in easy percolation of water.

(iii) **Vegetation** - Acid–producing vegetation such as coniferous pines is essential.

(iv) **Leaching and translocation of sesquioxide** - In the process of decomposition of organic matter various organic acids are produced. These organic acids act with sesquioxide and with the remaining clay minerals and form organic sesquioxide and organic clay complexes, which are water–soluble and move with the percolating water to the lower horizons (Bh, Bs).

Aluminium ions in a water solution hydrolyze and make the soil solution very acidic.

$$2Al + 6H_2O \rightarrow 2Al(OH)_3 + 6H^+$$

As iron and aluminium move about, the A horizon gives a bleached grey or ashy appearance.

The Russians used the term Podzols (*pod* means under, the *zola* means ash-like i.e., ash–like horizon appearing beneath the surface horizon) for such soils. To conclude, the Podzolization is a soil–forming process, which prevails in a cold and humid climate where coniferous and acid–forming vegetations dominate. The humus and sesquioxide become mobile and leach out from the upper horizons and get deposited in the lower horizon.

4. Laterization

The term 'laterite' is derived from the word *later* meaning brick or tile and was originally applied to a group of Indian soils with high clay content found in Malabar hills of Kerala, Tamil Nadu, Karnataka and Maharashtra. It specifically refers to a particular cemented horizon in certain soils, which when dried, become very hard such as a brick. Such soils (in tropics) when massively impregnated with sesquioxides (iron and aluminium oxides) to the extent of 70–80 per cent of the total mass, are called laterites or latosols (Oxisols). The soil forming process is called Laterization or Latozation.

Laterization is the process that removes silica instead of sesquioxides from the upper layers and thereby, leaving sesquioxides to concentrate in the solum. The process operates under the following conditions:

- **Climate** - Unlike podzolization, the process of laterization operates most favourably in warm and humid (tropical) climate with 2,000–2,500 mm rainfall and continuous high temperature (25°C) throughout the year.
- **Natural vegetation** - The rain forests of tropical areas are favourable for the process.
- **Parent Material** - Basic parent materials, having sufficient iron–bearing ferromagnesian minerals (Pyroxene, amphiboles, biotite and chlorite), which on weathering release iron, are congenial for the development of laterites.

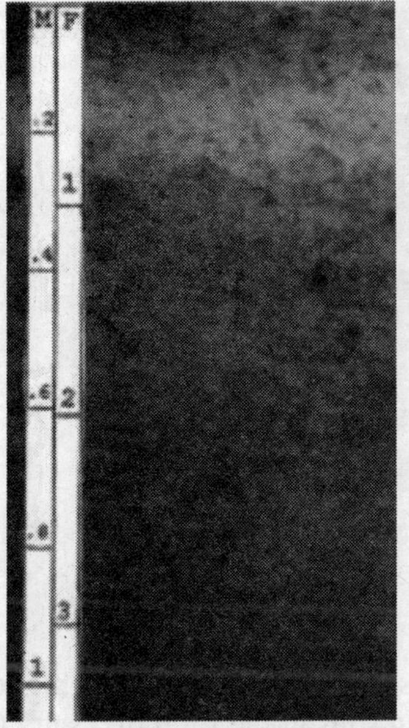

Fig. 15. A soil profile with Laterization

Fig. 16. Bricks cut from Laterite soil

5. Gleization

The term *glei* is of Russian origin meaning blue, grey or green clay. The Gleization is a process of soil formation resulting in the development of a glei (or gley horizon) in the lower part of the soil profile above the parent material due to poor drainage condition (lack of oxygen) and where waterlogged conditions prevail. Such soils are called hydromorphic soils.

The process is not particularly dependent on climate (high rainfall as in humid regions) but often on drainage conditions.

The poor drainage conditions result from:

- Lower topographic position such as depression land, where water stands continuously at or close to the surface

- Impervious soil parent material
- Lack of aeration

Under such conditions, iron compounds are reduced to soluble ferrous forms. The reduction of iron is primarily biological and requires both organic matter and microorganisms capable of respiring anaerobically. The solubility of Ca, Mg, Fe, and Mn is increased and most of the iron exists as Fe^{++} organo–complexes in solution or as mixed precipitate of ferric and ferrous hydroxides. This is responsible for the production of typical bluish to greyish horizon with mottling of yellow or reddish–brown colours.

Fig. 17. A gley horizon

6. Salinization

It is the process of accumulation of salts such as sulphates and chlorides of calcium, magnesium, sodium and potassium in soils in the form of a salty (salic) horizon. It is quite common in arid and semi–arid regions. It may also take place through capillary rise of saline groundwater and by inundation with seawater in marine and coastal soils. Salt accumulation may also result from irrigation or seepage in areas of impeded drainage.

7. Desalinization

It is the removal of excess soluble salts from horizons or soil profile (that contained enough soluble salts to impair the plant growth) through leaching by ponding water and by improving the drainage conditions using artificial drainage network.

8. Solonization or Alkalization

The process involves the accumulation of sodium ions on the exchange complex of the clay, resulting in the formation of sodic soils (Solonetz).

All cations in solution are engaged in a reversible reaction with the exchange sites on the clay and organic matter particles.

The reaction can be represented as:

$Ca.Mg.2NaX \rightarrow Ca^{++} + Mg^{++} + 2Na^{+} + x^{-6} + 3CO_3{}^{2-} \rightarrow Na_2CO_3 + MgCO_3 + CaCO_3$

(Where X represents clay or organic matter exchange sites)

9. Solodization or dealkalization

The process refers to the removal of Na^+ from the exchange sites. This process involves dispersion of clay. Dispersion occurs when Na^+ ions become hydrated.

Much of the dispersion can be eliminated if Ca^{++} and / or Mg^{++} ions are concentrated in the water, which is used to increase the rate of leaching process. These Ca and Mg ions can replace the Na on exchange complex, and the salts of sodium are leached out as:

$$2NaX + CaSO_4 \rightarrow \underset{\text{(leachable)}}{Na_2SO_4} + CaX$$

10. Pedoturbation

Another process that may be operative in soils is pedoturbation. It is the process of mixing of the soil. Mixing to a certain extent takes place in all soils. The most common types of pedoturbation are:

- **Faunal pedoturbation** - It is the mixing of soil by animals such as ants, earthworms, moles, rodents, and man himself
- **Floral pedoturbation** - It is the mixing of soil by plants as in tree tipping that forms pits and mounds
- **Argillic pedoturbation** - It is the mixing of materials in the solum by the churning process caused by swell shrink clays as observed in deep Black Cotton Soils.

5.2. SOIL PROFILE

The vertical section of the soil showing the various layers from the surface to the unaffected parent material is known as a soil profile. The various layers are known as horizons.

5.2.1. Soil Horizons (Layers)

Soil is made up of distinct horizontal layers; these layers are called horizons. They range from rich, organic upper layers (humus and topsoil) to underlying rocky layers (subsoil, regolith and bedrock). The master horizons are O, A, B and C horizons. Most soils have a distinct profile or sequence of horizontal layers. Generally, these horizons result from the processes of chemical weathering, eluviation, illuviation, and organic decomposition. Up to five layers can be present in a typical soil: O, A, B, C, and R horizons.

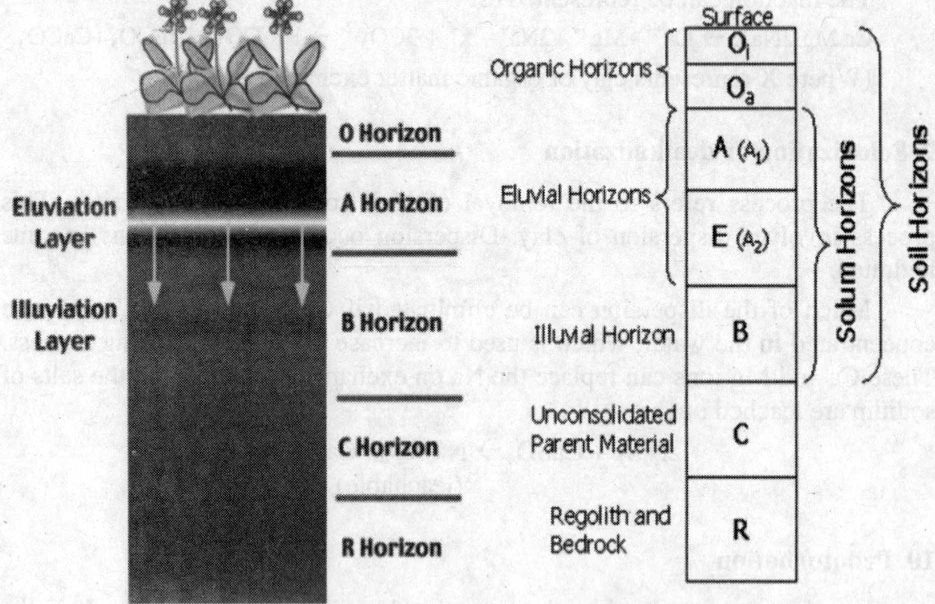

Fig. 18. Typical layers found in a soil profile.

1. O horizon

The O horizon is the topmost layer of most soils. It is mainly composed of plant litter at various levels of decomposition and humus.

2. A horizon

A horizon is found below the O layer. This layer is primarily composed of mineral particles, which has two characteristics - it is the layer in which humus and other organic materials are mixed with mineral particles, and it is a zone of translocation from which eluviation has removed finer particles and soluble substances, both of which may be deposited at a lower layer. Thus, the A horizon is dark in colour and usually light in texture and porous. The A horizon is commonly differentiated into a darker upper horizon or organic accumulation, and a lower horizon showing loss of material by eluviation.

3. B horizon

The B horizon is a mineral soil layer, which is strongly influenced by illuviation. Consequently, this layer reçeives material eluviated from the A horizon. The B horizon also has a higher bulk density than the A horizon due to its enrichment of clay particles. The B horizon may be coloured by oxides of iron and aluminium or by calcium carbonate illuviated from the A horizon.

4. C horizon

The C horizon is composed of weathered parent material. The texture of this material can be quite variable with particles ranging in size from clay to boulders. The C horizon is also not significantly influenced by the pedogenic processes, translocation, and/or organic modification.

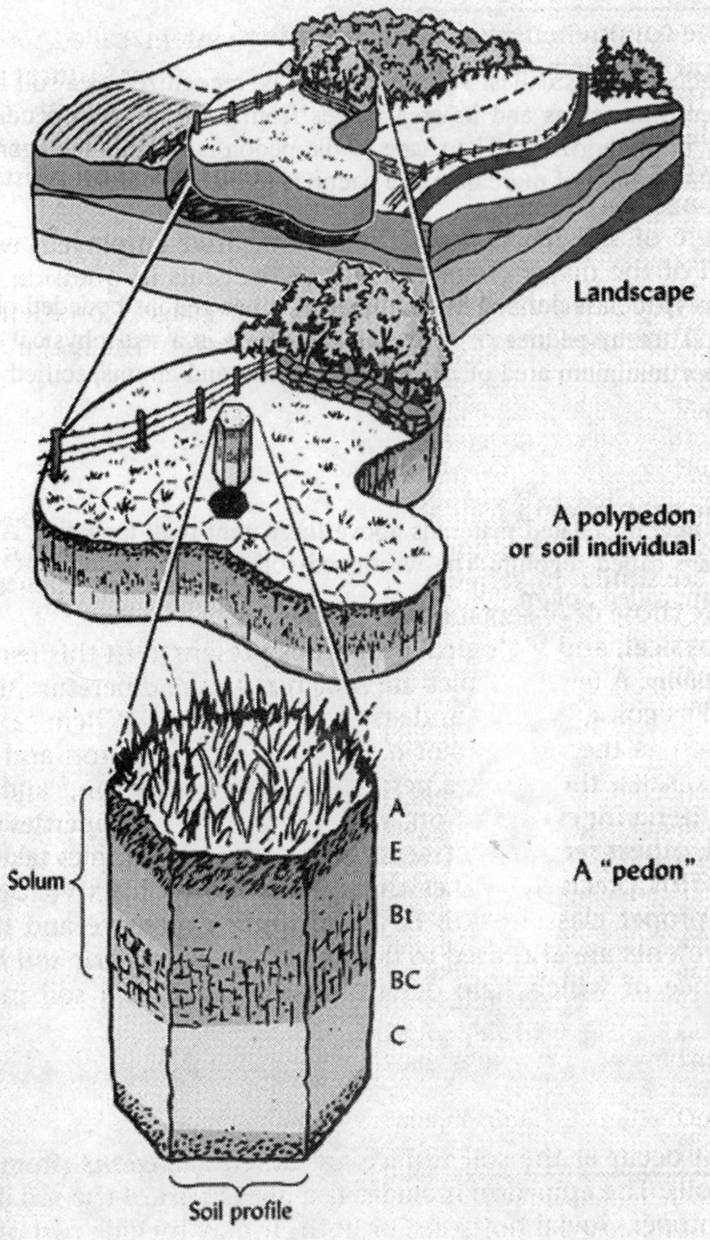

Fig. 19. A landscape showing a polypedon and a pedon

5. R horizon

The final layer in a typical soil profile is called the R horizon. This soil layer simply consists of unweathered bedrock.

Special Features

Pedon

A pedon is the smallest volume that can be recognized as a soil individual. It has three dimensions and its area ranges from 1–10 sq m, depending on the variability in the horizons. The shape of the pedon is roughly hexagonal. A soil volume that consists of more than one pedon is termed a polypedon.

Polypedon

A polypedon is defined as continuous similar pedons bounded on all sides by 'not-soil' or by pedons of unlike characters. It is a real physical soil body, which has a minimum area of more than 1 sq km and an unspecified maximum area.

Regolith

The unconsolidated materials above the parent rock including A, B and C horizons are called regolith. The weathered soil material including O, A and B horizons are called Solum.

6 SOIL CLASSIFICATION

Soils are a series of natural bodies on the Earth's surface and are evolved by the interactions of soil–forming factors and processes. A change in one such factor or process results in a different soil. When classifying the individuals of a large and wide population such as soils, it is necessary to group individuals into classes and further into higher classes. This kind of grouping is called multi–categorical or hierarchical system of classification. The soil individuals are grouped into classes of the lower category by adopting large number of differentiating characteristics. They are further grouped into classes of higher categories by means of a few differentiating characteristics.

6.1. PURPOSE OF CLASSIFICATION

Classification is grouping of objects into compartments in some orderly and logical manner. It is based on the properties of objects for the purpose of studying and identifying them. The properties are selected in accordance with the purpose of the classification. The various purposes of soil classification are:

- To organize our knowledge in such a way that it contributes to economy of thought.
- To remember properties of the objects classified.
- To learn new relationships and principles in the population being classified.
- To establish classes in a manner useful for practical and applied purposes to:
  - o Predict their behaviour
  - o Identify their best uses
  - o Estimate their productivity
  - o Provide objects for research
  - o Extrapolate research findings to other areas

6.2. SYSTEMS OF SOIL CLASSIFICATION

6.2.1. Early Systems of Soil Classification

Early systems of classifications, although having limited application, were quite simple and practical. The aim of this system was primarily utilitarian with little emphasis on the internal fabric and characteristics.

1. Economic classification

It is one of the earliest systems adopted by the revenue department for the purpose of taxation. Soil productivity is considered on the basis of grouping, which is judged with colour, texture and irrigation potentiality. The grouping was done as soils suitable for growing wheat, rice, groundnut etc.

2. Physical classification

It is one of the earliest systems and was based on soil texture. The soils were termed as loamy soils, sandy soils, clayey soils etc.

3. Chemical classification

Here soils were grouped as calcareous soils, gypriferous soils, acidic soils, alkaline soils etc., which are based on some chemical properties associated with soils.

4. Geological classification

This system is based on relationship recognized between soil and its parent rock.

a. Residual or sedentary soils: Developed *in–situ* from the underlying rocks such as basalt and limestone

b. Transported soils: Developed on unconsolidated sediments like alluvium, colluviums and Aeolian materials.

5. Physiographic classification

According to this system, the characteristics of the landscape were considered to classify the soils. e.g., level soils, basin soils, terrace soils, mountain soils, hilly soils, upland soils and low land soils. Classification based on only physiography may have limited scope as two or more soil groups with different properties may be classified in one group.

6. Others

(i) Based on organic matter

- Inorganic or mineral soils
- Organic soils

(ii) Based on soil structure

- Single–grained soils
- Aggregated soils

(iii) Based on moisture or humidity, vegetation and temperature

- Arid soils, humid soils, sub-humid soils, prairie soils, grass-land soils, forest soils etc.

6.2.2. Recent Systems of Soil Classification

1. Dakuchaiev's system (1900)

Soil differences observed on account of climate and vegetation by Dokuchaiev resulted in a classification system called 'Genetic system'. According to this system, soils were classified as:

- Normal (Zonal) soils - These are with fully–developed soil profiles and are in equilibrium with environmental conditions such as climate and vegetation. e.g., Sierozem, Chestnut, Podzols and Laterites.
- Abnormal (Azonal) soils - These are the soils where time has served as a limiting factor to produce horizonation. e.g., Alluvial soils and Regosols.
- Transitional (Intrazonal) soils -These are the soils that occur within the zonal areas and have characteristics that are determined by the local conditions such as topography, parent material etc. e.g., calcimorphic soils and hydromorphic soils.

2. Coffey's system (1912)

He emphasized that soils should be classified on the basis of their own properties. He proposed five major classes of soils under:

- Arid soils
- Dark–coloured prairie soils
- Light–coloured timbered soils
- Black–swamp soils
- Organic soils

Each class was further subdivided into series, on the basis of parent material, and the series were subdivided into types on the basis of surface soil texture.

3. Marbut's morphogenetic system (1935)

He accepted the concept of the Russian 'Soil type' but gave it the name 'Great soil groups'. His scheme of the soil classification was based on the iron alumina and lime content. At the highest category level, he divided zonal soils into two classes as pedalfers and pedocals. Pedalfers show accumulation of Fe and Al while pedocals show accumulation of calcium carbonate. Pedalfers occur in areas of high rainfall having surplus water for leaching. The pedocals occur in areas of high evaporation having real deficit of water. He also emphasized that classification should be based on morphology of soils. The major limitation of this system was that it was based, in part, on assumptions concerning soil genesis. Hence, many series could not find place in the Marbut's system.

4. Baldwin, Kellogg and Thorp (1938)

The morphogenetic system of Marbut was revised and elaborated by these authors. The system marked the start of truly comprehensive approach. The salient features of the system are:

- A return to the zonality concept of the Russian school
- The pedocal – pedalfer concept was de–emphasized
- More emphasis on soil as three–dimentional body and its characteristics

A new category viz., soil family was introduced between 'Great soil group' and 'soil series' but neither the soil families nor the higher categories were defined in terms of soil properties. Here also soil series did fit in any of the existing great soil groups.

5. Thorp and Smith (1949)

They revised the above system in which three new 'Great soil groups' were defined and three others grouped with existing 'Great soil groups'. According to the revised system, the soil was grouped in three orders viz., Zonal, Intrazonal and Azonal following the Russians zonality concept. The three orders were further subdivided into nine sub–orders on the basis of specific climate and vegetation. Each of the sub–orders, in turn, were divided into 'Great soil groups', which were an expression of more specific conditions. The Great groups are further subdivided into numerous soil families, series and soil types.

6.2.3. Limitations of Genetic Systems of Soil Classification

1. The two highest categories are defined in genetic terms and not on the basis of properties of soils.
2. The concepts and definitions of the order in terms of soil properties are not clear.
3. The great soil group concepts and definitions are based on environmental factors than on soil properties. The definitions are comparative and qualitative that led to difficulty in obtaining agreement among different workers.

4. Many of the soil properties are defined in terms of properties that were obvious under virgin conditions.
5. There is a tendency to attempt definition of units in lower categories in terms of too few properties. This poses problem in interpretation.
6. Nomenclature of highest category laid emphasis on colour and vegetation than soil properties.
7. Nomenclature was evolved from several languages and with the mixtures of nouns and adjectives and hence, it was difficult to name intergrades.

6.3. SOIL TAXONOMY

Soil Taxonomy (ST) is a basic system of soil classification in USA for making and interpreting soil surveys (ST USDA 1998). The primary objective of ST is to establish hierarchies of classes that permit understanding the relationship among soils and between soils and the soil–forming factors. A second objective is to provide a common language for soil science. Taxonomy is a system whereby, we name, describe, and categorize soil.

6.3.1. Soil Taxonomy (Or) Seventh Approximation (Or) Modern System of Soil Classification

Salient features

- Unlike the genetic system, the comprehensive system is based on measurable soil properties. Efforts have been made to define all classes in terms of soil properties that exist today.
- It considers all such properties, which affect soil genesis. Hence, soil genesis forms the backbone of the comprehensive system.
- The common definition of a class of taxonomic system is type or orthotype.
- The nomenclature, using coined words, is derived from Greek and Latin languages.
- A new category i.e., sub-group has been introduced to define the central concept of great groups and their intergrades.
- Unlike the genetic system, it is an orderly scheme without prejudices and facilitates easy remembering of the objects.

Structure

The system has six categories of classification. From the highest to the lowest levels of generalization, these are grouped under two broad categories *viz.*, higher and lower.

Higher Categories

1. Order - the highest category in the system
2. Sub–order - comparable to great soil groups of the genetic system
3. Great group - a basic category, based on diagnostic subsurface horizons

Lower categories

1. Sub–group - a new category designed to define the central concepts of great groups
2. Family - a practical category for making predictions and land–use plans
3. Series - the lowest and the most specific category

Application of differentiating criteria used

Order

The names of the orders are coined words and have a common ending, 'sol' (*L. solum*, soil) with the connecting vowel 'o' for Greek (or) 'i' for Latin. A formative element starts with first vowel (O) and ends with the last consonant (l) preceding the connecting vowel (i), e.g. oll from Mollisol.

Sub–order

Each sub–order consists of two syllables. The first is suggestive of a property of the class (i.e., sub–order) and second, the name of the order as reflected by the formative element.

For instance, in the names - Udoll, Xerert, Fluent, first part suggests the *property of the sub–order* (Ud from Udic, Xer from Xeric, flu from fluviatile) and second is the *name of the order* (oll from Mollisol, ert from Vertisol and ent from Entisol).

Great group

The names of great groups are coined by prefixing one additional prefix. For instance, Hapludoll, Chromustert. If a great group is distinguished from others within the same sub–order by a particular diagnostic horizon or property, the name of that property is used as prefix to form other great groups e.g., Argiudoll.

Sub–group

Name of the sub–group consists of the name of the approximate great group modified by one or more adjectives e.g., Typic Hapludoll, Typic Ustifluvent. The adjective *Typic* is used to typify the central concept of the great group.

The taxonomic name of a soil is given below for example. Various categories of soil taxonomy can be identified from this.

The fine silty–mixed calcareous mesic family of *Ustic Torri Orthents*

- *ent* – Entisol - order; *orthent* - Sub–order;
  Torri Orthents - Great group; *Ustic torri orthents* - Sub–group
- Calcareous mesic family of *Ustic Torri Orthents* - Family
- The fine silty–mixed calcareous mesic family of *Ustic Torri Orthents* - Soil texture with geographic name is series

Table 15. The nature and extent of differentiating characteristics used in the six categories

Category	Differentiating characteristics and description
1. Order	There are 12 orders (AVAAMIHOUSEG). Aridisol, Vertisol, Alfisol, Andisol, Mollisol, Inceptisol, Histosol, Oxisol, Ultisol, Spodosol, Entisol and Gelisol These are based on morphology, as produced by soil–forming processes and indicated by the presence or absence of major diagnostic horizons.
2. Sub–order	It emphasizes genetic homogenity, wetness, climatic environment, parent material and vegetation effects.
3. Great group	The major emphasis is on the diagnostic horizons (except in Entisols, which have no such horizon) and presence or absence of diagnostic layers (plinthite, fragipan, duripan), base status, soil temperature, and moisture regimes.
4. Subgroup	The Typic is used to define the central concept of a great group.
5. Family	The soil properties that are most important for plant growth (like texture, solum, mineralogical class, soil temperature class and pH) are used to differentiate families.
6. Series	It is the lowest category in the system. The series is a collection of soil individuals, essentially uniform in differentiating characteristics (like colour, texture, structure, consistency, pH and EC) and in arrangement of horizons. The series arc named after the geographic name of the place where it was first recognized.
7. Soil phase	It is based on differences in surface soil texture, solum thickness, percentage slope, the stoniness, the saltiness, the extent of erosion damage and other conditions.

6.3.2. Criticism and Appreciation of Soil Taxonomy

Criticism

1. The recently–developed system of soil classification viz., soil taxonomy (1975), apparently departs from the genetic approach.

2. The system does not have strong geographic bias towards the four orders, viz., Entisol, Vertisol, Inceptisol and Histosol.
3. The soils with a different genesis but, having identical properties, are classified within the same unit.
4. There is no particular order for the strictly hydromorphic and saline–sodic soils, as is the case in other systems.

Appreciation

1. It is the most elaborate system marked by great care and precision.
2. The primary basis for identifying different classes in the system is properties of soils as they exist in the field.
3. The nomenclature (with Latin and Greek origin) gives a definite connotation of the major soil characteristics.
4. The system has in–built mechanism to permit addition of new soil groups. e.g., the new orders Andisol and Gelisol have been included in the system recently. Other advantages of the system (Smith 1963) are:
   a. It permits classification of soils rather than soil–forming process
   b. It focuses on the soil
   c. It permits the classification of soils of unknown genesis

6.4. SOIL HORIZONS

A soil horizon is a layer that is approximately parallel to the soil surface and has properties that are produced by soil–forming processes. These properties, however, differ from those of the adjacent horizons in respect of characteristics such as colour, texture, structure, consistency etc. The horizons are identified partly by their morphology and partly by the properties of overlying and underlying horizons.

Horizon designation

Once horizons are recognized and demarcated, it becomes necessary to name each horizon. Capital letters, lower case letters, and Arabic numerals are used to designate horizons. The capital letters are used to designate master horizons, e.g., A, B, C. The lower case letters are used as suffixes to indicate specific characteristics of the master horizon e.g., A_p, B_t etc.

p - tillage or other disturbance by cultivation

t - Accumulation of silicate clay

Arabic numerals are used as suffixes to indicate additional vertical subdivisions within master horizon, e.g., B_t1, B_t2, B_t3 etc.

Lastly Arabic numerals are also used as prefixes to indicate discontinuities in a profile e.g., 2C, 2BC etc.

Table 16. Designations of master horizons in the new and old systems

Horizon Designation		Short description
New	*Old*	
O	A1	Organic horizon
A	A1	Mineral horizon
E	A2	Mineral horizon
C	C	Horizons or layers excluding hard bed rock
R	R	Hard bed rock

Diagnostic horizons

By definition, a diagnostic horizon is one, which is formed as a result of pedogenic processes and having distinct properties or features that can be measured in terms of measurable soil properties. The diagnostic horizons are largely used not only for identifying soils but also in classifying soils at various categorical levels, especially Great groups.

A number of diagnostic horizons have been defined in soil taxonomy. Based on location in soil profiles, these horizons have been divided into two categories - surface and sub–surface.

The diagnostic surface horizons are called 'epipedons'. The epipedons are simply the uppermost soil horizons and include the upper part of the soil darkened by organic matter. They are not synonymous with A - horizon.

Seven epipedons, viz., Mollic, Umbric, Anthropic, Ochric, Histic, Plaggen and Melanic are recognized but only three viz., Mollic, Ochric and Umbric are of importance in the soils of India.

The diagnostic sub –surface horizons are called 'endopedons'. The endopedons include the lower part of the soil where soil materials accumulate.

Nineteen endopedons, viz., Argillic, Natric, Agric, Spodic, Sombric, Cambic, Kandic, Oxic, Sulphuric, Salic, Placic, Albic, Glossic, Calcic, Gypsic, Duripan, Fragipan, Petrocalcic, Petrogypsic are recognized. Of these eight, viz., Argillic, Natric, Cambic, Kandic, Oxic, Salic, Calcic and Gypsic are commonly observed in India.

Table 17. Subordinate Distinctions within master horizons

New	Old	Short description
A	-	Highly decomposed organic material
B	b	Buried genetic horizon
C	cn	Concretions or nodules
D	d	Physical root restriction
E	-	Organic material of intermediate decomposition

F	f	Frozen soil
G	g	Strong gleying
H	h	Illuvial accumulation of organic matter
I	i	Slightly decomposed organic material
K	ca	Accumulation of carbonates
M	m	Cementation or induration
N	n	Accumulation of sodium
O	o	Residual accumulation of sesquioxides
P	p	Tillage or other disturbance by cultivation
Q	si	Accumulation of silica
R	r	Weathered or soft bed rock
S	ir	Illuvial accumulation of sesquioxides and organic matter
Ss	ss	Presence of silikensides
T	t	Accumulation of silicate clay
V	-	Plinthite
W	-	Development of colour or structure
X	-	Fragiapn
Y	cs	Accumulation of gypsum
Z	sa	Accumulation of salts

Diagnostic surface horizons (Epipedons)

Mollic epipedon - A thick, dark coloured, mineral horizon with high (>50%) base saturation and strong structure. It contains 1 per cent or more organic matter (when mixed to a depth of 17.5 cm) with colour values darker than 5.5 in dry and 3.5 in moist, and chroma less than 3.5 in moist conditions. Soil structure cannot be both massive and hard, very hard or extremely hard when dry. Base saturation is over 50 per cent and the epipedon is not naturally dry in all parts for more than nine months per year.

Anthropic epipedon - A surface horizon like the Mollic, but formed under long continued system of farming that involves large addition of organic matter (compost), containing more than 250 ppm of citric acid soluble phosphorus.

Umbric epipedon - A surface horizon like the Mollic, but is low (<50%) in base saturation (dominantly saturated with H^+) with high C:N ratio and is not naturally dry for more than three months per year.

Melanic epipedon - A thick black horizon at or near the soil surface, which contains 6 per cent or more organic carbon as weighted average, and/ or 4 per cent or more organic carbon in all the layers, usually associated with short-range-order minerals or alluvium–humus complex. Colour value, moist and chroma of 2 or less throughout and a Melanic index of 1.70 or less throughout.

Histic epipedon - A thin organic horizon (if virgin) or horizon with very high organic matter content depending on clay content, that remains saturated with water for 30 days at some season of the year unless artificially drained. It is thinner than 30 cm, if drained or 45 cm if not drained.

Ochric epipedon - A surface horizon that is light in colour (colour values >5.5 dry or >3.5 moist), contains less than 1 per cent organic matter, or is hard or very hard and massive when dry; or dry for more than three months per year.

Plaggen epipedon - A thick (>50 cm) man–made surface horizon, produced by a long and continued manuring with sod. In addition, the following designations are used for thick, sandy surface horizons.

Grossarenic horizon: A sandy (loamy fine sand or coarser) horizon, 100 cm, or more thick over an argillic horizon.

Diagnostic subsurface horizons (Endopedons)

Argillic horizon - A silicate clay–enriched horizon formed by illuviation of clay. The fine clay is carried downward by percolating water and deposited as clay skins or cutans on ped faces and on the walls of pores. In general, this is a B-horizon that has at least 1.2 times of clay content compared to horizons above or 3% more clay content (if the eluvial layer has <15% clay) or 8% more clay (if eluvial layer has >40% clay). It should be at least 1/10 th as thick as all overlying horizons or more than 15 cm whichever is thinner.

Natric horizon - A high sodium clay–enriched horizon with columnar or prismatic structure. The horizon meets all the requirements of an argillic but, has 15% or more of the exchange complex saturated with Na^+, or has more exchangeable Mg^{2+} plus Na^+ than Ca^{2+} plus exchange acidity at pH 8.2

Agric horizon - An illuvial horizon of clay, silt and humus formed directly under the plough due to long and continued cultivation.

Spodic horizon - A humus and/or sesquioxides–enriched horizon with or without iron. There are many specific limits dealing with AI, Fe, organic matter, and clay ratio, depending on whether the overlying horizon is virgin or cultivated. It is generally formed in cold humid regions.

Sombric horizon - A free-draining horizon located not under an *albic* horizon; has colours (darkness) and base status like an *umbric* epipedon, and has formed due to illuviation of humus and not of aluminium or sodium.

Cambic horizon - A colour or structural B-horizon formed due to alteration by physical movement or by chemical weathering. The pedogenic processes have altered the material enough to form structure (if the texture is suitable) to liberate free iron oxides to form silicate clays or both, and to obliterate most evidences of the original rock structure.

The alteration is not enough to completely destroy volcanic glass, allophane, glass, feldspars, and micas; similarly, illuviation of oxides, humus, or clay should not be enough that will qualify it for argillic or spodic horizon. The horizon is extremely variable in mineralogy because of its pedogenic youthness, occurs under widely differing environments, and may develop in the presence or absence of fluctuating ground water.

Kandic horizon - A horizon of low activity clays with or without clay–skins. It has ≤16 cmol (P^+) kg^{-1} clay CEC at pH 7; and ≤12 cmol (P^+) kg^{-1} clay CEC. It shows a clay content increase at its upper boundary of >1.2 x clay within a vertical distance of <15 cm, that is, abrupt or clear textural boundary.

Oxic horizon - A horizon enriched with Fe and Al–oxides with dominance of 1:1 type clay minerals and from where silica has leached. It is at least 30 cm thick and is sandy loam or finer in texture. It has dominance of low charge 1:1 clays with clay CEC at pH 7 of 16 or less cmol (p+) kg^{-1} and apparent CEC of 12 or less cmol (P+) kg^{-1} clay. The clay content increase is gradual than in *Kandic* horizon. With <10% weatherable minerals in sand fraction.

Sulphuric horizon - A mineral or organic soil horizon with pH of <3.5, is toxic to plant roots, and has yellow mottles of jarosite.

Salic horizon - A horizon with secondary accumulation of water soluble salts (NaCl, Na_2SO_4, etc.) at some depth in the soil profile. At least 15 cm thick with secondary soluble salt enrichment of over 2% and the multiple of thickness and salt is 60% or more.

Albic horizon - A bleached E horizon of podzols and planosols. It has typical colour values of ≥5 (dry) or ≤ 4 (moist).

Glossic horizon - It is a horizon in which albic horizon characteristics gradually intrudes into an argillic, a kandic, or a natric horizon. It may be 5 cm or more thick and consists of an eluvial part, which constitutes 15 to 85% (by volume) of the glossic horizon and an illuvial part (i.e., partially–altered argillic kandic or natric horizon).

Calcic horizon - A horizon with secondary Ca and/or Mg-carbonate enriched materials. It is 15 cm or more thick, has 15% or more of secondary accumulation of carbonate, and contains at least 5% more carbonate than any underlying layer horizon.

Gypsic horizon - It is a calcium and/ or magnesium sulphate-enriched horizon. It is more than 15 cm thick, and contains at least 5% more calcium sulphate than the underlying horizon/layer.

Petrocalcic horizon - An indurated calcic horizon that has hardness of three or more (Moh scale) and whose one-half or more of dry fragments break down in acid, but not in water.

Petrogypsic horizon - A strongly–cemented gypsic horizon whose dry fragments do not slake in water.

Placic horizon - A thin (2 to 10 mm thick), slowly permeable, dark reddish–brown to black–coloured iron or manganese pan that lies within 50 cm of the surface.

Salient characteristics and differentiating properties of different soil orders

In soil taxonomy, there are 12 soil orders including the newly proposed *Gelisol*. A basic system of soil classification for making and interpreting soil survey was published in 1975. The 1975 edition of soil taxonomy recognized 10 soil orders. The 1996 edition incorporated one more soil taxonomic order i.e., *Andisol* that included volcanic ash soils. In 1998, again taxonomy was revised to have 12 soil orders.

The 12th order covered soils that have a dark organic surface layer and mineral layers underlain by permafrost. These soils are commonly found in tundra regions and are termed as *Gelisols*.

Table 18. The salient characteristics of all the 12 orders are given below

Based on	Diagnostic property	Order
Nature of soil material	Very high content of organic matter	Histosols
	Cracking or swell-shrink clays	Vertisols
	Andic material > 35 cm soil	Andisols
	Dark organic surface layer underlain by permafrost	Gelisols
Presence or absence of diagnostic horizons	Spodic horizon	Spodosol
	Oxic horizon	Oxisols
	Argillic horizon but with low base saturation	Ultisols
	Argillic horizon with high base saturation	Alfisols
	Mollic epipedon and high base saturation	Mollisols
	Cambic horizon (or) umbric, plaggen epipedon	Inceptisols
	No distinct horizons	Entisols
Soil moisture regimes	Aridic or dry or saline	Aridisols

6.5. SOIL ORDERS (Simplified Key to Soil Orders)

Table 19. Formative elements of soil orders

Sr. No.	Soil Order	Derivation	Formative element
1	Alfisols	Nonsense symbol	alf
2	Andisols	Jap. *ando*, black soil	and
3	Aridisols	L. *aridus*, dry	id
4	Entisols	Nonsense symbol	ent
5	Gelisols	Gr. *gelid*, very cold	el
6	Histosols	Gr. *histos*, tissue	ist
7	Inceptisols	*L. inceptum*, beginning	ept
8	Mollisols	L. *mollis*, soft	oll
9	Oxisols	Fr. *oxide*, oxide	ox
10	Spodosols	Gr. *Spodos*, wood ash	od
11	Ultisols	*L. ultimus*, last	ult
12	Vertisols	*L. verto*, turn	ert

Table 20. Brief description of Soil Orders

Soil Order	General Features
Alfisols	Alfisols develop in humid and sub–humid climate with average annual precipitation of 500–1,300 mm. They are frequently under forest vegetation. Characteristic features: Clay accumulation in a Bt horizon, thick E horizon, available water for much of the growing season, slightly to moderately acid
Andisols	Andisols are soils with over 60% volcanic ejecta (ash, cinder, pumice, basalt) with bulk densities below 900 kg/m^3. Characteristic features: Dark A horizon, early–stage secondary minerals (allophane, imogolite, ferrihydrite clays), high adsorption and immobilization of phosphorus, very high cation exchange capacity
Aridisols	Aridisols exist in dry climates. Characteristic features: Horizons of lime or gypsum accumulation, salty layers, and/or A and Bt horizons.
Entisols	Entisols have no profile development except a shallow marginal A horizon. Many recent river floodplains, volcanic ash deposits, unconsolidated deposits with horizons eroded away, and sands are Entisols.
Gelisols	Gelisols are soils of very cold climates that contain permafrost within 2 m of the surface.
Histosols	Histosols are organic soils (peat and mucks) consisting of variable depths of accumulated plant remains in bogs, marshes, and swamps.
Inceptisols	Inceptisols, especially in humid regions, have weak to moderated horizon development. Horizon development has been retarded because of cold climate, waterlogged soils, or lack of time for stronger development. Characteristic feature: Texture has to be finer than loamy very fine sand.
Mollisols	Mollisols are frequently under grassland, but with some broadleaf forest–covered soils. Characteristic features: Deep, dark A horizons, they may have B horizons and lime accumulation.
Oxisols	Oxisols are excessively weathered, whereas few original minerals are left unweathered. They develop only in tropical and sub–tropical climates. Characteristic features: Often oxisols are over 3 m deep, have low fertility, have dominantly iron and aluminium clays, and are acidic.
Spodosols	Spodosols are typically the sandy, leached soils of cold coniferous forests. Characteristic features: O horizons, strongly acid profiles, well–leached E horizons, Bh or Bs horizons of accumulated organic material plus iron and aluminium oxides.
Ultisols	Ultisols are extensively–weathered soils of tropical and subtropical climates. Characteristic features: Thick A horizon, clay accumulation in a Bt, strongly acid.
Vertisols	Vertisols exist mostly in temperate to tropical climate with distinct wet and dry seasons. They have a high content of clays that swell when wetted and show cracks when dry. Characteristic feature: Deep self–mixed A horizon, top soil falls into cracks seasonally, gradually mixing the soil to the depth of the cracking.

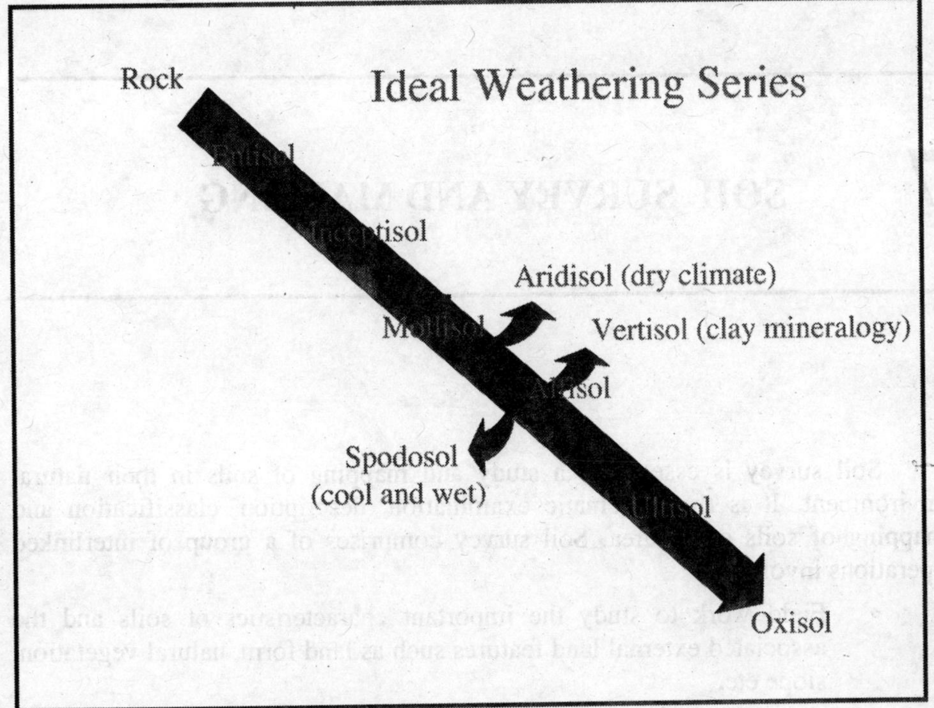

Fig. 20. Ideal Weathering Series.

7 SOIL SURVEY AND MAPPING

Soil survey is essentially a study and mapping of soils in their natural environment. It is the systematic examination, description, classification and mapping of soils of an area. Soil survey comprises of a group of interlinked operations involving:

- Field work to study the important characteristics of soils and the associated external land features such as land form, natural vegetation, slope etc.
- Laboratory analysis to support and supplement the field observations
- Correlation and classification of soils into defined taxonomic units
- Mapping of soils
- Soil survey interpretations i.e., predications about the potential of soils for alternate use and management.
- Transfer of technology from research stations to farmers' fields through soil taxa

7.1. Importance of soil survey

- Soil is the most valuable life supporting natural resource and must be properly used.
- Improper use results in degradation of soil; consequently food security is also questioned.
- Rational planning in land use ensures effective exploitation of soils on its potentialities and limitations.
- It is important to prepare an inventory of this resource so that optimum land use and conservation plans can be developed.
- Such plans will be useful while preparing national policies and their execution.

- It is also useful at other levels viz., State, town, farm, watershed development, Agricultural Research Station, Research Laboratories etc.

7.2. PURPOSES OF SOIL SURVEY

Fundamental

Soil survey helps in expanding our knowledge and understanding of different soils regarding their properties, genesis and classification for sustainable development.

Applied

- Soil survey and soil maps help in making predictions about the behaviour of different soils for agriculture, forestry, engineering, urban development etc.
- Transferring technology like correlating of characteristics of soils of known behaviour and predicting their adaptability to various uses and productivity under defined set of management practices.
- Soil survey provides information needed for developing optimum land–use plans and for bringing new areas under irrigation and drainage networks.
- Soil survey is useful in delineating the degraded soils such as saline–alkali, waterlogged or flood prone, eroded etc. and in suggesting soil and water conservation measures.
- Soil survey is also useful in delineating disease–infested areas and in providing indirect help in controlling the diseases.

Base maps

The base maps commonly used for soil survey are cadastral maps (1:5,000–1:8,000 scale), topographical maps (1:25,000–1:2,50,000 scale), aerial photographs (1:10,000–1:50,000 scale) or Indian remote sensing data (IRS) (1:25,000–1:2,50,000 scale).

Cadestral maps

It is prepared by the Department of Survey and Land Records. It shows the location of individual holdings, ponds, roads and streams. The physiographic features and contour lines are not available.

Aerial photographs

These are photographs of land surface taken from aircraft usually in vertical angle. For interpretation, aerial photos are viewed stereoscopically to give a three–dimensional view. Landforms, vegetation, land use, roads and tracks can be directly seen, while soil properties, geology and other land properties require indirect interpretation.

7.3. SOIL MAPPING UNITS

Soil mapping unit may be defined as the classification of areas having similar soil components. It differs from other mapping units in some of the soil characteristics. The area, which is completely circumscribed by a soil boundary on a map, is called 'soil delineation'. A soil mapping unit consists of the aggregate of all soil delineations, which are identified by a unique symbol, colour or name on a map. A soil mapping unit can be identified as a consociation (an area dominated by a soil of a single taxon such as soil series or a sub–order) or as a group (geographic mixture) of taxa such as association or complexes. A soil series has a much more narrowly–defined set of soil properties than a sub–order.

Depending upon the purpose, the following mapping units are being used in different kinds of soil survey.

7.3.1. Soil Series

It is a mapping unit consisting of a group of soils having similar pattern of pedon or profile characteristics. Basically, the morphological characters are taken into account for characterizing a soil series. The characters considered for characterization of soil series are:

- Arrangement of soil horizon
- Thickness of the individual horizon
- Solum depth
- Soil colour
- Soil structure
- Consistency
- Presence of clay films
- Soil pores
- Root distribution
- Development of surface cracks

By using profile characteristics we can differentiate the soils in the field. The physical and chemical characteristics are not taken for mapping. The environmental characters like slope, topography, rockiness, soil erosion and the surface layers are not taken into account for characterizing the soil unless they influence the arrangement of the soil horizons.

Each soil series is named after the name of a village, town or famous temples in which it is established. A minimum area of 2,000 acres is required to establish a soil series. The soil series is represented on the soil map by a three–letter symbol bearing the name of the soil series.

e.g.,

Plm - Pilamedu
Tyr - Turaiyur
Igr - Irugur
Klt - Kalathur
Mdk - Madukkur
Tlk - Thulakkanur
Vyg - Vylogam
Pvd - Palaviduthi
Klk - Kallakudi
Plr - Puvalur
Pkt - Pattukottai
Omd - Omandur

7.3.2. Soil Association

It is a mapping unit consisting of two or three soil series occurring in a repeated geographical pattern. The boundaries of soil series cannot be demarcated separately due to map scale limitation. The mapping unit proposed under these circumstances is soil association. e.g., Palathurai – Pichanur soil series occurs in association. The major soil series should be written first followed by the minor. Here also three–letter symbols are used.

Pth–Pch (Palathurai–Pichanur)
Pln–Pth–Pch (Pilamedu–Palathurai–Pichanur)

7.3.3. Soil Complex

It is a mapping unit consisting of more than three series occurring together in a repeated geographical pattern. Boundaries cannot be demarcated separately. It is a mapping unit in hilly terrains. It is indicated by Roman letters. The mapping units viz., soil series, soil association and soil complex are used in the Reconnaissance soil survey.

7.3.4. Soil Type

It refers to the surface texture of soil series. It is denoted by the name of the soil series followed by the texture of the surface layer e.g., Plm – Sandy loam.

7.3.5. Soil Phase

This is a mapping unit classified based on the environmental characters of the soil series viz., slope, erosion, stoniness etc.

Type and phase mapping units are used in detailed soil survey.

Miscellaneous land types

The areas, which have little or no soil to support any vegetation, are named as miscellaneous land types. These areas could be used only after proper reclamation. Mostly they are non–soil, however, they are complexes. Following are included as miscellaneous land types:

- Water bodies
- Salt pans
- Sand dunes
- Rubble lands
- Rockout crops
- Ravines

Inclusions

These are the soils not denoted in the soil map. They are so small to be delineated. Normally, a soil which occupies less than 20 per cent of the mapping area belongs to this type.

Variants

These are the soils of an established series differing in some properties of the series. They are indicated as variants of the series in Reconnaissance soil series. This variant will be proposed as a new series during detailed soil survey.

Based on the scale of mapping, the map is called as:

Large scale maps	– 1:5,000 and less
Medium scale maps	– 1:25,000 to 1:40,000
Small scale maps	– 1:50,000 to 1:2,50,000
Very small maps	– 1: million and above

Soil maps generally contain more than one mapping units. These units are arranged in many ways. An organized list of map units is called 'Legend'. There are several kinds of legends. Map units have symbols and names. The nomenclature may differ from one legend to another.

7.4. METHODS OF SOIL SURVEY

Soil surveys are being conducted for diagnosing the nature and extent of distribution of different soils including normal and problem soils. There are two methods of survey:

1. Grid survey
2. Free survey

7.4.1. Grid Survey

The procedure is adopted for mapping small areas such as a micro-watershed area or agriculture research station. The traverse lines are located on a grid pattern and the density of mapping is adjusted according to the area surveyed so that the number of observations per cm^2 of the final map is independent of the scale. Generally four or five observations per ha are recommended. Points of comparable observations are drawn. In this computer age, grid survey at geo reference points (with latitude and longitude) have been found to be of great value in digitizing of the database and generating several thematic maps of practical value. Such surveys are expensive.

7.4.2. Free Survey

In free survey, the surveyor chooses observation points on the assumption that changes in physiographic features observed in aerial photo or by satellite imagery interpretation, and other surface features such as colour, vegetation and land use, are indicative of differences in soil characteristics. The density of observations can be varied as the surveyor concentrates on confirming the inferred boundaries and checking the uniformity of the soil within each boundary. On small scales, the inferred boundaries are often accepted as soil boundaries. On a large scale, however, several new boundaries within the physiographic boundaries are recognized depending upon the scale of mapping.

7.5. TYPES OF SOIL SURVEY

Depending upon the objective, method, type of base map available and intensity of observations following types of soil surveys are recognized. They are:

- Reconnaissance
- Detailed
- Detailed – reconnaissance
- Semi-detailed
- Exploratory
- Rapid reconnaissance

7.5.1. Reconnaissance

The reconnaissance survey is undertaken to prepare resource inventory of large areas. It broadly identifies the kinds of soils and their extent of distribution. It enables to assess broad potentialities of soils, recognize areas of promise that are suitable for intensive agriculture and that are regaining priority treatment for amelioration. The survey relies on aerial photo or remotely–sensed data interpretation to develop soil physiographic relationship. The soil map provides information needed for broad or regional level planning in respect of land use. This kind of mapping precedes detailed soil survey of potential areas.

7.5.2. Detailed

This type of survey is undertaken in priority areas such as pilot projects, agricultural research stations, microwatersheds and areas in urban development. The tranverse lines are on grid pattern. The survey enables identification of soil units up to phases or series for planning the development, and provides sufficient information for interpretation of various kinds of soils and for understanding their pedogenic evolution. Such a survey is very time consuming and expensive. The detailed soil survey is of two types i.e., low and high intensity.

7.5.3. Detailed–reconnaissance

This kind of survey combines both the detailed and the reconnaissance survey, and is undertaken for understanding distribution of basic soil classes of series or families and their phases. The selected areas that have development potentials are surveyed in detail and the others at reconnaissance level.

7.5.4. Semi–detailed

This survey comprises very detailed study of some selected strops or bands cutting across many physiographic units for developing correlation between physiographic units and soils. Once correlation is developed and is found to be valid by random checking, the rest of the area is checked and soil boundaries based on physiographic units are delineated. This kind of survey provides sufficient information about various kinds of soils including problematic or degraded soils.

7.5.5. Exploratory and Rapid Reconnaissance

These lead to preparation of small scale soil maps that are needed for macro level planning for varied agro–based development programmes.

7.6. REMOTE SENSING

Remote sensing is a multi–disciplinary activity, which deals with the inventory, monitoring and assessment of natural resources through the analysis of data obtained by observations from a remote platform.

OR

Remote sensing is broadly defined as the process of collecting information about a subject without being in physical contact with the subject.

The term 'remote sensing' is used more commonly to denote identification of the Earth features by detecting the characteristic electromagnetic radiation that is reflected or emitted by the Earth surface. When electromagnetic energy is incident on any given feature of the Earth surface, fractions of the energy are reflected, absorbed and/or transmitted. It is important to note that the portion of energy reflected, absorbed, and transmitted will vary for different Earth features depending on their material types and condition.

These differences permit us to distinguish between different features of an image. Many remote sensing systems operate in the wavelength regions in which reflected energy predominates and the reflectance properties of the Earth features are very important. The reflectance/emittance of any object at different wavelengths follow a pattern, which is characteristic of that object, and is referred as the 'Spectral signature' of that object.

Drawbacks of traditional methods

Although first generation realistic and factual data can be collected through traditional ground survey methods, these techniques suffer from following drawbacks:

- Data collection cannot be done throughout the year due to unfavourable weather conditions.
- Data collection from inaccessible areas is not possible.
- Data collection through traditional methods is time consuming and there is a wide gap between data collection and its possible utilization.

Advantages of remote sensing

Remote sensing technique provides the synoptic view of large areas.

- The data is collected on a permanent basis.
- The data is an unbiased record of the objects.
- The data is amenable for multi–disciplinary use i.e., the same data can be used for studies in forestry, soil science, hydrology and geology.
- The process of data acquisition and analysis is faster.
- Satellite data is received periodically and it helps in updating the information and monitoring the changes at shorter intervals.
- This has unique capability of recording data in visible as well as invisible parts of electromagnetic spectrum.

7.7. SOIL SURVEY INTERPRETATION

The term 'soil survey interpretation' comprises of two parts, namely soil survey and interpretation. The meaning of the word interpretation is to explain meaning, to show the purpose, or to translate something, which is under consideration, into familiar terms.

Soil survey comprises of collection of wide variety of scientific information about the soils of the area, classification of soils into different categories and showing their distribution on suitable scaled base maps. Any data without interpretation is of limited value to land users, planners, and decision makers.

The most widely–adopted land evaluation methods (Interpretations) are:

- Thematic maps - single theme interpretative

- Land capability classification
- Land irrigability classification
- Land suitability classification
- Productivity classification

7.7.1. Thematic Interpretative Classification

Interpretations of soil depth, soil surface texture, soil drainage, available water capacity, soil erosion, soil gravelliness and soil calcareousness based on site characteristics recorded during soil survey mapping, help in identifying areas that have specific problems or have high potential for agriculture.

1. Soil depth

Depth of soil determines the effective rooting depth for plants, and in soils of the same textural class, the capacity of the soil column to hold water.

Table 21. Soil depth classes

Class	Solum depth (cm)	Description
d_1	0–25	Very shallow
d_2	25–50	Shallow
d_3	50–100	Medium Deep
d_4	100–150	Deep
d_5	150–200	Very deep
d_6	> 200	Extremely deep

2. Surface soil texture

The surface layer of soils to a depth of about 25 cm is the portion of the soil column most used by crop plants. The texture of the surface soil is an important factor in evaluating the suitability of soils for the production of specific crops.

Table 22. Soil textural classes

Class	Surface texture	Description
1.	s/ ls	Sandy
2.	sl/ scl	Loam
3.	sc/ cl/ sicl/ c	Clayey (clay 35–60%)
4.	c/sic	Fine clayey (Clay 60% and above)

3. Soil drainage

Soil drainage refers to the rate of removal of added water. The class of soil drainage is indicated by the morphology of the soil in relation to slope and texture.

Table 23. Soil drainage classes

Class	Description
1.	Very poorly–drained
2.	Poorly–drained
3.	Imperfectly–drained
4.	Moderately well–drained
5.	Well–drained
6.	Somewhat excessively–drained
7.	Excessively–drained

4. Available water capacity (AWC)

Classes of AWC are based on the ability of the soil column to retain water between the tensions of 0.33 and 15 bar in a depth of 100 cm. AWC of the soils composing the mapping unit was estimated from a consideration of texture, mineralogy, depth, and gravel content.

Table 24. Soil AWC classes

Class	Water (mm)	Description
1	< 50	Very low
2	50–100	Low
3	100–150	Medium
4	150–200	High
5	> 200	Very high

5. Soil erosion

Erosion broadly means removal of Earth's surface layer by water and wind. In agriculture, the term soil erosion is commonly used for accelerated erosion caused by disturbance of the landscape, usually by man. The erosion classes showing an estimate of the current erosion status is judged from the field observations.

Table 25. Soil erosion classes

Class	Description
e_1	Slight erosion
e_2	Medium erosion
e_3	Severe erosion
e_4	Very severe erosion

6. Soil gravelliness

Gravel is the name given to soil particles of diameter larger than 2 mm. The materials that constitute gravel are usually rock fragments both fresh and partly weathered. The presence of gravel in the soil hinders plant growth directly by impeding root development and seedling emergence, and indirectly through reducing the volume of soil that can be drawn upon by plants for water and nutrients.

Table 26. Soil gravel classes

Class	% Gravel in profile	Description
1	< 15	Non–gravelly
2	15–35	Gravelly
3	35–80	Skeletal

7. Soil calcareousness

Soil calcareousness is the term used to indicate the content of calcium carbonate in the soil.

Table 27. Soil calcareousness

Class	Description
1	Slight effervescence
2	Medium effervescence
3	Strong effervescence

$CaCO_3$ content is estimated in the field by observation of the degree of effervescence given by the soil when it is moistened with dilute HCl.

7.7.2. Land Capability Classification (LCC)

LCC is an interpretative grouping of soils based on the factors that limit the land use for Agriculture. The factors taken into consideration for LCC are as follows:

Table 28. Factors for LCC

1. Inherent soil characters	2. External land features	3. Environmental factors
a. Depth	Slope	a. Climate
b. Colour	Erosion	b. Vegetation
c. Texture	Stoniness	
d. Structure	Higher water table	
e. Consistency	Hard clay pan	

f. Permeability	Salinity and alkalinity
g. Reaction (pH)	
h. Parent material	

A standard soil survey provides information on the first two items i.e., inherent soil characters and external land features. The environmental factors such as climate and vegetation can be collected from other sources.

Capability system

The capability classification consists of three categories:

- Capability class
- Capability sub–class
- Capability units

Capability class

There are eight capability classes recognized and they are designated by Roman letters I to VIII. The capability classes show the suitability of the soils for cultivation. Classes V, VI, VII are not suitable for cultivation but suitable for pasture, forestry; and class VIII are the lands not suitable for cultivation, pasture, and forestry but suitable for wild life, water supply, or recreational purposes.

Table 29. Capability class

Class I	Soils have few limitations that restrict their use
Class II	Soils have moderate limitations that reduce the choice of crops, require moderate conservation practices
Class III	Soils have severe limitations that reduce the choice of crops, require special conservation practices
Class IV	Soils have very severe limitations that reduce the choice of crops, require very careful management
Class V	Soils not suitable for farming, but highly suitable for grazing
Class VI	Non–arable land, unsuitable for cultivation; well–suited for grazing or woodland, have moderate limitations like steep slope, severe erosion
Class VII	Soils have very severe limitations viz., very steep or very shallow soils, and stony soils having not enough soil moisture. Largely suitable for forestry with limited use for pasture.
Class VIII	Soils and land forms have limitations that prevent their use for commercial plant production like forestry, and soil is restricted to the use of recreation, wild life or to aesthetic purposes.

Table 30. Map legend for land capability

Capability class	Colour
I	Green
II	Yellow
III	Brown
IV	Pink
V	Dark grey
VI	Orange
VII	Red
VIII	Purple

Land capability classification is shown on the map by standard colours fixed for each of the classes.

Capability sub–classes

The capability sub–classes are based on kinds of dominant limitation such as wetness or excess water (w), climate (c), root zone limitation (s), and erosion (e). Those sub–classes are added to capability class as subscripts.

e.g., IIe, IIIw, hence, sub–class indicates both degree and kind of limitations.

Capability unit

These are further sub divisions of capability sub–classes. A capability unit includes soils, which is sufficiently uniform in its characteristics, potentials, and limitations and require fairly uniform management practices.

The soils in a capability unit are sufficiently uniform to:

- Produce similar kinds of cultivated crops with similar management practices
- Require similar conservation treatments
- Have comparable potential productivity

7.7.3. Land Irrigability Classification

The stability of land for irrigation depends upon physical and socio–economic factors in addition to the soil irrigability class. Among the more important considerations, in addition to soil suitability, in deciding suitability of land for irrigation are the following:

Quality and quantity of water

- Equilibrium salinity level
- Equilibrium exchangeable sodium percentage levels

- Availability of water to the land in relation to water requirement of crops

Drainage requirements

- Permeability of substrata and feasibility of providing needed drainage
- Cost of drainage measures

Other economic considerations

- Production costs and yield potential
- Land development costs
- Other factors affecting benefit–cost ratio

Based on the above factors, lands are grouped into six irrigability classes.

Table 31. Land irrigability classes

Class	Description	
Class 1	Lands that have few limitations for sustained use under irrigation. The soils in this class are nearly levelled, have deeper rooting zones, have favourable permeability, texture, available moisture holding capacity.	
Class 2	Lands that have moderate limitations for sustained use under irrigation. Limitations may be single or combinations of factors such as:	
	1. Very gentle slopes	3. Moderate salinity or alkali
	2. Less than ideal depth, texture permeability	4. Somewhat unfavourable topography or drainage
Class 3	Lands that have severe limitations for sustained use under irrigation. Limitations are:	
	1. Gentle slopes	3. Moderately severe salinity or alkalinity
	2. Unfavourable soil depth, texture, permeability	4. Unfavourable topography
Class 4	Lands that are marginal for sustained use under irrigation because of very severe limitations. Limitations are permeability	
	1. Moderately steep slopes	2. Very unfavourable soil depth, texture
	3. Severe salinity / alkali when in equilibrium with irrigation water	4. Very unfavourable topography
Class 5	Lands that are temporarily classed as not suitable for sustained use under irrigation pending further investigations.	
Class 6	Lands not suitable for sustained use under irrigation.	

Land irrigability sub–classes

These are groups of land irrigability units that have same kinds of dominant limitations for sustained use under irrigation when lands are placed in a class lower than 1. The reasons should be indicated whether limitation is due to soil (s), topography (t), or drainage (d). Lands with one or more limitations may also be indicated with relevant letters after the class e.g., class 2 st - It means land irrigability class 2 limitations are soil and topography.

Land irrigability units

Land irrigability units are grouping of lands that are nearly alike in suitability for irrigation and having similar crop adaptation, yield potential, and soil and water management needs. Detailed information regarding crops adopted (a), water requirement (b), method of irrigation (c), cost and benefits of land development (d), and land treatment (t) are to be suggested in formulating irrigation projects.

Irrigability units are designated by letter in brackets appended to the class, subclass, and irrigability unit.

e.g., 2s (a) - It means Class 2, Subclass = limitation in soil (s) Unit (a) = adopted crops

7.7.4. Soil Irrigability Classification

For irrigation projects, special interpretations and classification of soils for sustained use under irrigation are often required. The soils are first grouped into soil irrigability classes, according to their limitations for sustained use under irrigation.

Soil irrigability classes are groups of similar soil types, which have their similar soil limitations for sustained use under irrigation. Soil irrigability classes are established with regards to availability of irrigation water, water quality, land preparation costs, availability of drainage, and other non–soil related factors.

The soil properties used as criteria for classifying soil into irrigability classes in semi-arid and arid regions are:

- Effective soil depth
- Surface soil texture (30 cm)
- Soil permeability
- Available water holding capacity to a depth of 90 cm
- Coarse fragments (> 75 mm diameter)
- Gravel and kankar (25-75 mm diameter)
- Rock out crops
- Salinity of soil
- Salt affected (visual) (percentage of area affected)
- Severity of alkali problem

Criteria for classes are quantitatively defined and soil is classified in one irrigability class. The most limiting property is determined in classification e.g., a soil may have all the properties of the most desirable class except one, but based on this one undesirable property, it is assigned to a lower class.

There are five soil irrigability classes:

Table 32. Soil irrigability classes

Class A	None to slight soil limitation for sustained use under irrigation
Class B	Moderate soil limitations for sustained use under irrigation
Class C	Severe soil limitations for sustained use under irrigation
Class D	Very severe soil limitations for sustained use under irrigation
Class E	Not suited for irrigation (or non–irrigable soil class)

7.7.5. Crop Suitability Classification

Each plant species requires definite soil and site conditions for its optimum growth. Although some plants may be found to grow under different soils and extreme agro–ecological conditions, yet not all the plants can grow in the same soil and under the same environment. The conspicuous absence of *Pinus* species in tropical and of *Eucalyptus* in the temperate (cool/cold) regions may be illustrative.

Since the availability of water and plant nutrients is largely controlled by the physico–chemical and microenvironment of soils, the success and failure of any plant species, in a particular area, is largely determined by these factors.

Several soil–site studies for different plant species have been reported in the literature. These illustrate how soil depth, (sub) soil texture, salinity, and drainage conditions are related to site quality. The objectives of various soil–site suitability evaluation studies have been to predict and classify land for plant growth. Observations on growth–inhibiting factors for certain species and tolerance of others to extremely adverse conditions have been evaluated by many soil scientists.

Suitability criteria

Most of the plant species need well drained, moderately fine to medium textured soil, free of salinity and having optimum physical environment. Once the suitability criteria is established, soil resource maps, based on several parameters, can predict the behaviour and suitability of soils for growing crops and forest or other plantation crops. Several systems of land evaluation have been proposed for use in different regions; the important being that of Storie (1954), and Riquier and coworkers (1970).

The FAO panel for land evaluation (1976) defined the concept of land utilization types and suggested the classification of land for specific use. The classification itself is presented in different categories viz., orders, classes, subclasses, and units.

There are two orders (S, N), which reflect the kind of suitability (S for suitable and N for unsuitable). There are three classes under the order S and two classes under the order N.

Table 33. Suitability orders

S_1	Highly suitable with no limitations
S_2	Moderately suitable with moderate limitations
S_3	Marginally suitable with severe limitations
N_1	Currently not suitable
N_2	Permanently not suitable

The subclasses reflect the kind(s) of limitations or the main kinds of improvement measures required within a class. The following subclasses have been defined:

Table 34. Suitability Subclasses

c	Climate
t	Topography
w	Wetness
n	Salinity / sodicity
f	Soil fertility (limitation not readily correctable)
s	Physical (influencing soil / water relationship and management)

The land suitability unit is used to identify land development unit having minor differences in management requirements. This suggests the relative importance of land improvement works. It is indicated by Arabic numerals, enclosed in parenthesis, following the subclass symbol e.g., Total unit

The whole unit is indicated by a symbol as in figure:

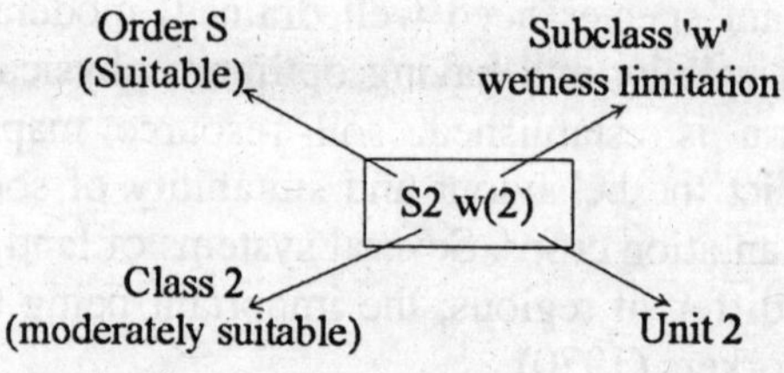

Fig. 21. Symbol of land suitability unit

Sys and Verhese (1975) proposed the following capability index (Ci) based on nine parameters for crop production in the arid and semi–arid regions:

Ci = ABCDEFGHI

Where, A to I represent: soil texture, calcium carbonate, gypsum, salinity, sodium saturation, drainage, soil depth, epipedon and weathering stage / profile development respectively.

Sys (1976) proposed the following scheme for evaluating the degree of limitation ranging from 0 to 4.

Table 35. Degree of limitations

No limitation (0)	The characteristics (quality) are optimal for plant growth
Slight limitation (1)	The characteristics are nearly optimal for land utilization type and affect productivity for not more than 20 per cent to optimal yield (ci - 60–80)
Moderate limitation (2)	The characteristics have moderate influence on crop yield decline, however benefits can still be made and the land use remains economical (ci - 45 to 60)
Severe limitation (3)	The characteristics have such an effect on productivity that the use becomes uneconomical for the considered land utilization type (ci – 30–45)
Very severe limitation (4)	Such limitation does not only decrease the yields below the profitable level, but may inhibit the use of the soil for the considered land utilization type (ci < 30)

The soil site parameters considered for evaluating land for agriculture, forestry and for plantation crops and for defining suitability classes are given in the following Table 36.

Table 36. Soil site parameters considered for evaluation of land

Climate (C)	
Topography and landscape (t)	
Wetness (w) conditions	• Drainage
	• Flooding
Physical conditions of soil	• Texture
	• Gravel/stoniness (Surface and Subsoil)
	• Depth
	• $CaCO_3$
	• Gypsum
Soil fertility (f)	• Organic matter
(not readily correctable)	• Cation exchange capacity (CEC)
	• Base saturation

	• Nutrient availability
Salinity and Alkalinity	• Salinity
	• Ground water depth and its quality
	• Alkalinity / sodicity

7.8. SOIL– PRODUCTIVITY RATING

Soil productivity is the capability of soils to produce a certain amount of crop yield per unit area per annum.

Potential productivity

It is nothing but the productivity of soil when all possible improvements have been made. OR

It is the future productivity of the soil after eliminating all the constraints, which restricts the yield potential.

Determination of productivity index

Productivity is a function of the intrinsic properties of soil; firstly as determined in the process of describing the soil profile *in situ* and secondly by laboratory analysis. Soil moisture and temperature are included for determination; and environmental factors such as slope, vegetation and climate are used to determine what set of management practices are necessary.

From the countless characteristics that influence the soil productivity, following nine factors have been selected:

Moisture (H)	Drainage (D)	Effective depth (P),
Texture / structure (T),	Base saturation (N),	Soluble salt concentration (S),
Organic matter content (O),	Mineral exchange capacity / nature of clay (H)	Mineral reserves (M).

Therefore,

Productivity = H x D x P x T x N (or) S x O x H x M

A soil is considered more fertile if the soil is deep (depth: P), richer in bases (base saturation, N) and contains more water and nutrients, and facilitates better root penetration (Texture and structure, T). The other factors are organic matter, availability of nutrients, greater cation exchange capacity (40 - more nutrients are retained). Greater the mineral reserves, the more will be the nutrient replacement (M). Finally, soil moisture factor (H), water factor (D) also affect productivity.

Each factor is rated on the scale from 0–100. The actual percentage of each factor is multiplied by the percentage of each other. The resultant index of productivity lying between 0% and 100% is set against a scale placing soil in one or other of following five productivity classes.

Table 37. Soil Productivity Classes

Sr. No.	Class	Rating (%)
1.	Excellent	65–100
2.	Good	35–64
3.	Average	20–34
4.	Poor	8–19
5.	Extremely poor	0–7

Riquier index

This is another way of assessing productivity index of a soil, when all possible improvement has been made. It is the future productivity of that soil taking into account of physical and chemical characteristics as modified by conservation practices or improvements.

First, we have to determine which management practices are necessary and secondly assess what will be their impact on potentiality.

1. Each limiting factor requiring soil management

Table 38. Limiting factors that require management

H (dryness)	Requires irrigation
D (Poor drainage)	Requires drainage
P (Shallowness)	Deepening
T (Poor structure or texture)	Improvement of structure / texture
N (Low nutrient content)	Requires fertility management
C (Salinity)	Requires desalting
O (Low organic matter)	Requires application of organic matter

2. Other types of improvement required

J (Wind erosion); K and L (Water erosion); M (Land clearance)

The potentiality index is calculated from the improved characteristics and percentages of increased productivity resulting from management. The soil can be classified against the potentiality scale of 1–5 productivity classes as given earlier.

Coefficient of improvement

It is the ratio of the indices of productivity (P) and potential productivity (P^1).

$$\text{Coefficient of improvement} = \frac{\text{Potential productivity index } (P^1)}{\text{Productivity index } (P)}$$

This ratio reveals the amount of improvement possible in quantitative terms.

Storie Index Rating Or Soil Suitability Rating

Storie index of soil rating is a function of four major factors viz., A,B,C,X

A - Rating of character of physical profile

B - Rating on the basis of surface texture

C - Rating on the basis of slope

X - Rating of conditions other than those in factors A, B and C like drainage nutrient (fertility) levels and erosion etc.

The index rating in per cent has been worked out by taking all the factors in multiplication pattern as under:

$$\text{Soil index rating} = \frac{A}{100} \times \frac{B}{100} \times \frac{C}{100} \times \frac{X}{100} \times 100$$

Soil grading: Six soil grades as per Storie Index Rating.

Table 39. Soil grading

Grade	Description
Grade 1 (excellent)	Soils that rate between 80 per cent and 100 per cent and which are suitable for wide range of crops.
Grade 2 (good)	Soils that rate between 60 per cent and 79 per cent and which are suitable for most crops. Yields are generally good to excellent.
Grade 3 (fair)	Soils that rate between 40 per cent and 59 per cent and which are generally of fair quality with less wide range of suitability than grades 1 and 2. Soils in this grade may give good results with certain specialized crops.
Grade 4 (poor)	Soils that rate between 20 per cent and 39 per cent and which have a narrow range in their agricultural possibilities. e.g., a few soils in this grade may be good for rice, but opt for many other crops.
Grade 5 (very poor)	Soils that rate between 10 per cent and 19 per cent are of very limited use except for pasture, because of adverse conditions such as shallowness, roughness, and alkali content.
Grade 6 (non–agricultural)	Soils that rate less than 10 per cent, which include tidelands, river wash soils of high alkali content and steep broken land.

8 SOILS OF INDIA

8.1. Indian Production System

A Production System is defined by the key elements in its farming systems. These include commodity production trends and key socioeconomic variables. A Production System is also defined by environmental resources, geography, and important issues, or constraints, and opportunities for improving productivity and sustainable agriculture (ICRISAT 1994). Of the 12 ICRISAT (International Crops Research Institute for the Semi-Arid Tropics) Production Systems in Asia, 10 occur in India.

In India, out of the 12 soil orders of soil taxonomy, seven occur in these 10 Production Systems in India. The seven orders and their 14 suborders are shown in Table 40. The Entisols (suborder Orthents) are the most pervasive of all the soils, and occur in all the Production Systems (Table 40). Alfisols (suborder Ustalfs) and Vertisols (suborder Usterts) are found in eight of the 10 Productions Systems but Alfisols occupy a total area of 6,15,016 km^2 and Vertisols 4,70,148 km^2 in all the Production Systems.

Table 40. Soil order and sub order in the production systems of India

Order	Suborder	Order	Suborder
Alfisols	Aqualfs	Inceptisols	Aquepts
	Udalfs		Ochrepts
	Ustalfs		Tropepts
Aridisols	Orthids	Mollisols	Aquolls
Entisols	Aquents	Ultisols	Udults
	Fluvents		
	Orthents		
	Psamments	Vertisols	Usterts

8.2. MAJOR GROUPS OF SOILS

Indian soils have been divided into eight major groups. They cover the largest areas. They are composed of sediments deposited by rivers and the waves. Their chemical composition makes them one of the most fertile in the world. Usually deficient in nitrogen and humus (thus fertilizers are needed), they occupy the plains (from Punjab to Assam) and also occur in the valleys of Narmada and Tapti in Madhya Pradesh and Gujarat, Mahanadi in the Madhya Pradesh and Orissa, Godavari in Andhra Pradesh and Cauvery in Tamil Nadu. They can be divided into Khadar (new) and Bhangar (older, more clayey and kankary) alluvium.

Black soils

It is also called Regur and is ideal for cotton crop. These soils have been formed due to the solidification of lava spread over large areas during volcanic activity in the Deccan Plateau, thousands of years ago. These soils are black due to compounds of iron and aluminium (also because of titaniferous magnetite). Mainly found in Deccan Plateau – Maharashtra, Gujarat, Madhya Pradesh, Karnataka, Andhra Pradesh, and Tamil Nadu. Apart from cotton cultivation, these fertile soils are suitable for growing cereals, oilseeds, citrus fruits and vegetables, tobacco and sugarcane. They have high moisture–retention level, but lack in phosphorus, nitrogen and organic matter.

Red soils

These soils are mainly formed due to the decomposition of ancient crystalline rocks like granites and gneisses, and from rock types rich in minerals such as iron and magnesium. The term 'red soil' is due to the wide diffusion of iron oxides through the materials of the soil. Red soils cover almost the whole of Tamil Nadu, Karnataka, Andhra Pradesh, Maharashtra, Chhattisgarh, parts of Orissa, Jharkhand and Bundelkhand. They are generally deficient in nitrogen, humus and phosphorus, but rich in potash. These soils are suitable for rice, millets, tobacco and vegetables (also groundnuts and potatoes at higher elevations).

Laterite soils

Laterite soils are found in typical monsoon conditions i.e., under conditions of high temperature and heavy rainfall with alternate wet and dry periods. The alterations of wet and dry season leads to the leaching away of siliceous matter and lime of the rocks, and a soil rich in oxides of iron and aluminium compounds is left behind. Found in parts of Western Ghats, Eastern Ghats, Rajmahal hills, Maharashtra, Karnataka, Kerala, Orissa, West Bengal, Assam, Tamil Nadu, etc. These are poor in nitrogen and minerals. Best for tea, coffee, rubber, cinchona, coconut and suitable for rice and millet cultivation, if manured.

Forest and mountain soils

Such soils are mainly found on the hill slopes covered by forests. The formation of these soils is mainly governed by the characteristic deposition of organic matter derived from forest growth. In the Himalayan region, such soils are mainly found in valley basins, depressions and less steeply inclined slopes. Apart from the Himalayan region, the forest soils occur in higher hills in the South and the peninsular region. It is very rich in humus, but deficient in potash, phosphorous and lime, and needs fertilizers. It is suitable for plantation of tea, coffee, spices and tropical fruits.

Arid and desert soils

A large part of the arid and semi–arid region in Rajasthan and adjoining areas of Punjab and Haryana, lying between the Indus and the Aravallis receiving less than 50 cm of annual rainfall, is affected by desert conditions. This area is covered by a mantle of sand, which inhibits soil growth. The phosphate content of these soils is as high as in normal alluvial soils. Nitrogen is originally low but its deficiency is made up to some extent by the availability of nitrogen in the form of nitrates. Thus, the presence of phosphates and nitrates make them fertile soils wherever moisture is available. The changes in the cropping pattern in the Indira Gandhi Canal Command Area are a living example of the utility of the desert soils.

Saline and alkaline soils

In the drier parts of Bihar, Uttar Pradesh, Haryana, Punjab, Rajasthan and Maharashtra, there are salt-impregnated or alkaline soils. They are known by different names such as Reh, kallar, USAR, etc. Some of the salts are transported in solution by the rivers and canals, which percolates in the sub–soils of the plains. The accumulation of salts makes the soil infertile and renders it unfit for agriculture.

Peaty and marshy soils

They have originated in the humid regions as a result of accumulation of large amounts of organic matter in the soil. They contain considerable amounts of soluble salts and 10–40% of organic matter. Peaty soils are found in Kottayam and Alappuzha districts of Kerala, where it is called Kari. Marshy soils, high in vegetable matter, are found in northern Bihar, coastal parts of Orissa, Tamil Nadu and West Bengal, and parts of Uttar Pradesh.

8.3. LAND–USE CLASSIFICATION

India has diverse agro-climatic conditions, topography and soil types; on the basis of which it has been categorized into various regions. Major part of the country is rainfed. Rainfall, therefore, constitutes an important parameter in the classification of the country into various regions for the purpose of planning.

8.3.1. Agro-climatic regions

India has been divided into 15 agro-climatic zones on the basis of climate, in combination with soil and other factors that affect the agriculture in the region. This classification was done in 1979 by the Indian Council of Agricultural Research (ICAR) through the National Agricultural Research Project (NARP). These are:

1. Western Himalayan Region
2. Eastern Himalayan Region
3. Lower Gangetic Plains Region
4. Middle Gangetic Plains Region
5. Upper Gangetic Plains Region
6. Trans Gangetic Plains Region
7. Eastern Plateau and Hill Region
8. Central Plateau and Hill Region
9. Western Plateau and Hill Region
10. Southern Plateau and Hill Region
11. East Coast Plains and Hill Region
12. West Coast Plains and Ghat Region
13. Gujarat Plains and Ghat Region
14. Western Dry Region
15. Island Region

Agro–ecological regions (AERs) and sub–regions (AESRs)

Agro–ecological regions

The country has also been categorized into 20 agro–ecological regions on a 1:4 million scale map, based on physiography, soils, climate, growing period and also available water capacity of the soil, etc. The mapping and classification of the various parts of the country for generation of agro–ecological regions involved the superimposition of four base maps, namely physiography, soils, bioclimate and length of growing period, and these have been used for resource planning at national level.

Agro–ecological sub–regions (AESRs)

The agro–ecological regions were subsequently refined to prepare a 60 Agro–Ecological Sub–Regions (AESRs) Map for regional level planning using the detailed soil information at subgroup level, physiography at land form level, and bioclimate (refined limits of arid, semi–arid and sub–humid bioclimate) types and length of growing period (LGP) at 30 day class interval. The AESR map is useful for regional level planning and resource allocation. The information of length of growing period as well as agro–ecological zoning, supported by moisture availability index, can act as an excellent base for crop modelling and crop suitability evaluation.

8.3.2. Bioclimatic Regions

The agro–ecological regions fall into six major climatic regions as given below:

1. Arid
2. Semi–arid
3. Dry Sub–humid
4. Moist Sub–humid
5. Humid
6. Perhumid

The classification was laid down by the National Bureau of Soil Science and Land Use Planning (NBSSLUP) in the publication "Agro-Ecological Sub-regions of India for Planning and Development" (Velayutham, 1999). It is important to note that the Thornthwaite system of classification is used for agricultural planning in the country considering the annual rainfall, moisture index and other aspects such as the growing period of crops.

Dryland (arid, semi–arid, dry sub–humid) regions of India

According to NBSSLUP, Nagpur, the arid, semi–arid and dry sub–humid regions constituting the 'drylands' (as per Thornthwaite Classification) cover about 228.3 mha (69.6%) of the total land area (328 mha) of the country. A large number of States fall within the drylands. However, almost the entire north–eastern region covering the States of Assam, Meghalaya, Nagaland, Tripura, Manipur, Mizoram, Sikkim and Arunachal Pradesh, and the State of Uttaranchal in North India do not fall within the dryland region. Also, parts of Jammu & Kashmir, parts of Himachal Pradesh, coastal areas of Karnataka, Maharashtra, Goa and major parts of the States of Kerala, Orissa, West Bengal, and the Islands of Andaman & Nicobar and Lakshwadeep and Minicoy, which are predominantly humid or perhumid, do not fall within the dryland regions.

It is, however, important to note that land degradation is a serious problem even in the humid or perhumid regions of the country, particularly in the hilly regions, where the main process of land degradation is water erosion resulting in high losses of top soil and fertility. Major part of the dryland regions in the country are rainfed, while some are irrigated. Therefore, the country's programmes are not targeted to address land degradation in the drylands alone, but also special emphasis is given to rainfed regions.

1. Arid Region

According to NBSSLUP, about 50.8 mha (15.8 per cent) of the country's geographical area is arid (Table 41).

Hot Arid Region - These include the hot arid regions that occupy major parts of Rajasthan (western), Gujarat, southern parts of Punjab and Haryana and a small portion of Deccan Penninsula in the States of Andhra Pradesh, Karnataka and Maharashtra. Roughly, three-fourths of the State of Rajasthan, comprising of 12 western districts falls within the hot arid zone. These are the districts of Barmar, Bikaner, Churu, Sri Ganganagar, Hanumangarh, Jaisalmer, Jalore, Jhunjhunu, Jodhpur, Nagaur, Pali and Sikar (Faroda, 1999). The hot arid region suffers low and erratic rainfall, frequent droughts, high evaporation, intense heat and high winds.

The soils are not conducive to intensive crop production. The nomadic lifestyle has, over the years, being replaced by agrarian in regions not suitable for arable cropping systems, which has further degraded the fragile ecosystem. In arid areas, the growing season being very short, millets and short duration pulses dominate the cropping systems. Livestock farming forms an integral part of this ecosystem. The 'Thar Desert', also known as the Great Indian Desert, lies in western Rajasthan and comprises of an area of 1,96,150 sq km.

Cold Arid Region - An area of about 15.2 mha of cold desert is located in Jammu and Kashmir and the Lahul-Spiti region in Himachal Pradesh.

2. Semi–Arid

About 123.4 mha (37.6%) of the country's geographical area consists of the semi–arid region. (NBSSLUP, 2001). The Semi–Arid Tropical (SAT) areas can be further classified into dry and wet. In SAT areas, the crops and cropping systems are quite diverse depending on the soil type and the length of growing season. Sorghum, cotton, soyabean, groundnut and pulses are the major crops grown in this zone.

3. Dry sub-humid

About 54.1 mha (16.5 per cent) of the country's geographical area falls within the dry sub–humid region. The dry sub-humid region receives fairly high rainfall providing ample opportunities for water harvesting. This can be effectively integrated with the safe disposal of excess runoff to overcome water congestion of soils for crops (other than rice). Rainfed rice is the predominant crop followed by pulses, oilseeds and to some extent, vegetables. Fruit crops, particularly in Orissa, are also an important component of the production system.

8.4. LAND–USE PATTERN IN INDIA

Land–use pattern in India is mainly determined by economic, institutional and physical structure. Land area in India is ample, but it needs proper care to survive and sustain.

Land–use pattern in India relates to the physical characteristics of land, the institutional and other resource frameworks like labour, capital etc. All these aspects are associated with the economic development. India has a total land area of approximately 328 million hectares. Mostly, land utilization statistics are obtainable for almost 93 per cent of the entire area that is around 306 million hectares. It is considerable to note that every forefather over the past 8,000 years or so have been successful in harbouring nearly 140 million hectares of land from the natural ecosystem to agriculture. From the time of independence, people have been successful to add another 22 million hectares. As a result, 162 million hectares of land excels as the net sown area at present. It forms a stupendous percentage of as high as 51%. No other large country is as fortunate as India in this regard.

The following are the nine–fold classification followed in India since independence:

1. Forests
2. Area under non–agricultural uses
3. Barren and uncultivable land
4. Permanent pastures and other grazing land
5. Land under miscellaneous tree crops
6. Culturable waste land
7. Fallow land other than current fallows
8. Current fallows
9. Net area sown

Table 41. Land Use Classification in India since Independence (in million hectares)

Sr. No.	Classification	Land use 2007–2008 (in million hectares)
1	Total geographical area	328.37
2	Reporting area for land use statistics	305.67
3	Forest	69.63 (22.8)
4	Not available for cultivation	43.22
5	(a) Non-Agricultural uses	25.92 (8.5)
6	(b) Barren and uncultivable land	17.29 (5.7)
7	Other uncultivated land excluding fallow land	26.82
8	(a) Permanent pastures and other grazing land	10.39 (3.4)

9	(b) Miscellaneous tree crops and groves	3.31 (1.1)
10	(c) Culturable wasteland	13.12 (4.3)
11	Fallow land	25.15
12	(a) Fallow land other than current fallow	10.34 (3.4)
13	(b) Current fallow	14.81 (4.8)
14	Net area sown	140.86 (46.1)

9 PHASES OF SOILS

Soil is a heterogeneous, multiphase, disperse and porous system. It is the most common media for plant growth.

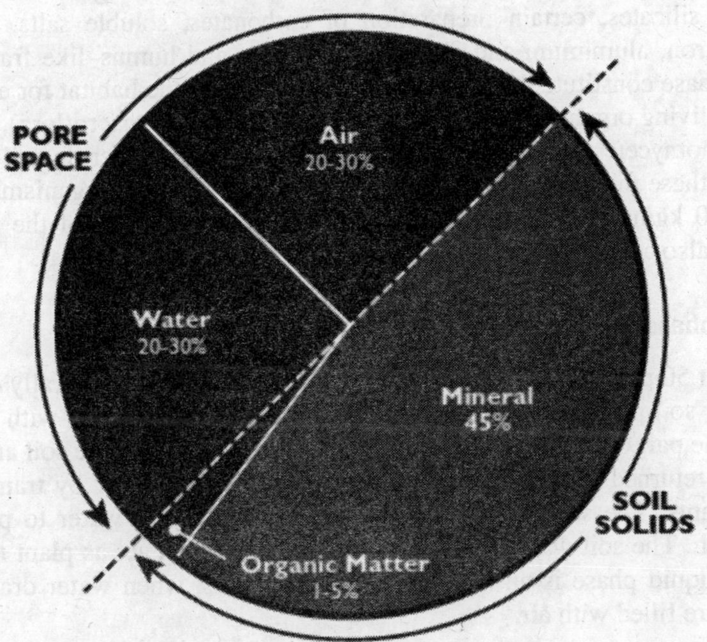

Fig. 22. Phases of soil

9.1. Soil - A three phase system

Minerals soils consist of four major components: Mineral materials, organic matter (OM), water and air in various proportions. Approximately 50 per cent of the total volume of the surface horizon of many soils is made up of inorganic materials (mineral matter) and OM (5 per cent) and the remaining volume is pore space between the soil particles. Water and air occupy these pore spaces in various proportions. The proportion of air and water varies from one season to another. At optimum moisture for plant growth, the 50 per cent of pore space possessed is divided roughly into half i.e., 25 per cent of water space and 25 per cent of air. The soil may be described as the three phase system: Solid, Liquid and Gaseous phase.

1. Solid phase

The soil material less than 2 mm size constitutes the soil sample. It is broadly composed of inorganic and organic constitutes. Soils having more than 20 per cent of organic constitutes are arbitrarily–designated organic soils. Where inorganic constituents dominate, they are called mineral soils. The majority of the soils of India are mineral soils. It accounts for nearly 50 per cent of the total volume, and 95% of the total weight of the solid phase is made up of inorganic or mineral matter. The remaining 5% weight comprises of OM which is mainly derived from dead parts of the vegetation and animals. Inorganic constituents consist of silicates, certain preparation of carbonates, soluble salts, and free oxides of iron, aluminium and silicon. The humus and humus–like fractions of the solid phase constitute the soil organic matter. Soil is the habitat for enormous number of living organisms like roots of higher plants (Soil macroflora), bacteria, fungi, actinomycetes and algae (Soil microflora). A gram of fertile soil contains billions of these microorganisms. The live weight of the microorganisms may be about 4,000 kg/ha and may constitute about 0.01–0.4 per cent of the total soil mass. Soil also consists of protozoa and nematodes (Soil microfauna).

2. Liquid phase

About 50 per cent of the bulk volume of the soil body is generally occupied by voids or soil pores, which may be completely or partially filled with water. A considerable part of the rain, which falls on soil is absorbed by the soil and stored in it to be returned to the atmosphere by direct evaporation or by transpiration through plants. The soil acts as the reservoir for supplying water to plants for their growth. The soil water keeps salts in solution, which act as plant nutrients. Thus, the liquid phase is an aqueous solution of salts, when water drains from soil pores are filled with air.

3. Gaseous phase

The air–filled pores constitute the gaseous phase of soil system and are dependent on the liquid phase. The N and O_2 contents of soil air are almost equal to that of the atmospheric air, but the concentration of CO_2 is much higher (8–10 times more), which may be toxic to plant roots. This phase supplies O_2 and thereby, prevents CO_2 toxicity. The three phases of the soil system have definite roles to play. The solid phase provides mechanical support and nutrients to the plants. The liquid phase supplies water and along with it dissolved nutrients to plant roots. The gaseous phase satisfies the aeration (O_2) need of the plants.

Irrigation and drainage are concerned with maintaining the optimum balance between soil water and soil air as given below:

- Too much water - Shallow root growth; roots may rot; anaerobic reactions produce toxic by–products that reduce growth
- Too little water - Limited nutrient supply retards plant growth; wilting occurs; reduced yields; if conditions are dry enough, plants can die.

Some plants are adapted for growth in extremely wet or in extremely dry conditions. These plants have efficient mechanisms for supplying oxygen or for conserving water.

9.2. PHASE RELATIONSHIPS OF SOILS

The soil is particulate in nature as opposed to the rock that is more or less solid, from which it comes. Any sample of soil obtained in the field has more than just soil particles in it. Every sample of soil has voids between the particles because soil cannot be compacted into a solid mass and still remain soil! The voids between the soil particles can contain only two types of substances beside soil particles: liquids (usually water) and gas (usually air) as illustrated in Fig. 30.

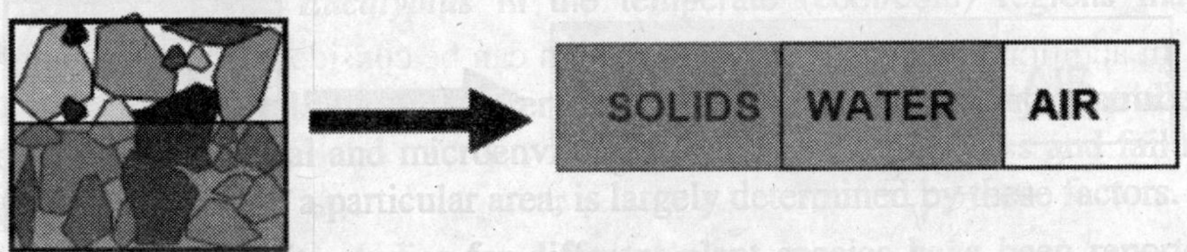

Fig. 23. Phases of Elements in a Soil Sample

If the different elements in the soil sample on the left are rearranged and grouped together, the three different phases of matter in the sample appear as on the right side. This is the basic structure of a Phase Diagram, which is used to express the relationships that exists between the volumes and masses of each element of the soil sample. The volumes of each component of the diagram are shown on the top of the diagram and the masses are shown on the bottom as illustrated in Fig. 31. The volumes of each component of the diagram are shown on the top of the diagram and the masses are shown on the bottom as illustrated in Fig. 31.

Where,

V = total volume of sample | M = total mass of sample

V_s = Volume of solids | M_s = Dry mass or mass of solids

V_w = Volume of water | M_w = Mass of water

V_a = Volume of air | V_v = Volume of voids

The mass of air compared to the other elements is negligible and so is never included. From Fig. 31, it is easy to see the following relations:

$$V = V_s + V_v$$

$$V_v = V_w + V_a$$

$$M = M_s + M_w$$

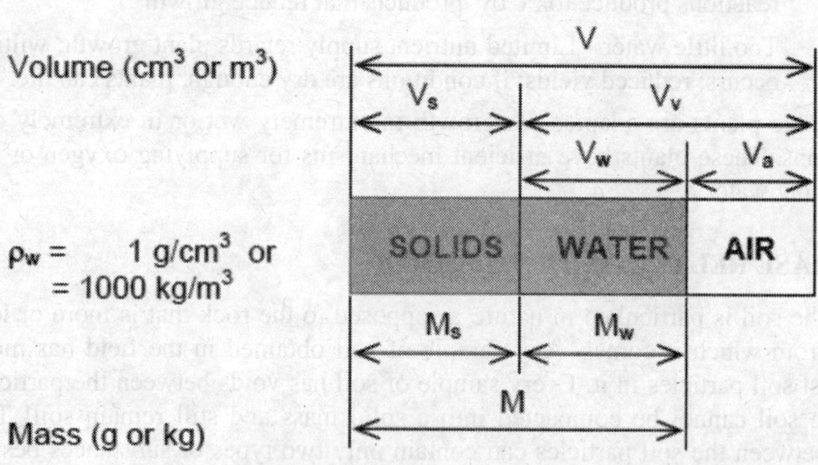

Fig. 24. Phase Diagram of Soil Sample Elements

In addition, the density of water, which can be considered as a constant for most calculations, can also be expressed as:

$$\rho_W = \frac{M_W}{V_W}$$

It is absolutely essential that the units used in the diagram should be expressed explicitly as shown. Masses should be in grams (g) or kilograms (kg). If grams are used for masses then cubic centimetres (cm^3) MUST be used for volumes. If kilograms are used for masses, then volumes MUST be expressed in cubicmetres (m^3). These conditions are non-negotiable!

There are also a number of phase–related properties that can be determined once all volumes and masses on the phase diagram are known:

Density (Also known as wet density, bulk density; g/cm3 or kg/m3)

$$\rho = \frac{M}{V}$$

Dry Density (g/cm^3 or kg/m^3)

$$\rho_d = \frac{M_s}{V}$$

Void Ratio (decimal, no units)

$$e = \frac{V_v}{V_s}$$

Porosity (usually expressed as percentage)

$$n = \frac{V_v}{V}$$

Saturation Rate (usually expressed as percentage)

$$S = \frac{V_w}{V_v}$$

Air Content (usually expressed as percentage)

$$A = \frac{V_a}{V}$$

Water Content (expressed as percentage)

$$W = \frac{M_w}{M_s}$$

Specific Gravity of Soil Solids (dimensionless)

$$G_s = \frac{M_s}{V_s \rho_w}$$

10 PHYSICAL PROPERTIES OF SOIL

Physical properties (mechanical behaviour) of a soil greatly influence its use and behaviour towards plant growth. The plant support, root penetration, drainage, aeration, retention of moisture and plant nutrients are linked with the physical condition of the soil. Physical properties also influence the chemical and biological behaviour of soil. The physical properties of a soil depend on the amount, size, shape, arrangement and mineral composition of its particles. These properties also depend on organic matter content and pore spaces.

10.1. IMPORTANT PHYSICAL PROPERTIES OF SOILS

1. Soil texture
2. Soil structure
3. Surface area
4. Soil density
5. Soil porosity
6. Soil colour
7. Soil consistence

Table 42. Soil properties that change over time

Soil Properties That Change Over Time		
Properties that change over minutes or hours	**Properties that change over months or years**	**Properties that change over hundreds and thousands of years**
Temperature Moisture content Composition of air in soil pores	Soil pH Soil colour Soil structure Soil organic matter content Soil fertility Microorganisms Density	Kinds of minerals Particle size distribution Horizon formation

11 SOIL TEXTURE

11.1. SOIL SEPARATES

Soil is composed of solid particles (minerals and organic matter) and pore space (air and water). The characteristics of the soil are determined by the size, distribution and shape of the solid particles in addition to the size and number of pore spaces. Texture of soil refers to the size of mineral particles, ranging in size from fine to coarse. There are three categories of sizes - clays, silt and sand. The proportion of sand, silt and clay particles in the soil determines whether a soil is classified as sandy, silty or clayey. A balanced soil (loam) contains about 40 per cent sand, 40 per cent silt and 20 per cent clay, and is preferred for growing crops. Other soil textural classes include:

- Sandy loam
- Loam
- Clay loam
- Silty clay loam

Sand particles are the only particles which may be large enough to be seen with the naked eye. Predominantly sandy soil has a gritty feel (coarse texture) when rubbed between the fingers.

Silt particles are smaller than sand particles. Predominantly, silty soils feel powdery (like flour) and do not hold together well when wet, though they are more cohesive than sandy soils.

Clay particles are the smallest soil particles with many small pore spaces. Soils with a high number of clay particles have a very high water–holding capacity and are very fine–textured, it becomes smooth and sticky (like soap) when wet.

To emphasize more on the relative sizes of sand, silt and clay particles, imagine you are holding a clay particle of the size of a grain of salt in your hand. A silt particle next to it would then be the size of a candy and a sand particle would be the size of a stove! (The ratio is about 1:26:1025).

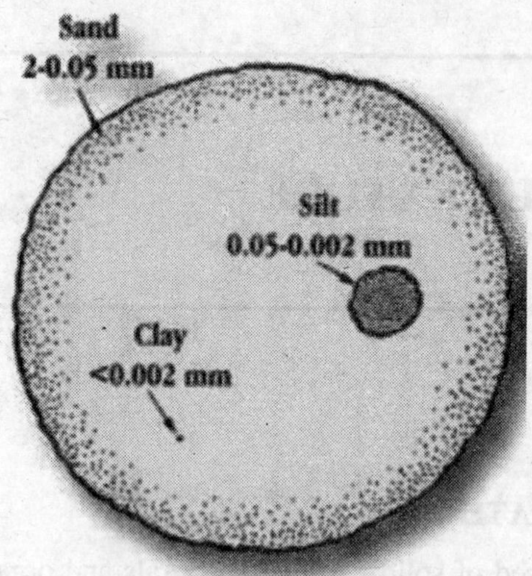

Fig. 25. Sizes of soil particles

Definition - Soil texture refers to the relative proportion of particles or it is the relative percentage by weight of the three soil separates viz., sand, silt and clay or simply refers to the size of soil particles.

The proportion of each size group in a given soil (the texture) cannot be easily altered and it is considered as a basic property of a soil. The soil separates are defined in terms of diameter in millimetres of the particles. Soil particles more than 2 mm in diameter are excluded from soil textural determinations. Stones and gravels may influence the use and management of land because of tillage difficulties but these larger particles make little or no contribution to soil properties such as water holding capacity (WHC) and the capacity to store plant nutrients and their supply.

Gravels -	2 – 4 mm
Pebbles -	4 – 64 mm
Cobbles -	64 – 256 mm
Boulders -	> 256 mm

The particles less than 2 mm are called fine earth, normally considered in chemical and mechanical analysis.

11.2. SOIL TEXTURAL CLASSIFICATION

The components of fine earth are sand, silt and clay (Soil separates). The limits for the sizes of these fractions have been established by various organizations. There are a number of systems of naming soil separates:

- The American System Developed by USDA
- The English System or British System (BSI)
- The International System (ISSS)

- European System
- Indian Standard Classification (I.S.)

Table 43. USDA system

Sr. No.	Soil separates	Diameter (mm)
1	Clay	< 0.002
2	Silt	0.002 – 0.05
3	Very Fine Sand	0.05 – 0.10
4	Fine Sand	0.10 – 0.25
5	Medium Sand	0.25 - 0.50
6	Coarse Sand	0.50 - 1.00
7	Very Coarse Sand	1.00 – 2.00

Table 44. BSI system

Sr. No.	Soil separates	Diameter (mm)
1	Clay	< 0.002
2	Fine silt	0.002–0.01
3	Medium silt	0.01–0.04
4	Coarse silt	0.04–0.06
5	Fine sand	0.06–0.20
6	Medium sand	0.20–1.00
7	Coarse sand	1.00–2.00

Table 45. ISSS system

Sr. No.	Soil separates	Diameter (mm)
1	Clay	< 0.002
2	Silt	0.002–0.02
3	Fine Sand	0.02–0.2
4	Coarse Sand	0.2–2.0

Table 46. European system

Sr. No.	Soil separates	Diameter (mm)
1	Fine clay	< 0.0002
2	Medium clay	0.0002–0.0006
3	Coarse clay	0.0006–0.002
4	Fine silt	0.002–0.006
5	Medium silt	0.006–0.02

6	Coarse silt	0.02–0.06
7	Fine sand	0.06–0.20
8	Medium sand	0.20–0.60
9	Coarse sand	0. 60–2.00

As per I.S. Classification (IS: 1498–1970), the soil is divided into six groups as follows:

Table 47. Indian standard classification

Sr. No.	Soil separates	Diameter (mm)
1	Clay (size)	< 0.002 (2 μm)
2	Silt (size)	0.002–0.075
3	Sand	0.075– 4.75
4	Gravel	4.75–80
5	Cobble	80–300
6	Boulders	>300

Sand

- Usually consists of quartz but may also contain fragments of feldspar, mica and occasionally heavy minerals viz., zircon, tourmaline and hornblende
- Has uniform dimensions
- Can be represented as spherical
- Not necessarily smooth and has jagged surface

Silt

- Particle size - intermediate between sand and clay
- Since the size is smaller, the surface area is more
- Coated with clay
- To a limited extent, has the physico–chemical properties as that of clay
- Sand and silt forms the skeleton

Clay

- Particle size - less than 0.002 mm
- Plate–like or needle–like shape
- Belong to alumina silicate group of minerals

- Sometimes considerable concentration of fine particles, which does not belong to alumino silicates, e.g., iron oxide and $CaCO_3$
- These are secondary minerals derived from primary minerals in the rock.
- Flesh of the soil

Knowledge of soil texture is important

It is a guide to the value of the land. Land use capability and methods of soil management depend on soil texture.

11.3. DETERMINATION OF PARTICLE SIZE DISTRIBUTION

The determination of relative distribution of the ultimate or individual soil particles below 2 mm diameter is called as Particle size analysis or Mechanical analysis.

Two steps are involved:

1. Separation of all the particles from each other i.e., complete dispersion into ultimate particles
2. Measuring the amount of each group

Table 48. Separation methods

S.N.	Aggregating agents	Dispersion method
1	Lime and Oxides of Fe and Al	Dissolving in HCl
2	Organic matter	Oxidizes with H_2O_2
3	High concentration of electr-olytes (soluble salts)	Precipitate and decant or filter with suction
4	Surface tension	Elimination of air by stirring with water or boiling

After removing the cementing agents, disperse the soil particles by adding NaOH.

Measurement

Measurement can be done once the soil particles are dispersed into ultimate particles.

1. Coarser fractions

By sieving - Sieves are used in the mechanical analysis correspond to the desired particle size separation. Sieves with circular holes are used for 2 mm, 1 mm and 0.5 mm sizes. For smaller sizes, wire mesh screens are used (screening).

2. Finer fractions

By settling in a medium - The settling or the velocity of the fall of particles is influenced by viscosity of the medium and difference in density between the medium and falling particles.

Size and shape of object

Stokes' law

Particle size analysis is based on a simple principle i.e., when soil particles are suspended in water, they tend to sink. Their velocity of settling (V) is proportional to the square of the radius (r) of each particle because there is little variation in the density of most soil particles.

Thus, $V = kr^2$, where k is a constant

This equation is referred to as Stokes' law. Stokes (1851) was the first to suggest the relationship between the radius of the particles and its rate of fall in a liquid. He stated that the velocity of a falling particle is proportional to the square of the radius and not to its surface. The relation between the diameter of a particle and its settling velocity is governed by Stokes' law:

$$V = \frac{2}{9} gr^2 \frac{(ds - dw)}{n};$$

Where,

V = Velocity of settling particle (cm/sec)

g = Acceleration due to gravity cm/ sec^2 (981)

ds = Density of soil particle (2.65)

dw = Density of water (1)

n = Coefficient of viscosity of water (0.0015 at 4°C)

r = Radius of spherical particles (cm).

Assumptions and limitations of Stokes' law

Particles are rigid and spherical / smooth. This requirement is very difficult to fulfil, because the particles are not spherical and irregularly–shaped such as plate and other shapes.

The particles are large in comparison with the molecules of the liquid so that in comparison with the particle the medium can be considered as homogenous. The particles must be big enough to avoid Brownian movement. The particles less than 0.0002 mm exhibit this movement so that the rate of falling is varied.

The fall of the particles is not hindered or affected by the proximity (very near) of the wall of the vessel or of the adjacent particles. Many fast falling particles may drag finer particles down along with them.

The density of the particles and water as well as the viscosity of the medium remains constant. But this is usually not so because of their different chemical and mineralogical composition.

The suspension must be still. Any movement in the suspension can alter the velocity of fall and such movement is brought by the sedimentation of larger particles (> 0.08 mm). They settle fast and create turbulence in the medium. The temperature should be kept constant so that convection currents are not set up.

11.4. METHODS OF SOIL TEXTURAL DETERMINATION

Numerous methods for lab and field–use have been developed:

- Elutriation method - Water and air
- Pipette method
- Decantation/beaker method
- Test tube shaking method
- Feel method – Applicable to the field – quick method – by feeling the soil between thumb and fingers

Feel method

The texture is evaluated by attempting to squeeze the moistened soil into a thin ribbon as it is pressed with rolling motion between thumb and index finger or alternately to roll the soil into a thin wire.

Four aspects to be seen – i) Feel by fingers, ii) Ball formation, iii) Stickiness, iv) Ribbon formation

Soil textural classes

To convey an idea of the textural make up of soils and to give an indication of their physical properties, the names of soil textural class are used. These are grouped into three main fractions viz., sand, silt and clay. According to the proportion of these three fractions, a soil is given a name to indicate its textural composition. Such a name gives an idea not only of the textural composition of a soil but also of its various properties in general.

On this basis, soils are classified into various textural classes such as sands, clays, silts, loams etc.

Sands - The sand group includes all soils in which the sand separates make up at least 70 per cent and the clay separate 15 per cent or less of the material by weight. Therefore, the properties of such soils are characteristically those of sand in contrast to the stickier nature of clays. Two specific textural classes are recognized in this group - sandy and loamy sand, although in practice two subclasses are also used i.e., loamy fine sand and loamy very fine sand.

Silt - The silt group includes soils with at least 80 per cent silt and 12 per cent or less clay. Naturally, the properties of this group are dominated by those of silt. Only one textural class is included in this group i.e., silt.

Clays - To be designated as clay, a soil must contain at least 35 per cent of the clay separate and in most cases not less than 40 per cent. In such soils, the characteristics of the clay separates are distinctly dominant, and the class names are clay, sandy clay and silty clay. Sandy clays may contain more sand than clay. Similarly, the silt content of silty clays usually exceeds the clay fraction.

Loams - The loam group, which contains many sub–divisions, is a more complicated soil textural class. An ideal loam may be defined as a mixture of sand, silt and clay particles that exhibits the properties of those separates in about equal proportions. Loam soils do not exhibit dominant physical properties of sand, silt or clay. Loam soil does not contain equal percentage of sand, silt and clay, however, exhibit approximately equal properties of sand, silt and clay.

Determination of textural class

A system for determining the soil textural class known as the "American system" has been developed by the United States Department of Agriculture. This system proposes twelve textural classes.

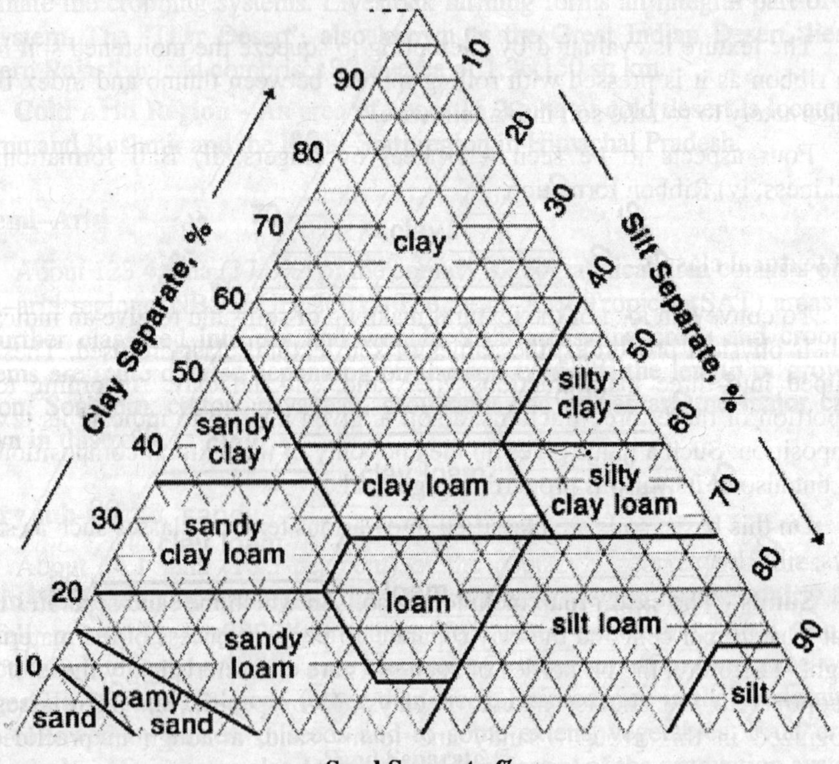

Fig. 26. Soil textural triangle

The textural triangle

It is used to determine the soil textural name after the percentages of sand, silt, and clay are determined from a laboratory analysis. Since the soil's textural classification includes only mineral particles and those of less than 2 mm diameter, the sand plus silt plus clay percentages equal 100 per cent. (Note that organic matter is not included). The amount of any two fractions automatically gives the percentage of the third one.

While using the diagram, first find out the percentage of clay and project inward parallel to sand line. Similarly, find out the silt percentage and project inward parallel to clay line, and for sand project inward parallel to silt line. The point at which the projections intersect, gives the class name.

Sometimes, the intersecting point exactly falls on the line between the textural classes. In such condition, it is customary to use the name of the finer fraction, e.g., soil containing 40 per cent clay, 30 per cent sand and 30 per cent silt is called as clay rather than clay loam.

- Soils with more than 30 per cent clay are clays
- Soils with 20–30 per cent clay are clay loams
- Soils with less than 20 per cent clay are loams unless they have more than 80 per cent sand
- Soils with more than 80 per cent sand are sands
- Soils with more than 50 per cent sand are sandy
- Soils with more than 50 per cent silt are silty

Importance of soil texture

Presence of each type of soil particles makes its contribution to the nature and properties of soil as a whole. Texture has good effect on management and productivity of soil.

1. Sandy soils

- Sandy soils are of open character, usually loose and friable.
- Such type of the texture is easy to handle in tillage operations.
- Sand facilitates drainage and aeration. It allows rapid evaporation and percolation.
- Sandy soils have very little water holding capacity. Such soils cannot stand drought and are unsuitable for dry farming.
- Sandy soils are poor storehouse of plant nutrients.
- In sandy soils, organic matter content is low and leaching of applied nutrients is very high.

- In sandy soils, only a few crops can be grown such as potato, groundnut and cucumbers.

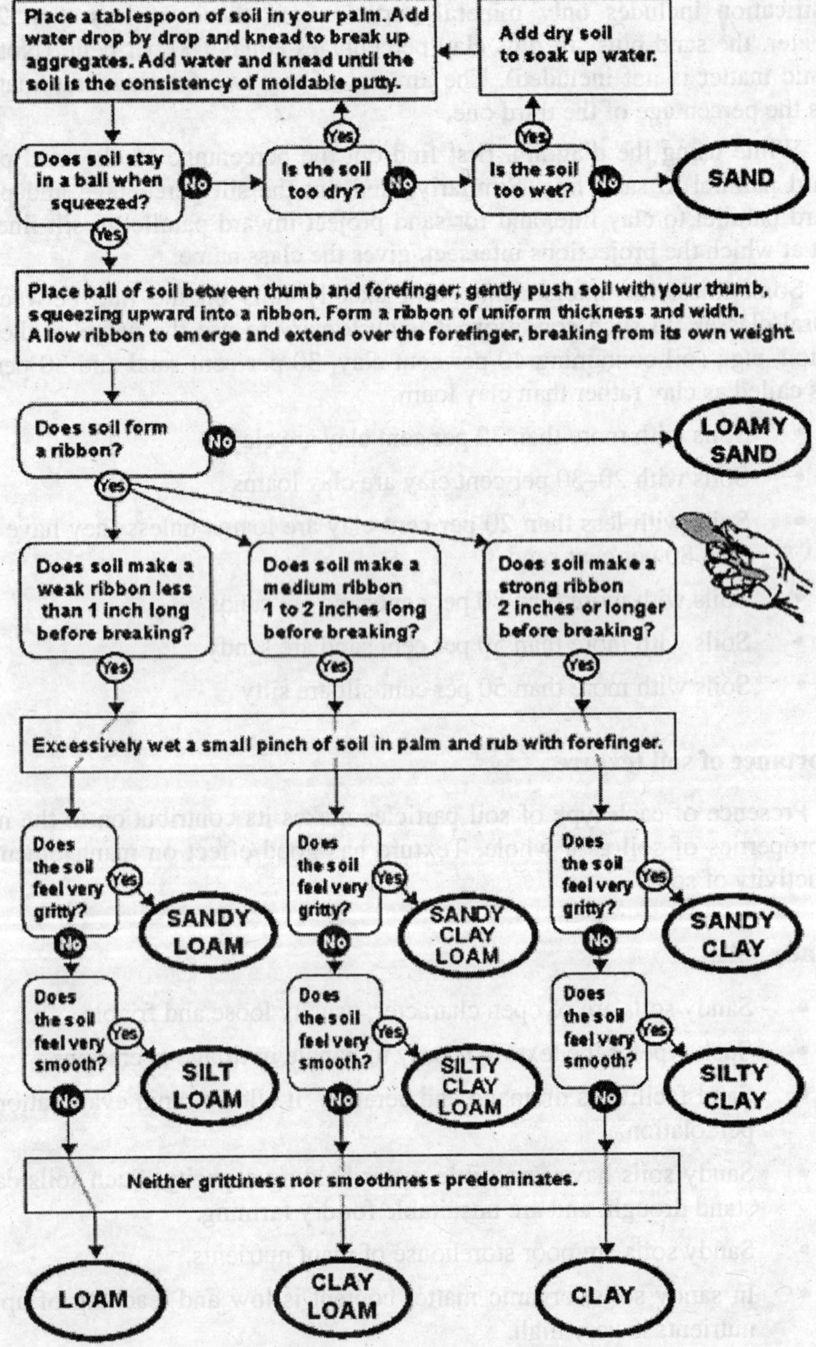

Fig. 27. Determining soil texture by the feel method

2. Clayey soils

- Clay particles play a very important role in soil fertility.
- Clayey soils are difficult to till and require skill in handling. If moist clayey soils are exceedingly sticky, after drying they become very hard and difficult to break.
- They have fine pores, and are poor in drainage and aeration.
- They have a high water–holding capacity and poor percolation, which usually results in waterlogging.
- They are generally very fertile soils, in respect of plant nutrient content. Rice, jute, sugarcane can be grown very successfully in these soils.
- Loam and silt loam soils are highly desirable for cultivation.
- Generally, the best agriculture soils are those, which contain 10–20 per cent clay, 5–10 per cent organic matter and the rest equally shared by silt and sand.

12 SOIL STRUCTURE

Soil conditions and characteristics such as water movement, heat transfer, aeration and porosity are much influenced by soil structure. In fact, the important physical changes imposed by the farmer in ploughing, cultivating, draining, liming, and manuring his land are structural rather than textural.

Definition

The arrangement and organization of primary and secondary particles in a soil mass is known as soil structure. Soil structure controls the amount of water and air present in soil. Plant roots and germinating seeds require sufficient air and oxygen for respiration. Bacterial activities also depend upon the supply of water and air in the soil.

12.1. FORMATION OF SOIL STRUCTURE

Soil particles may be present either as single individual grains or as aggregate i.e., group of particles bound together into granules or compound particles. These granules or compound particles are known as secondary particles. A majority of particles in a sandy or silty soil are present as single individual grains while in clayey soil they are present in granulated condition. The individual particles are usually solid, while the aggregates are not solid but they possess a porous or spongy character. Most soils are mixture of single grain and compound particle. Soils, which predominate with single grains, are said to be structureless, while those possessing majority of secondary particles are said to be aggregate, granulated or crumb structured.

Mechanism of aggregate formation

The bonding of the soil particles into a structural unit is the genesis of soil structure. The bonding between individual particles in the structural units is generally considered to be stronger than the bonding between the structural units themselves.

In aggregate formation, a number of primary particles such as sand, silt and clay are brought together by the cementing or binding effect of soil colloids. The cementing materials taking part in aggregate formation are colloidal clay, iron and aluminium hydroxides and decomposing organic matter. Whatever may be the cementing material, it is ultimately the dehydration of colloidal matter accompanied with pressure that completes the process of aggregation.

Colloidal clay

By virtue of high surface area and surface charge, clay particles play a key role in the formation of soil aggregates. Sand and silt particles cannot form aggregates as they do not possess the power of adhesion and cohesion. These particles usually carry a coating of clay particles and are enmeshed in the aggregates formed by the adhering clay particles. Colloidal particles form aggregates only when they are flocculated. There is a vast difference between flocculation and aggregation.

Flocculation is brought about by coalescence of colloidal particles and is the first step in aggregation.

Aggregation is something more than flocculation involving a combination of different factors such as hydration, pressure, dehydration etc. Aggregation requires cementation of flocculated particles. The cementation may be caused by cations, oxides of Fe and Al, humus substances and products of microbial excretion and synthesis. Clay particles form aggregates only if they are wetted by a liquid such as water whose molecules possess an appreciable dipole moment.

$$\text{Clay}^{-} - {}^{+}\text{Water}^{-} - {}^{+}\text{Cation}^{+} - {}^{-}\text{Clay}^{-} - {}^{+}\text{Water}^{-} - {}^{+}\text{Cation}^{+} - {}^{-}\text{Clay}^{-}$$

The aggregation also depends upon the nature of clay particles, size and amount of clay particles, dehydration of clay particles, cations such as calcium and anions such as phosphate.

Fe and Al oxides - The colloidal Fe oxides act as cementing agent in aggregation. Al oxides bind the sand and silt particles. These act in two ways. A part of the hydroxides acts as a flocculating agent and the rest as a cementing agent.

Organic matter - It also plays an important role in forming soil aggregates.

- During decomposition, cellulosic substances produce a sticky material very much resembling mucus or mucilage. The sticky property may be due to the presence of humic or humic acid or related compounds produced.
- Certain polysaccharides formed during decomposition.
- Some fungi and bacteria have cementing effect probably due to the presence of slimes and gums on the surface of the living organisms produced as a result of the microbial activity

12.2. CLASSIFICATION

The primary particles, sand, silt and clay, are usually grouped together in the form of aggregates. Natural aggregates are called peds, whereas clod is an artificially–formed soil mass. Structure is studied in the field under natural conditions and it is described under three categories:

1. Type - Shape or form and arrangement pattern of peds
2. Class - Size of peds
3. Grade - Degree of distinctness of peds

12.3. TYPES OF STRUCTURE

There are four principal forms of soil structure:

1. Plate–like (Platy) - In this type, the aggregates are arranged in relatively thin horizontal plates or leaflets. The horizontal axis or dimensions are larger than the vertical axis. When the units or layers are thick, they are called platy. When they are thin, they are laminar. Platy structure is most noticeable in the surface layers of virgin soils but may be present in the sub–soil. This type is inherited from the parent material, especially by the action of water or ice.

2. Prism–like - The vertical axis is more developed than horizontal, giving a pillar–like shape. It varies in length from 1–10 cm and commonly occur in sub–soil horizons of Arid and Semi–arid regions. When the tops are rounded, the structure is termed as columnar and when the tops are flat or plane, levelled and clear cut, it is termed as prismatic.

3. Block–like - In this type, all the three dimensions are of same size. The aggregates have been reduced to blocks. The soil blocks are irregularly six faced with their three dimensions more or less equal. When the faces are flat and distinct, and the edges are sharp angular, the structure is named as angular blocky. When the faces and edges are rounded, it is called sub– angular blocky. Usually, these types are confined to the sub–soil and characteristics have much to do with soil drainage, aeration and root penetration.

4. Spheroidal (Sphere–like) - All rounded aggregates (peds) may be placed in this category. Not exceeding an inch in diameter, usually these rounded complexes are loosely arranged and readily separated. When wetted, the intervening spaces generally are not closed so readily by swelling as may be the case with a blocky structural condition. Therefore, in sphere–like structure, infiltration, percolation and aeration are not affected by wetting of soil. The aggregates of this group are usually termed as granular, which are relatively less porous. When the granules are very porous, it is termed as crumb. This is specific to surface soil particularly high in organic matter or to grass land soils.

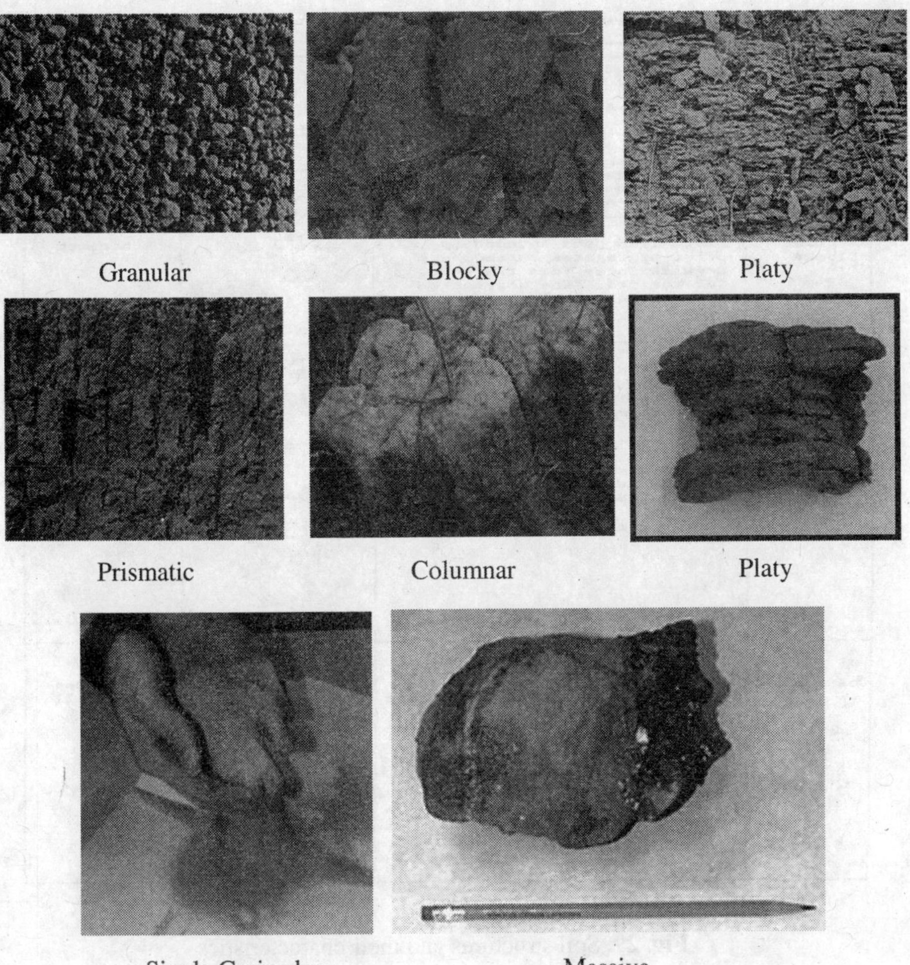

Fig. 28. Different soil structures

12.4. CLASSES OF STRUCTURE

Each primary structural type of soil is differentiated into five classes depending upon the size of the individual ped.

The terms commonly used for the size classes are:

1. Very fine or very thin
2. Fine or thin
3. Medium
4. Coarse or thick
5. Very coarse or very thick

The terms thin and thick are used for platy types, while the terms fine and coarse are used for other structural types.

Structure type	Aggregate description	Diagrammatic aggregate	Usual location
Granular	Relatively nonporous, small and spheroidal peds; not fitted to adjoining aggregates.		A horizon
Platy	Aggregates are platelike. Plates often overlap and impair permeability.		E horizon or in compacted soils
Angular Blocky	Block-like peds bounded by other aggregates whose sharp angular faces form the cast for the ped. The aggregates often break into smaller blocky peds.	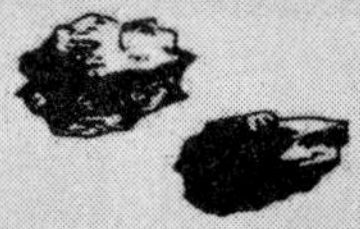	B horizon
Subangular Blocky	Block-like peds bounded by other aggregates whose rounded subangular faces form the cast for the ped.		B horizon
Prismatic	Column-like peds without rounded caps. Other prismatic aggregates form the cast for the ped. Some prismatic aggregates break into smaller blocky peds.		B horizon
Columnar	Column-like peds with rounded caps bounded laterally by other columnar aggregates which form the cast for the peds.		B horizon in alkali soils

Fig. 29. Soil structures and their characteristics

12.5. GRADES OF STRUCTURE

Grades indicate the degree of distinctness of the individual ped. It is determined by the stability of the aggregates. Grade of structure is influenced by the moisture content of the soil. Grade also depends on organic matter, texture etc. Four terms commonly used to describe the grade of soil structure are:

1. **Structureless** - There is no noticeable aggregation such as conditions exhibited by loose sand.
2. **Weak Structure** - Poorly formed, indistinct formation of peds, which are not durable and much unaggregated material.
3. **Moderate structure** - Moderately well–developed peds, which are fairly durable and distinct.
4. **Strong structure** - Very well–formed peds, which are quite durable and distinct.

Structure naming - For naming a soil structure the sequence followed is grade, class and type, e.g., strong coarse angular blocky, moderate thin platy, weak fine prismatic.

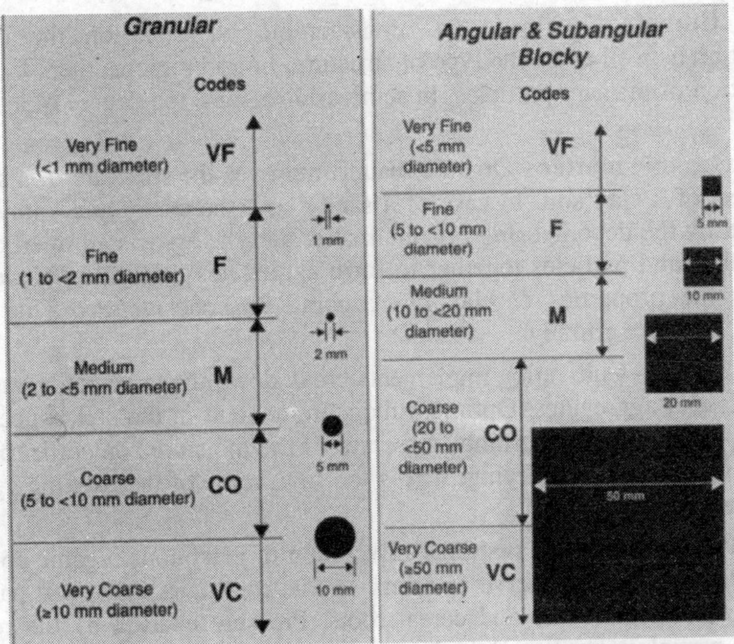

Fig. 30. Soil structure size (Granular and Blocky)

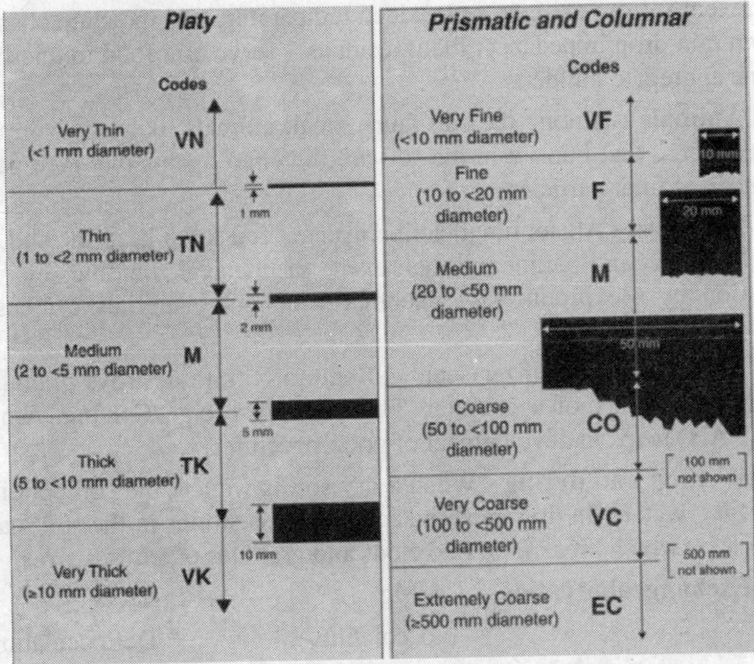

Fig. 31. Soil structure size (Platy and Prismatic)

12.6. Factors Affecting Soil Structure

The development of structure in arable soil depends on the following factors:

1. Climate - Climate has considerable influence on the degree of aggregation as well as on the type of structure. In arid regions, there is very little aggregation of primary particles. In semi–arid regions, the degree of aggregation is greater.

2. Organic matter - Organic matter improves the structure of a sandy soil as well as of a clay soil. In case of a sandy soil, the sticky and slimy material produced by the decomposing organic matter and the associated microorganism, cement the sand particles together to form aggregates. In case of clayey soil, it modifies the properties of clay by reducing its cohesiveness. This helps in making clay more crumby.

3. Tillage - Cultivation implements break down the large clods into smaller fragments and aggregates. Optimum moisture content in the soil is necessary to obtain good granular and crumby structure. If the moisture content is too high, it will form large clods on drying. If it is too low, some of the existing aggregates will be broken down.

4. Plants, roots and residues - Excretion of gelatinous organic compounds and exudates from roots serve as a link so that root hairs make soil particles to cling together e.g., Grass and cereal roots. Pressure exerted by the roots also holds the particles together. Dehydration of soil - strains the soil due to shrinkage → result in cracks → leads to aggregation. Plant tops and residues – shade the soil – prevent it from extreme and sudden temperature and moisture changes, and also from rain drop impedance. Plant residues – serve as a food to microbes –are the prime aggregate builders.

5. Animals - Among the soil fauna small animals like earthworms, moles and insects etc., that burrow in the soil are the chief agents that take part in the aggregation of finer particles.

6. Microbes - Algae, fungi, actinomycetes and fungi keep the soil particles together. Fungi and actinomycetes exert mechanical binding by mycelia, cementation by the products of decomposition and materials synthesized by bacteria.

7. Fertilizers - Fertilizer such as Sodium Nitrate destroys granulation by reducing the stability of aggregates. Few fertilizers, e.g., Calcium Ammonium Nitrate (CAN) helps in development of good structures.

8. Wetting and drying - When a dry soil is wetted, the soil colloids swell on absorbing water. On drying, shrinkage produces strains in the soil mass gives rise to cracks, which break it up into clods and granules of various sizes.

9. Exchangeable cations - Ca, Mg ------------ H, Na

Flocculating Deflocculating

Good structure Poor structure

10. Inorganic cements - $CaCO_3$ and Sesquioxides

11. Clay: The content and type of clay particles in soil will influence the aggregate stability and formation of soil structures.

12. Water: The soil moisture will aid in the formation of soil aggregates.

12.7. EFFECT OF SOIL STRUCTURE ON OTHER PHYSICAL PROPERTIES

Porosity - Porosity of a soil is easily changed. In plate–like structure, pore spaces are less whereas in crumby structure pore spaces are more.

Temperature - Crumby structure provides good aeration and percolation of water in the soil. Thus, these characteristics help in keeping optimum temperature in comparison to plate–like structure.

Density - Bulk density varies with the total pore space present in the soil. Structure chiefly influences pore spaces. Platy structure with less total pore spaces has high bulk density whereas crumby structure with more total pore spaces has low bulk density.

Consistence - Consistence of soil also depends on structure. Plate–like structure exhibits strong plasticity.

Colour - Bluish and greenish colours of soil are generally due to poor drainage of soil. Platy structure normally hinders free drainage.

Importance of structure

Soil structure, rather indirectly, influences by the formation of an array of pores of various shapes and sizes. These pores are controlling factors governing water, air and temperature in soil.

The role of soil structure in relation to plant growth

- Soil structure influences the amount and nature of porosity.
- Structure controls the amount of water and air present in the soil. Not only the amount of water and air is dependent on soil structure, but their movement and circulation are also controlled by soil structure.
- It affects tillage practices.
- Structure controls runoff and erosion.
- Platy structure normally hinders free drainage whereas sphere–like structure (granular and crumby) helps in drainage.
- Crumby and granular structure provides optimum infiltration, water holding capacity, aeration and drainage. It also provides good habitat for microorganisms and supply of nutrients.

Table 49. Classes of soil structure as differentiated by size of soil peds

Class	Platy	Prismatic	Columnar	Blocky	Sub angular Blocky	Granular	Crumb
Very fine or Very thin	<1	<10	<10	<5	<5	<1	<1
Fine or Thin	1–2	10–20	10–20	5–10	5–10	1–2	1–2
Medium	2–5	20–50	20–50	10–20	10–20	2–5	2–5
Coarser or thick	5–10	50–100	50–100	20–50	20–50	5–10	-
Very coarse or Very thick	>10	>100	>100	>50	>50	>10	-

13 SOIL CONSISTENCY

Soil consistency is also an important physical and dynamic property, which varies with the variation of soil moisture and applied stress.

Consistency is the behaviour of soil under stress. This stress is commonly evidenced by feeling the soil, manipulating it by hand, or by tillage operation. Soil consistency is considered as a combination of soil properties dependent upon the forces of attraction between soil particles influenced by soil moisture.

Soil consistency is defined as the manifestations of the physical forces of cohesion and adhesion acting within the soil at various moisture constants. These manifestations include:

- The behaviour towards gravity, pressure, thrust and pull
- The tendency of the soil mass to adhere to foreign bodies or substances
- The sensations which are evidenced as felt by the fingers of the observer

Cohesion refers to the attraction of substances of similar characteristics such as the attraction of one water molecule for another. **Adhesion** is the attraction of different materials i.e., attraction of water molecule for the soil particle.

Soil consistency depends on the texture, nature and amount of inorganic and organic colloids, structure and moisture content etc. With decreasing moisture content, the soils lose their stickiness and plasticity, and become friable and soft, and finally, when dry become hard and coherent.

13.1. CONSISTENCY IN RELATION TO DIFFERENT MOISTURE REGIMES

Soil consistency is described on the basis of three moisture levels - wet, moist and dry. In addition, soil consistency is also expressed based on the degree of cementation.

13.1.1. Wet Consistency

Stickiness

It is the capacity of soil to adhere to other objects. It is estimated at moisture content that displays maximum adherence between thumb and fore finger

Stickiness classes

Non–sticky - After release of pressure, practically no soil material adheres to thumb and finger.

Slightly sticky - After pressure, soil material adheres to both thumb and finger but comes off one or the other rather cleanly. It is not appreciably stretched when the digits are separated.

Sticky - After pressure, soil material adheres to both thumb and finger and tends to stretch somewhat and pull apart rather than pulling free from either digit.

Very sticky - After pressure, soil material adheres strongly to both thumb and finger and is decidedly stretched when they are separated.

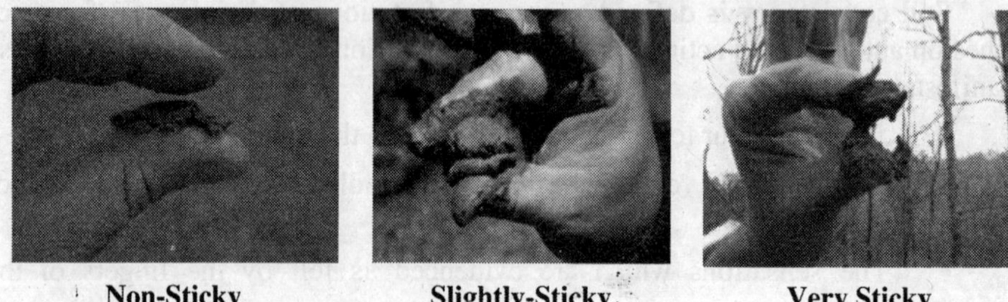

Non-Sticky **Slightly-Sticky** **Very Sticky**

Fig. 32. Soil stickiness classes

Plasticity

It is a degree to which a soil can be moulded or reworked causing permanent deformation without rupturing.

Plasticity classes

Plasticity is the ability to change shape continuously under the influence of an applied stress and retain the impressed shape on removal of the stress.

Non–plastic - The soil is not able to form a wire–like shape (6 mm diameter, 4 cm length), or if it is formed, it cannot support itself.

Slightly plastic - The soil is able to form a wire–like shape (6 mm diameter, 4 cm length) and it supports itself. If wire of 4 mm diameter and 4 cm length is formed, it cannot support itself.

Moderately plastic - The soil is able to form a wire–like shape (4 mm diameter and 4 cm length) and it supports itself. If wire of 2 mm diameter and 4 cm length is formed, it cannot support itself.

Very plastic - The soil is able to form a wire–like shape (2 mm diameter and 4 cm length) and it supports itself.

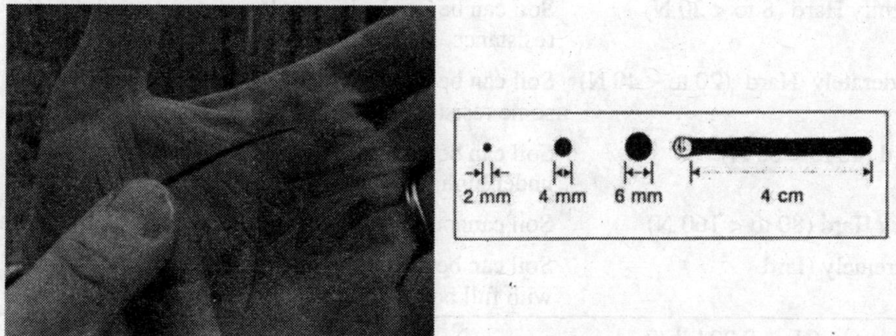

Fig. 33. Soil plasticity classes

13.1.2. Moist Consistency

When soil is moist, moist consistency is determined at moisture content approximately midway between air–dry and field capacity. The consistency of moist soils is most important since it best describes the condition of soils when they are tilled in the field. Consistency of a moist soil is described starting from the least coherent material to the strongly adherent material.

Table 49. Moist consistency

Class	Description
Loose	Non–coherent
Very friable	Soil material can be crushed easily under very gentle pressure. The soil again coheres when pressed together.
Friable	Soil material can be crushes easily under some pressure.
Firm	Soil material can be crushed under moderate pressure between thumb and forefinger. The resistance is distinctly noticeable.
Very firm	Soil can be barely crushed between thumb and forefinger.
Extremely firm	Soil cannot be crushed between thumb and forefinger.

13.1.3. Dry Consistency

The consistency of soil material when dry is characterized by rigidity, brittleness, maximum resistance to pressure, tendency to crush to a powder and inability of crushed material to cohere again when pressed together.

Table 49. Dry consistency

Class	Description
Loose	Non–coherent
Soft (<8 N)	Soil can be crushed under very low pressure.
Slightly Hard (8 to < 20 N)	Soil can be crushed under low pressure with little resistance.
Moderately Hard (20 to < 40 N)	Soil can be crushed under moderate pressure with some resistance.
Hard (40 to < 80 N)	Soil can be crushed between thumb and forefinger under high pressure.
Very Hard (80 to < 160 N)	Soil cannot be crushed between thumb and forefinger.
Extremely Hard	Soil can be crushed after applying pressure by foot with full body.

1 Newton (N) = 0.224 lb/ft

Cementation

Some soil horizons exhibit cementation quite independent of soil moisture level. The cementing agents are generally oxides of iron and aluminium, calcium carbonate and silica.

Table 50. Consistency of soil horizons due to cementation

Class	Description
Weakly cemented	Cemented units can be broken by hands.
Strongly cemented	Cemented units cannot be broken by hands but can be broken easily with a hammer
Indurated	Cemented units can be broken only with sharp blows of a hammer.

Scouring point

Another important character of soil consistency is the scouring point. This point represents that moisture content at which the soil no longer sticks to a foreign object. In highly plastic soils, the scouring point is slightly below the liquid limit, and in slightly plastic soils, it occurs above the liquid limit.

13.2. ATTERBERG LIMITS

Atterberg, a well–known scientist, suggested three values, viz., upper plastic limit, lower plastic limit and plasticity number or plasticity index. These values are called Atterberg limits. Atterberg limits are used extensively by soil scientists.

Liquid limit (LL) is defined as the moisture content at which soil begins to behave as a liquid material and begins to flow. Liquid limit of a fine-grained soil gives the moisture content at which the shear strength of the soil is approximately 2.5 kN/m^2.

Plastic limit (PL) is defined as the moisture content at which soil begins to behave as a plastic material.

Shrinkage limit (SL) is defined as the moisture content at which no further volume change occurs with further reduction in moisture content. SL represents the amount of water required to fully saturate the soil (100% saturation).

Atterberg limits are the limits of water content used to define soil behaviour. The consistency of soils according to Atterberg limits gives the following diagram:

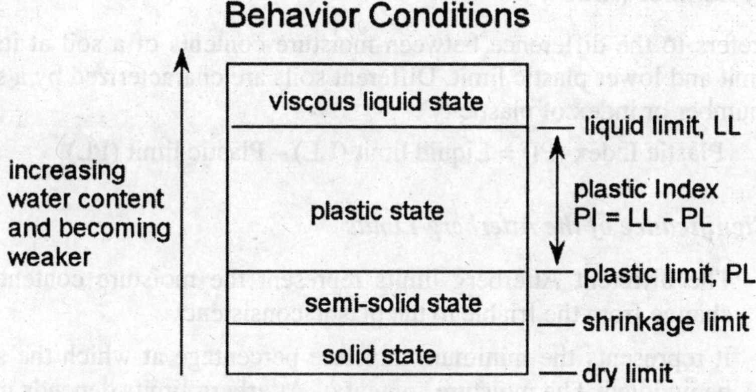

Fig. 34. Behaviour Conditions of soil at different moisture contents

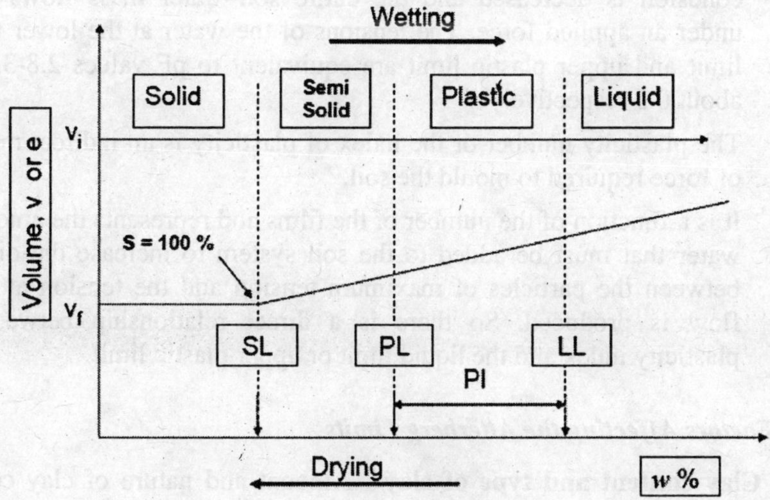

Fig. 35. Atterberg limits

Upper plastic limit

It is also called liquid limit. It represents the moisture content of soil at a point where the soil–water mass just flows under an applied force and fails to retain its shape.

Lower plastic limit

Lower plastic limit refers to the moisture content of a soil at a point where its consistency changes from plastic to friable. Soil–water mass is unable to change shape continuously under the influence of an applied force, and ultimately the mass breaks into fragments.

Plasticity number (Index)

If refers to the difference between moisture contents of a soil at its upper plastic limit and lower plastic limit. Different soils are characterized by a specific 'plastic number or index of plasticity'.

Plastic Index (PI) = Liquid limit (LL) – Plastic limit (PL)

13.2.1. Significance of the Atterberg Limits

- The different Atterberg limits represent the moisture content of the change from the friable to the plastic consistency.
- It represents the minimum moisture percentage at which the soil can be puddled. The moisture content of Atterberg limits depends upon the amount and nature of the soil colloids present. In the upper plastic limit, the film of oriented water molecules becomes so thick that cohesion is decreased and the entire soil–water mass flows freely under an applied force. The tensions of the water at the lower plastic limit and upper plastic limit are equivalent to pF values 2.8-3.3 and about 0.5 respectively.
- The plasticity number or the index of plasticity is an indirect measure of force required to mould the soil.
- It is a function of the number of the films and represents the amount of water that must be added to the soil system to increase the distance between the particles of maximum tension and the tension at which flow is produced. So there is a direct relationship between the plasticity index and the liquid limit or upper plastic limit.

13.2.2. Factors Affecting the Atterberg Limits

1. Clay content and type of clay - Amount and nature of clay colloids greatly influence the plasticity. An increase in the percentage of clay causes plastic limits to be higher with the moisture content and increases the plasticity number or index. With the decrease in clay content in the soil, the upper plastic limit decreases and thereby, decreases the plasticity number or the index of plasticity. Atterberg limits are raised as the surface is increased due to higher amount of clay present in the soil.

The clay content, therefore, determines the amount of surface available for water adsorption. For a particular clay mineral, the amount of absorbed water required at the plastic limit will increase with the amount and size of the particles present. Plastic limits of different types of silicate clays are given in Table 51.

Table 51. Plastic limits of different types of silicate clays

Types of Clay	Calcium saturated (%)		Sodium saturated (%)	
	Lower limit	Upper limit	Lower limit	Upper limit
- Montmorillonite	63	177	97	700
- Illite	40	90	34	61
- Kaolinite	36	73	26	52

Table 52. Consistency and mineralogy

Wet Consistence	Moist Consistence	Moist Matrix Color	Estimated Mineralogy
Very sticky Very plastic	Firm to Very Firm or greater	10YR 2.5YR 5Y	2:1 Smectite
Slightly Sticky to Sticky Slightly Plastic to Plastic	Friable to Firm	2.5YR or Redder (e.g. 10R, 7.5R)	1:1 Kaolinite
Sticky Plastic	Firm	5YR 7.5YR 10 YR	Mixed, 1:1 and 2:1

So, the type of clay mineral has a tremendous influence on the adsorption of water by the colloidal system. The plasticity of the montmorillonite group is high because of greater surface hydration, which results from more orientation of water and more of interlattice swelling as the thickness of the water films increases.

2. Nature of exchangeable cations - The exchangeable cations have considerable influence on soil plasticity. The Na–saturated montmorillonite has the highest plastic limit as compared to the K, Ca and Mg–saturated montmorillonite clay. The effect of these ions on non–expanding clays is quite different from the effect on the montmorillonite group. The higher hydration energies of the divalent cations cause a raising of the Atterberg limits.

3. Organic matter - Organic matter exhibits an interesting effect upon soil plasticity. It has been found that the plasticity limits decreased due to oxidation of organic matter with hydrogen peroxide. Organic matter has a high adsorptive capacity for water. Hydration of organic matter must be fairly complete before sufficient water is available for film formation around the mineral particles. Consequently, the plastic limit occurs at relatively high moisture contents. Thus, the addition of organic matter to soil may be expected to extend the zone of friability to fairly high moisture contents.

14 DENSITIES OF SOIL

Density is the weight per unit volume of a substance. It is expressed in gram per cubic centimetre or pound per cubic foot. Two density measurements - particle density and bulk density, are common for soils.

$$\text{Density (D)} = \frac{\text{Mass (M)}}{\text{Volume (V)}} \text{ gm / cc}$$

14.1. PARTICLE DENSITY

The weight per unit volume of the solid portion of soil excluding pore space is called particle density. It is also termed as 'true density'. It is expressed in gm / cm^3 (CGS system). It depends upon the accumulative densities of the individual inorganic and organic constituents of the soil. Generally, in the normal soils, the particle density is 2.65 grams per cubic centimetre. The particle density is higher if large amounts of heavy minerals such as magnetite, limonite, hematite and zircon are present. With increase in organic matter of the soil, the particle density decreases. When particle density is divided by density of water, a relative weight number is obtained, which is called 'specific gravity'. Water is the reference; its density is 1 gm / cm^3.

Table 53. Densities of different material

Material	Density (g cm^{-3})
Water	1.0
Pine wood	0.7
Loose sand	1.6
Soil minerals, rocks	2.65
Steel	7.7
Lead	11.3

14.2. BULK DENSITY

Bulk density is defined as the mass (weight) per unit volume of a dry soil (volume of solid and pore spaces). It is also expressed in gm/cc (CGS system). The bulk density of a soil is always smaller than its particle density. Loose and porous soils have low weights per unit volume and compact soils have high values. Volcanic ash also has a very low bulk density. The bulk density of sand–dominated soils is about 1.7 gm/cc, whereas in organic peat soils the value of bulk density is about 0.5 gm/cc. Bulk density normally decreases as mineral soils become finer in texture. Bulk density is of greater importance than particle density in understanding the physical behaviour of soils. Generally, soils having low and high bulk densities exhibit favourable and poor physical conditions respectively. Generally, in normal soils bulk density ranges from 1–1.6 gm/cc except in very compact sub-soils where bulk density is 2 gm/cc.

14.3. FACTORS AFFECTING BULK DENSITY

- **Amount of pore space** - If the soil contains more pore spaces than solid spares per unit volume, the value of bulk density is very low.
- **Compactness of the soil** - In highly compacted soil or waterlogged soils the bulk density is more.
- **Texture of the soil** - Textural variations influence the value of bulk density in soils. e.g., clay, clay loam and silt loam surface soils show low bulk density as compared to sands and sandy loam soils, which show high bulk density value.
- **Organic matter content** - Soils containing high organic matter show lower value of bulk density.
- **Soil structure** - Soil structure affects bulk density by influencing the porosity of the soil. e.g., crumb soil structure shows low bulk density than that of platy soil structure.

14.4. BULK DENSITY (BD) DETERMINATION

Farmers often refer to it as 'heavy', but they mean that it is difficult to plow, not that it has a high bulk density. Clay soils generally will have a lower bulk density than sandy soils (**clay=1.3 g/cc vs sand=1.6g/cc**). Sandy soils have a higher bulk density, but are easier to plow since they have weaker consistence. Thus, they are often referred to as 'light soils'. The lower B.D. of clays is due to their better aggregation.

- Clod Method of BD Determination

  BD = Mass of Clod ÷ Volume of water displaced

  Volume of water displaced = weight of water or the (gms clod in air) - (gms clod in water)

- Cylinder method of BD Determination

BD = Mass of oven dry soil (gms) ÷ total volume of soil (cm^3)

- Soil core method of BD Determination

Another method of determining bulk density is by using **soil cores**. These are obtained with a **probe** that forces a cylinder into the soil. The soil can be removed from this cylinder and cut into sections and placed in another cylinder for transportation back to the laboratory. Care needs to be taken to avoid compactness of the soil during and after obtaining the core.

14.4.1. Core Method

BD of soil = {(Mass of soil + Mass of core) – (Mass of core} ÷ (Volume of core)}

Volume of core = Π (3.1416) $r^2 \times h$

r = Radius of core

H = Height of core

For the cores in the lab,

r = 2.45 cm (½ of Diameter); h = 5 cm

$V = 3.1416 \times 2.45^2 \times 5 = 94.3\ cm^3$

Bulk density is a fairly easy determination that yields information about the soil. This information is significant for determining the soil's potential for plant growth or a building foundation.

Bulk density can be used to calculate soil mass to a given depth, e.g.,

1 ha = 10,000 m^2 = 10,00,00,000 (10^8) cm^2

Let $\rho_B = 1.3\ g\ cm^{-3}$ or 1,300 kg m^{-3} and soil depth = 15 cm

Soil volume = (0.15 m) x (10,000 m^2) = 1,500 m^3

Soil mass = (1,500 m^3) x (1,300 kg m^{-3}) = 19,50,000 kg

So, we have 19,50,000 kg ha^{-1}

This is often used in agriculture and ecology to calculate water and nutrient storage per ha for plant growth or hydrology.

14.5. POROSITY OF SOIL

Pore spaces (also called voids) in a soil consist of that portion of the soil volume not occupied by solids, either mineral or organic. The pore spaces under field conditions are always occupied by air and water. Pore spaces directly control the amount of water and air in the soil and indirectly influence the plant growth and crop production. In general, there are broadly two types of pores in soils i.e., Macropores and Micro or capillary pores.

1. **Macropores** - Large-sized pores (dimension is larger than about 0.08 mm) are referred to as macropores, which readily allow air and water movement. Sands and sandy soils have a large number of macropores. It is found in between the granules.

2. **Micro or capillary pores** - Smaller–sized pores (dimension is smaller than about 0.08 mm) are generally referred to as a micro or capillary pores in which movement of air and water is restricted to some extent. Clays and clayey soils have a greater number of micro or capillary pores. It has got more importance in the plant growth relationship. It is found within the granules.

Biopores

Macropores created by roots, earthworms and other organisms constitute a very important type of pores termed as biopores. Soil pores are divided into following four categories based on the size groupings of soil separates:

1. **Coarse pores** - Average diameter is greater than 0.2 mm or 200 microns (0.008 inch). 1 micron = 1 millionth of a metre
2. **Size of medium sands** - These are medium pores. Average diameter is 0.2–0.02 mm or 200–20 microns (0.008–0.0008 inch).
3. **Size of coarse silt particles** - These are fine pores. Average diameter is 0.02–0.002 mm or 20–2 microns (0.0008 inch).
4. **Size of fine silt particles** - These are very fine pores. Average diameter is less than 2 microns (0.00008 inch). Large clay particles are 2 microns in average diameter.

Porosity refers to the percentage of soil volume occupied by pore spaces. Size of individual pores, rather than total pore space in a soil, is more significant in its plant growth relationship. For optimum growth of the plant, approximately equal amount of macro and micropores are required, which influence aeration, permeability, drainage and water retention favourably. Porosity of a soil can be easily changed.

14.6. FACTORS AFFECTING POROSITY OF SOIL

Wide difference in the total pore space of various soils occurs depending upon the following factors:

1. **Soil structure** - A soil having granular and crumb structure contains more pore spaces than that of prismatic and platy soil structure. So, well–aggregated soil structure has greater pore spaces as compared to structureless or single grain soil.
2. **Soil texture** - In sandy soils the total pore space is small whereas in fine textured clay and clayey loam soils total pore space is high and there is a possibility of more granulation in clay soils.
3. **Arrangement of soil particles** - When the sphere–like particles are arranged in columnar form (i.e., one after another on the surface forming column like shape), it gives the most open packing system resulting into very low amount of pore spaces. When such particles

are arranged in the pyramidal form it gives the closest packing system resulting into high amount of pore spaces.

4. **Organic matter** - Soil containing high organic matter possesses high porosity because of well–aggregated formation.
5. **Macroorganisms** - Macroorganisms such as earthworm, rodents, insects etc. increase macropores in the soil.
6. **Depth of soil** - With the increase in depth of soil, the porosity decreases because of compactness in the sub–soil.
7. **Cropping** - Intensive crop cultivation tends to lower the porosity of soil as compared to fallow soils. The decrease in porosity may be due to reduction in organic matter content.
8. **Puddling** - Due to puddling under sufficient soil moisture, the soil surface layer is made dense and compact. Eventually, the porosity of this surface soil is reduced by the infiltration of muddy surface materials.

14.7. RELATIONSHIP BETWEEN POROSITY & DENSITIES OF SOIL

$$\% \text{ solid space} = \frac{\text{bulk density}}{\text{particle density}} \times 100$$

Since, % pore space + % solid space = 100

or $\quad \% \text{ pore space} = 100 - \% \text{ solid space}$

or $\quad \% \text{ pore space} = 100 - \dfrac{\text{bulk density}}{\text{particle density}} \times 100$

or $\quad \% \text{ pore space} = 100\left(1 - \dfrac{\text{bulk density}}{\text{particle density}}\right)$

Table 54. General relationship among texture, bulk density and porosity of soils

Textural Class	Bulk Density (Mg/m³)	Porosity (%)
Sand	1.55	42
Sandy loam	1.40	48
Fine sandy loam	1.30	51
Loam	1.20	55
Silt loam	1.15	56
Clay loam	1.10	59
Clay	1.05	60
Aggregated clay	1.00	62

15 SOIL COLOUR

Soil colour indicates many soil features. Colour is produced by the spectral reflectance of light by the soil. It is an indirect measure of many important soil properties. A change in soil colour from the adjacent soils indicates a difference in the soil's mineral origin (parent) material or in the soil development. Soil colour varies among different kinds as well as within the soil profile of the same kind of soil. It is an important soil property through which its description and classification can be made.

Kinds of Soil Colour

Soil colour is inherited from its parent material and that is referred to as lithocromic e.g., red soils developed from red sandstone. Besides soil colour also develops during soil formation through different soil–forming processes and that is referred to as acquired or pedochromic colour e.g., red soils developed from granite geneiss or schist.

15.1. FACTORS AFFECTING SOIL COLOUR

There are various factors or soil constituents that influence the soil colour which are as follows:

1. Organic matter - Soils containing high amount of organic matter show the colour variation from black to dark brown colour.

Fig. 36. Soil rich in organic matter

2. Iron compounds - Soils containing higher amount of iron compounds generally impart red, brown and yellow–tinge colour.

Fig. 37. Soil rich in iron compounds

3. Silica, lime and other salts - Sometimes soils contain either large amounts of silica and lime or both. Due to presence of such materials in the soil the colour of the soil appears like white or light–coloured.

4. Mixture of organic matter and iron oxides - Very often soil contains certain amount of organic matter and iron oxides. As a result of their existence in the soil, the most common soil colour found is brown.

5. Alternate wetting and drying condition - During monsoon period due to heavy rain the reduction of soil occurs and during dry period the oxidation of soil takes place. Due to development of such alternating oxidation and reduction condition, the colour of soil in different horizons of the soil profile is variegated or mottled. This mottled colour is due to residual products of this process especially iron and manganese compounds.

6. Oxidation–reduction conditions - The permanent reduced condition develops when soils are waterlogged for a longer period. The presence of ferrous compounds resulting from the reducing condition in waterlogged soils impart bluish and greenish colour.

Therefore, it may be concluded that the soil colour is indirectly indicative of many other important soil properties. Besides soil colour directly modifies the soil temperature e.g., dark–coloured soils absorb more heat than light–coloured soils.

15.2. DETERMINATION OF SOIL COLOUR

Munsell colour chart is used for measuring soil colour. This chart consists of 196 different coloured chips systematically arranged on cards according to the Munsell notation. The soil colours are best determined by the comparison with the Munsell colour. The Munsell colour notations are systematic numerical and letter designations of each of the three variables. These three variables of colour are Hue, Value and Chroma e.g., the numerical notation 2.5 YR 5/6 suggests a hue of 2.5 YR, value of 5 and chroma of 6. The equivalent or parallel soil colour name for this Munsell notation is 'red'.

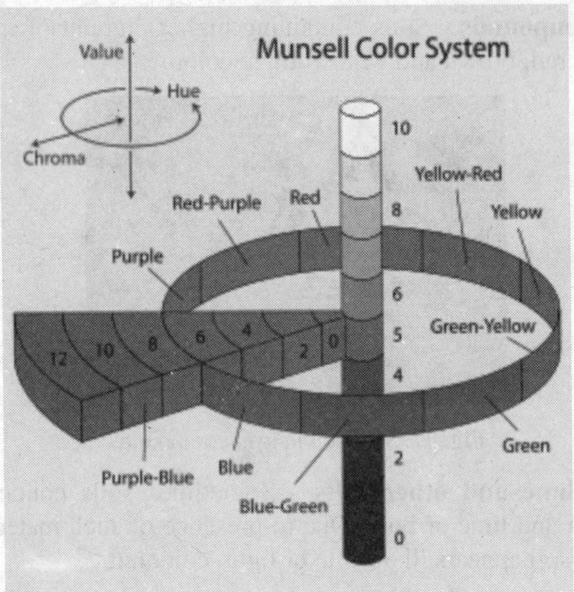

Fig. 38. Munsell colour notations

Hue is the dominant spectral colour (It denotes the dominant spectral colour - red, yellow, blue and green). In the colour chart hue is arranged radially from one card to next. The symbol is the letter of abbreviation of colour of rainbow. R stands for red, YR for yellow red, and Y for yellow preceded by numbers from 0-10.

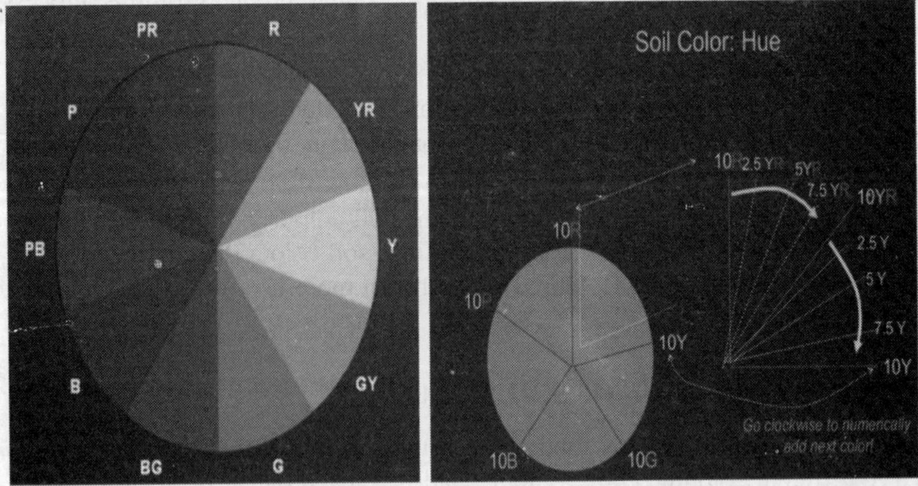

Fig. 39. Dominant spectral colour (Values of Hue)

Value is the intensity of light i.e., the lightness or darkness of a colour (the amount of reflected light). The notations for value consist of numbers from 0 for absolute black to 10 for absolute white. Value is vertically arranged in the colour chart. The colour becomes successively lighter from the bottom of the card to the top and the value increases. Numbers ranging above 5 indicates lighter colours while numbers below 5 indicates dark colours.

0 0

Fig. 40. Intensity of light

Chroma is the relative purity of light i.e., the purity of the colour (strength of the colour). The notation for chroma consists of numbers beginning at zero for grey and increases with decreasing greyness. In writing the Munsell notation, the order is hue, value and chroma with a space between the hue letter and the value number and a virgule between the two numbers for value and chroma e.g., in 10YR 5/6 (10YR is hue, 5 is the value and 6 is the chroma).

0 8

Fig. 41. Purity of the colour (Values of Chroma)

The colour of the soil is determined by holding the soil sample below the colour chip being compared. The colour is determined, under dry and moist conditions. If the soil is moist, indicate the same by letter 'M' after the colour notation. e.g., 10YR 5/6 M. If the soil sample has more than one distinct colour, record both the colours. The nomenclature for soil colour consists of two complementary systems - colour name and Munsell notation of the colour, the former for describing the soil and the latter for the purposes of international soil correlation.

Table 55. Colours of common minerals

Material	Chemical Composition	Colour
Manganese	MnO_2	Purplish black
Hematite	Fe_2O_3	Red
Geothite	FeOOH	Yellow
Hydrated Ferric Oxide	$Fe(OH)3*nH_2O$	Red brown
Calcite	$CaCO_3$	Whitish
Glauconite	$KMg\ (Fe,Al)(SiO_3)_6\ 3H_2O$	Greenish

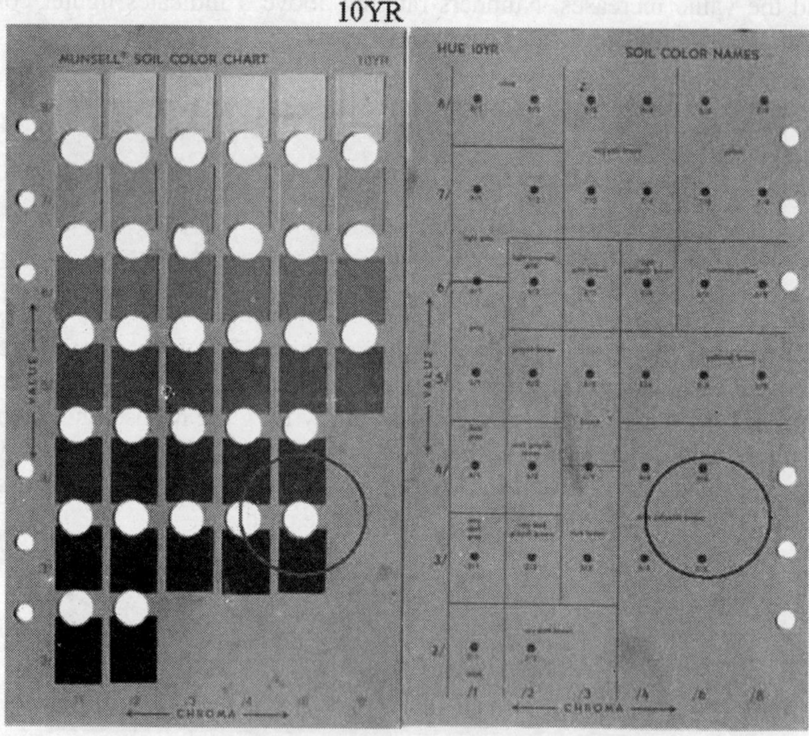

Fig. 42. 10YR 3/6 – dark yellowish brown

Colour Patterns-Mottling

- Matrix colour is the dominant colour in the soil.
- Mottling is spots or blotches of colour in the soil that differ from the matrix colour. The pattern may relate to the aeration or drainage of the soil.
- Gleyed colours are low chroma matrix colours with or without mottles. If the soil is gleyed, it is likely to be reduced and wet for much of the year.
- Mottling associated with periods of saturation and anaerobic conditions are known as redoximorphic features.
- Note the soil moisture condition when describing. Soils that contain reduced iron may appear light in place, but become darker as the reduced iron begins to oxidize.

16 SOIL TEMPERATURE

There are various sources of soil heat such as solar radiation, biochemical reactions, conduction, precipitation, exposure and vegetation etc.

16.1. SOURCES OF SOIL HEAT

1. Solar radiation

- Radiant energy from the Sun
- The power source that determines the thermal regime of the soil and the growth of plants
- Agriculture is the exploitation of solar energy in the presence of an adequate water supply and sufficient plant nutrients to maintain plant growth.
- Much of the radiation from the Sun is dissipated and distributed by components of the Earth' atmosphere as it passes downward towards the Earth.
- The heat absorbed by the surface of the Earth from the solar radiation is affected by several parameters such as latitude, distribution of land and water, slope of the land etc.

2. Bio-chemical reactions

In the soil atmosphere, a variety of chemical reactions are going on and during such reaction, liberation of large amount of heat in the soil environment occurs.

Besides, decomposition of organic matter and other crop residues in the soil, and other microbial processes liberate large amounts of heat in the soil and thus, contribute towards soil heat..

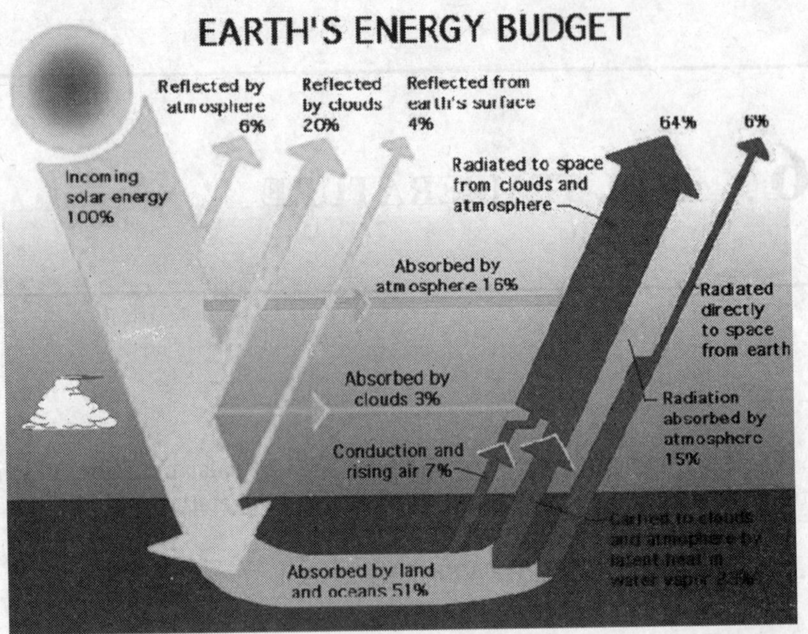

Fig. 43. Earth's Energy Budget

3. Conduction

The inner part of the Earth is very hot, and the conduction of heat in the soil environment is very slow. Generally, during night, the uppermost surface soil becomes cooler than sub–surface soil. Thus, heat flows from the regions of sub–soil to the regions of surface soil (cooler soil layer).

4. Precipitation

During the winter season precipitation increases soil heat because of the higher specific heat of precipitation.

5. Vegetation

Vegetation plays a significant role in reducing soil heat because of the insulating properties of plant cover. Bare soil is unprotected from the direct rays of the Sun and becomes very warm during the hottest part of the day.

16.2. LOSS OF SOIL HEAT

There are various factors that influence the loss of soil heat such as radiation, conduction, evaporation and precipitation.

1. Radiation - The quantity of heat absorbed by the surface soil does not remain constant. Some portion of the absorbed soil heat is lost to the atmospheric environment by radiation.

2. Conduction - The conduction process means the transmission of anything from one point to the other point. Here heat is transmitted to the sub–surface horizon of the soil from the surface soil layer by the conduction process and thereby resulting in the loss of soil heat from the surface soil.

3. Evaporation - The major portion of the global radiation in humid climates is used in the process of evapo–transpiration. This heat energy is lost to the soil and ultimately returned to the atmospheric air resulting into loss from the soil mass. This results into a cooling effect especially at the surface. The potential evapo–transpiration process may be defined as the amount of water lost by evaporation and transpiration from a surface that is completely covered with vegetation, if there is sufficient water in the soil at all times for the use of the vegetation.

4. Precipitation - During summer, precipitation has generally a cooling action in soil because rainwater usually has a lower temperature than soil.

16.3. FACTORS INFLUENCING SOIL TEMPERATURES

There are various factors that influence the soil temperature, which are as follows:

1. Composition of the soil

Since the soil is a granular medium consisting of solid, liquid and gaseous phases, the thermal regimes of the soil depends upon its volumetric composition, the size and arrangement of the solid particles and also interface relationship between the solid and liquid phases. Among the solid materials, the soil is composed of partly mineral matter and partly organic matter. The specific heat of mineral matter such as iron (Specific heat 0.11) is less than that of organic matter, humus (Specific heat 0.44). So, the soils containing much more mineral matter get heated very easily than those soils containing higher amount of organic matter.

2. Soil structure

Soil structure also influence soil temperature by controlling pore spaces resulting from the different types of arrangement of soil particles such as open and close systems of packing etc. Soils having spheroidal type of structure, warm up more quickly because there is no prevailing condition of waterlogging.

3. Soil texture

Light–textured sandy soil, in general, absorbs heat very quickly than heavy–textured clayey soil. Heavy soil carries a greater quantity of water and due to this reason it warms up very slowly.

4. Soil moisture

Soil moisture plays a vital role in controlling its thermal regime. The specific heat of water is higher than the soil. Moist soils have a higher specific heat than dry soils. Consequently, a moist soil has a lower temperature than dry soil. Moist soil gets heated very slowly and it gets cooler faster than the dry soils.

5. Soil colour

In a dark soil, where colour is caused by large amounts of humus, the larger amounts of water held by the humus may offset the increased heat absorption due to the dark colour.

6. Vegetation

Soils covered with different types of vegetation absorb less heat as compared to soils without vegetation cover. Vegetation acts as an interceptor of heat on the soil surface and thereby, reduces the thermal regime of the soil.

7. Irrigation and drainage

Irrigation raises the humidity of the air, lowers the air temperature over the soil and reduces the daily soil temperature variations. Drainage decreases the heat capacity of wet soils, which raises the soil temperature. This plays an important role in warming up the soil in the spring.

8. Topography

The slope of the land, especially ground slope, also influences the soil temperature. The temperature of the ridged fields is higher than those that are levelled.

9. Compactness of the soil

Compactness of the soil surface increases the density and the thermal conductivity. Compacted soils have better heat conductivity than loosened soil. Tillage, on the other hand, cereals can act as a surface mulch which reduces heat flux from the surface to the sub–surface layers.

10. Climate

Climate has a profound influence on the thermal regime of the soil. Soils in temperate climate are cooler than those in tropical climate.

11. Season

Soil temperature differs with the seasonal variations of the year. The summer months (June to July) in the northern hemisphere, especially midday, represent the peak of the global radiation and the maximum temperatures. The winter months have a similar effect to night temperatures, which is much lower. The temperature of the surface soil is always higher than the air temperature.

16.4. SPECIFIC HEAT, HEAT CAPACITY, THERMAL, CONDUCTIVITY AND THERMAL DIFFUSIVITY

16.4.1. Specific Heat

Specific heat may be defined as the amount of heat required to raise the temperature of one gram of a substance by 1°C.

The specific heat of dry soil (0.2 cal/g) is only about one–fifth that of water (1 cal/g). Hence, moist soils are cooler due to their high specific heat and also due to the heat energy spent in evaporation of soil moisture.

16.4.2. Heat Capacity

The heat capacity of a given material is equal to its specific heat multiplied by its mass. The specific heat of soil varies with various soil constituents. It is found that among the major soil constituents, quartz has the lowest specific heat and humus has the highest specific heat, except water. It is also evident that the heat capacity is influenced by humus and water content of the soil.

16.4.3. Thermal Conductivity and Diffusivity

Once the soil gains the heat, the soil temperature is governed by the thermal characteristics of the soil i.e., heat capacity, thermal conductivity and thermal diffusivity. The amount of heat required to produce a given change in the soil temperature depends upon the mass and the nature of the soil.

The change of heat *dQ/dt* flowing into or out of the soil depends on the temperature gradient *dT/dZ* in the soil and the thermal conductivity K. For any given soil volume, the change of heat flow is given by:

$$\frac{dQ}{dt} = -k\,\frac{dT}{dz}$$

where, *z* is measured along the gradient.

k is the thermal conductivity of soils expressed in J/°C /cm/sec and it depends on the physical properties of the soil viz., porosity, air, moisture and soil minerals. The more compact and wetter the soil, the greater is its conductivity.

The effect of a heat flow on the rate of change of temperature at a given depth depends on the difference between the heat flow into and out of a small volume of soil, at that depth; and is given by:

$$\frac{dT}{dt} = \frac{k}{\rho c}\,\frac{d^2T}{dz^2}$$

where, p is the density of the soil and c is the specific heat.

k/pc is the thermal diffusivity of the soil expressed in cm^2/sec.

The thermal diffusivity of soil may either increase or decrease with increasing moisture content since *k*, p and c all increase with it, but in general, starting with a dry soil, it first increases and after reaching a peak, may decrease slightly.

16.5. INFLUENCE OF SOIL TEMPERATURE ON PLANT GROWTH AND NUTRITION

There are various factors that influence the soil temperature which affects soil fertility by changing the soil environment, which in turn, affects plant growth.

1. Germination of seeds - The germination of different seeds is greatly influenced by the variation in soil temperature. If the temperature is too low, the seed fails to germinate or germinates at a slow rate. On the other hand, seeds may be injured if the temperature is very high.

2. Physical properties of soil - Soil structure is greatly influenced by the temperature. The temperature has a great influence on the aggregation of the soil as well as on the binding materials present in it.

3. Microbial activity - The activity of microorganisms having thermophobic and thermophilic nature is influenced by the variation in soil temperature. A certain amount of heat is necessary for the proper functioning of various types of soil microorganisms. Various microbiological processes like mineralization of nitrogen, nitrogen fixation, pesticide degradation etc. are influenced by the temperature.

The activity of microorganisms is lowest when soil temperature is below 5°C and above 54°C. The optimum temperature for the activity of most of the microorganisms is in the range of 25–35°C.

4. Decomposition of organic matter in soil - Various important functions of organic matter depend on the process of decomposition, which is largely influenced by the soil temperature. The decomposed products vary with the rate of decomposition of organic matter. At low temperature the rate of organic matter decomposition is low resulting into various toxic organic substances in soil and at high temperature the rate of decomposition is very fast resulting into beneficial products of organic matter decomposition, and hence influences the plant growth.

5. Absorption of water - Variation in soil temperature (very low to high temperature) affects the absorption of soil water by the plant roots. The effect of temperature in reducing the rate of water uptake may be important in spring time resulting into strong transpiration.

6. Availability of nutrients - Various physico–chemical and chemical reactions are greatly influenced by soil temperature. Temperature influences the solubility reactions of different nutrients and releases larger amount of nutrient elements in the soil solution at higher temperatures. The rate of nutrient uptake, however, is almost different for each crop, but it appears to be low on the relative effect of temperature on the uptake of nutrients.

7. Root growth - Soil temperature affects the growth of different types of root. Low temperature encourages white succulent roots with little branching, while high temperatures encourages a browner, finer and much more freely branching root system. However, the soil temperature at which *roots* grow fastest is usually higher than the temperature, which encourages the most extensive root system.

8. Plant diseases - Development of various diseases are also related to the soil temperature. At low temperatures, many parasitic fungi grow actively and very rapidly in the soil and can kill the seedlings.

On the other hands, seedlings of the temperate zone cereals, which are adapted to grow actively at lower temperatures, are relatively resistant to their attack.

16.6. MANAGEMENT OF SOIL TEMPERATURE

The thermal regimes favourable for plant growth as well as different microorganisms can be maintained through different management practices under field condition. Soil temperature can be maintained by regulating the amount of water either by draining under waterlogged condition or by adding water under dry condition in the field. The soil temperature under waterlogged soil condition is very low and to increase the soil temperature under such condition, it is essential to drain out the excess water from the field. The removal of water in this way lowers its specific heat and thus, increases the soil temperature.

The colour of the surface soil also influences the thermal regime either favourably or adversely e.g., a soil surface containing black substances viz., charcoal powder maintain high temperature while a soil surface containing white substances such as chalk and other light substances shows low soil temperature because of more heat absorption of the former than that of the latter where more reflection of radiant energy occurs.

Soil temperature can also be maintained by providing different types of mulches on the soil surface. Mulches applied to the surface of a soil affect the amount of heat received and the way it is dissipated. A surface mulch with a layer of dead vegetable waste material has two types of effect on the soil - (i) a characteristic effect on the soil surface, and (ii) supply of different plant nutrients after decomposition through ploughing into the soil. The mulched soil is much cooler during the heat of the day and rather warmer during the night.

Due to organic matter application, dark–coloured soils are not warmer than that of light–coloured soils because of high water content in the dark–coloured soils, which helps in lowering down the temperature.

17 SOIL AIR

A part of the soil volume that is not occupied by soil particles is known as pore space. Pore spaces are filled partly with soil water and partly with soil air. As the pore space is occupied by both water and air, the volume of air varies inversely with that of water. As the moisture content of the soil increases, the air content decreases and vice–versa.

Definition of soil aeration

The constant movement of air in the soil mass resulting in the renewal of gases is known as soil aeration. Thus, a well–aerated soil is one in which gases are available to growing aerobic organisms (including higher plants) in sufficient quantities and in the proper proportions to encourage optimum rates of the essential metabolic processes of these organisms.

17.1. COMPOSITION OF SOIL AIR

The soil air contains a number of gases of which nitrogen, oxygen, carbon dioxide and water vapour are the most important. Soil air constantly moves from the soil pores into the atmosphere and from the atmosphere into the pore space. Although soil air and atmospheric air differ in the compositions, soil air contains a much greater proportion of carbon dioxide and a lesser amount of oxygen than atmospheric air. At the same time, soil air contains a far greater amount of water vapour than atmospheric air. The amount of nitrogen in soil air is almost the same as in the atmosphere.

Table 56. Composition of soil and atmospheric air

Air	Percentage by Volume		
	Nitrogen	Oxygen	Carbon dioxide
Soil air	79.2	20.6	0.25
Atmospheric air	79.0	20.97	0.03

Soil atmosphere

Inverse relationship between O_2 and CO_2

– O_2 ~ 20 per cent at surface to < 5 per cent in lower horizons

– No O_2, anaerobic (typical of wet soils)

– Carbon dioxide levels often 0.35 per cent

Other gases are H_2O vapour (typically 100 per cent relative humidity). In strongly reduced soils: methane (CH_4), ethylene (C_2H_4), and hydrogen sulfide (H_2S) (toxic to plants if air exchange is too slow)

17.2. FACTORS AFFECTING THE COMPOSITION OF SOIL AIR

The composition of soil air is influenced by a number of factors such as nature of soil, soil condition, type of crop, microbial activity, season etc.

1. Oxygen

The quantity of oxygen in soil air is less than that in atmospheric air. Plant roots and various microorganisms require oxygen which they take from the soil air, thus, depleting the concentration of oxygen in the soil air. The amount of oxygen also depends upon the soil depth. The oxygen content of the air in lower layer is usually less than that of the surface soil. This is possibly due to the diffusion of oxygen from the atmosphere into the surface soil than in the sub–soil.

The quantity of oxygen is usually higher in dry season than during the monsoon. Because soils are normally drier during the summer, opportunity for gaseous exchange is greater during this period. This results in relatively high O_2 and low CO_2 levels. Light–textured soil i.e., sandy soil contains much higher oxygen percentage than heavy soil.

2. Carbon dioxide

Decomposition of organic matter produces CO_2. Hence, soils rich in organic matter contain higher percentage of carbon dioxide. Production of CO_2 is associated with microbial activity. CO_2 increases with the increasing number and activity of microorganism. Temperature and season also influence the carbon dioxide content in the soil air. High temperature during summer season encourages microorganism activity, which results in higher production of CO_2. Soils on which crops are grown contain more CO_2 than fallow lands. The amount of CO_2 is usually much greater near the roots of plants than further away. It may be due to respiration by the roots. The concentration of CO_2 is usually greater in sub–soil probably due to more sluggish aeration in lower layer than in the surface soil.

3. Water vapour

Soil air, however, contains much more water vapour than atmospheric air. Capillary water in the soil is used to saturate the soil air with water vapour. If the soil moisture falls below hygroscopic coefficient, the water vapour begins to decrease. During crop–growing period, when soil remains moist, the amount of water vapour in the soil air would be more.

17.3. EXCHANGE OF GASES BETWEEN SOIL AND ATMOSPHERE

The exchange of gases between the soil and the atmosphere is facilitated by two mechanisms:

1. Mass flow

Mass flow of air is apparently due to the pressure difference between the atmosphere and the soil air. With every rain or irrigation, a part of the soil air moves out into the atmosphere as it is displaced by the incoming water. The atmospheric air enters the soil pores as and when moisture is lost by evaporation and transpiration. The variations in soil temperature cause changes in the temperature of soil air. As the soil air gets heated during the day, it expands, and the expanded air moves out into the atmosphere; when the soil begins to cool, the soil air contracts and the atmospheric air is drawn in.

2. Diffusion

Most of the gaseous interchange in soils occurs by diffusion. Atmospheric air and soil air contain a number of gases such as nitrogen, oxygen, carbon dioxide etc. Each of these gases exerts its own partial pressure in proportion to its concentration. The movement of each gas is regulated by the partial pressure under which it exists. If the partial pressure of one of the gases (i.e., carbon dioxide) is greater in the soil air than in the atmospheric air, it (CO_2) moves out into the atmosphere. Hence, the concentration of CO_2 is more in soil air.

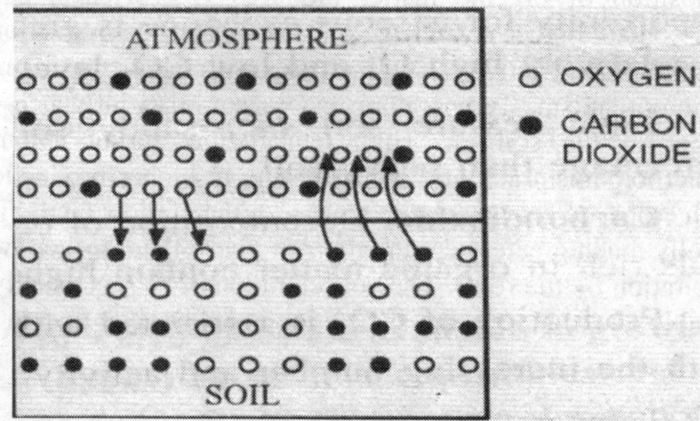

Fig. 44. Gaseous exchange between soil and atmosphere (diffusion)

On the other hand, partial pressure of oxygen that is low in the soil air is consumed as a result of biological activities. The oxygen present in the atmospheric air (partial pressure of O_2 is greater), therefore, diffuses into the soil air till an equilibrium is established, Thus, diffusion allows extensive movement and continual change of gases between the soil air and the atmosphere. Oxygen and carbon dioxide are the two important gases that take part in diffusion.

17.4. IMPORTANCE OF SOIL AERATION

The role of soil air in relation to fertility of soil and plant growth is given below:

1. Plant and root growth

Soil aeration is an important factor in the normal growth of plants. The supply of oxygen to roots in adequate quantities and the removal of CO_2 from the soil atmosphere are very essential for healthy plant growth. When the supply of oxygen is inadequate plant growth either retards or ceases completely as the accumulated CO_2 hampers the growth of plant roots.

The abnormal effect of insufficient aeration on root development has been observed. On the root crops such as carrots and sugar beets, the influence is most noticeable. Abnormally–shaped roots of these plants are common on compact, poorly–aerated soils. The penetration and development of root are poor so such undeveloped root system cannot absorb sufficient moisture and nutrients.

2. Microorganism population and activity

The microorganisms living in the soil also require oxygen for respiration and metabolism. Some of the important microbial activities such as the decomposition of organic matter, nitrification and sulphur oxidation depend upon oxygen present in the soil air. The deficiency of air (oxygen) in soil slows down the rate of microbial activity e.g., the decomposition of organic matter is retarded and nitrification is arrested. The microorganism population is also drastically affected by poor aeration.

3. Formation of toxic material

Poor aeration results in the development of toxin and other injurious substances such as ferrous oxide, H_2S gas and CO_2 gas in the soil.

4. Water and nutrients absorption

A deficiency of oxygen has been found to check the nutrient and water absorption by plants. The energy of respiration is utilized in absorption of water and nutrients. Under poor aeration condition (this condition may arise when waterlogged), plants exhibit water and nutrient deficiency.

5. Development of plant disease

Insufficient aeration of the soil also leads to the development of diseases e.g., wilt of arhar (red gram) and gram, dieback of citrus and peach.

18 SOIL MOISTURE

Water contained in the soil is called soil moisture. The water is held within the soil pores. Soil water is the major component of the soil in relation to plant growth. If the moisture content of a soil is optimum for plant growth, plants can readily absorb soil water. Not all the water held in the soil is available to plants. Much of the water remains in the soil as a thin film. Soil water dissolves salts and makes up the soil solution, which is important as a medium for supply of nutrients to growing plants.

Importance of soil water

- Soil water serves as a solvent and carrier of food nutrients for plant growth.
- Yield of crop is more often determined by the amount of water available rather than the deficiency of other food nutrients.
- Soil water acts as a nutrient itself.
- Soil water regulates soil temperature.
- Soil–forming processes and weathering depend on water.
- Microorganisms require water for their metabolic activities.
- Soil water helps in chemical and biological activities of the soil.
- It is a principal constituent of the growing plant.
- Water is essential for photosynthesis.

18.1. RETENTION OF WATER BY SOIL

The soils hold water (moisture) due to their colloidal properties and aggregation qualities. The water is held on the surface of the colloids and other particles, and in the pores. The forces responsible for retention of water in the soil after the drainage has stopped are due to surface tension and surface attraction, and are called surface moisture tension. This refers to the energy concept in moisture retention relationships. The force with which water is held is also termed as suction. The water is retained in the soil in the following ways:

1. Cohesion and adhesion forces - These two basic forces are responsible for water retention in the soil. One is the attraction of molecules for each other i.e., cohesion, and the other is the attraction of water molecules for the solid surface of soil i.e., adhesion. By adhesion, solids (soil) hold water molecules rigidly at their soil-water interfaces. These water molecules, in turn, hold together by cohesion. Together, these forces make it possible for the soil solids to retain water.

2. Surface tension - This phenomenon is commonly evidenced at water–air interfaces. Water behaves as if its surface is covered with a stretched elastic membrane. At the surface, the attraction of the air for the water molecules is much less than that of water molecules for each other. Consequently, there is a net downward force on the surface molecules, resulting in a sort of a compressed film (membrane) at the surface. This phenomenon is called surface tension.

3. Polarity or dipole character - The retention of water molecules on the surface of clay micelle is based on the dipole character of the molecule of water. The water molecules are held by electrostatic force that exists on the surface of colloidal particles. By virtue of their dipole character and under the influence of electrostatic forces, the molecules of water get oriented (arranged) on the surface of the clay particles in a particular manner.

Each water molecule carries both negative and positive charges. The clay particle is negatively charged. The positive end of water molecule gets attached to the negatively–charged surface of clay, leaving its negative end outward. The water molecules attached to the clay surface in this way present a layer of negative charges to which another layer of oriented water molecules is attached. The number of successive molecular layers goes on increasing as long as the water molecules get oriented. As the molecular layer gets thicker, orientation becomes weaker, and at a certain distance from the particle surface the water molecules cease to orientate and capillary water (liquid water) begins to appear. Due to the forces of adsorption (attraction) exerted by the surface of soil particles, water gets attached on the soil surface. The force of gravity also acts simultaneously, which tries to pull it downwards. The surface force is far greater than the force of gravity so water may remain attached to the soil particle. The water remains attached to the soil particle or may move downward into the lower layers, depending on the magnitude of the resultant force.

Factors affecting soil water

- **Texture** - Finer the texture, more is the pore space and also the surface area, and greater is the retention of water.
- **Structure** - Well–aggregated porous structure favours better porosity, which in turn enhances water retention.

- **Organic matter** - Higher the organic matter, more is the water retention in the soil.
- **Density of soil** - Higher the density of soil, lower is the moisture content.
- **Temperature** - Cooler the temperature, higher is the moisture retention.
- **Salt content** - More the salt content in the soil, less is the water available to the plant.
- **Depth of soil** - More the depth of soil, more is the water available to the plant.
- **Type of clay** - The 2:1 type of clay increases the water retention in the soil.

18.2. SOIL WATER POTENTIAL

The retention and movement of water in soils, its uptake and translocation in plants and its loss to the atmosphere are all energy related phenomenon. The more strongly water is held in the soil the greater is the heat (energy) required. In other words, if water is to be removed from a moist soil, work has to be done against adsorptive forces. Conversely, when water is adsorbed by the soil, a negative amount of work is done. The movement is from a zone where the free energy of water is high (standing water table) to the zone where the free energy is low (a dry soil). This is called 'soil water energy concept'.

Soil water potential is a measure of the energy status of the soil water. It is important because it reflects how hard plants must work to extract water. Units of measure are normally bars or atmospheres. Soil water potentials are negative pressures (tension or suction). Water flows from a higher (less negative) potential to a lower (more negative) potential.

Free energy of soil solids for water is affected by

- Matric (solid) force (matric potential) i.e., the attraction of the soil solids for water (adsorption), which markedly reduces the free energy (movement) of the adsorbed water molecules.
- Osmotic force (osmotic potential) i.e., the attraction of ions and other solutes in water to reduce the free energy of soil solution.
- Matric and osmotic potentials are negative and reduce the free energy level of the soil water. These negative potentials are referred to as suction or tension.

Force of gravity (gravitational potential)

This acts on soil water. The attraction is towards the Earth's centre, which tends to pull the water downward. This force is always positive. The difference between the energy states of soil water and pure free water is known as soil water potential. Total water potential (Pt) is the sum of the contributions of gravitational potential (Pg), matric potential (Pm) and the osmotic potential or solute potential (Po).

$$Pt = Pg + Pm + Po$$

Potential represents the difference in free energy levels of pure water and of soil water. The soil water is affected by the force of gravity, presence of soil solid (matric) and of solutes.

Another way of expression

$$\psi_t = \psi_g + \psi_m + \psi_o$$

ψt = Total soil water potential

ψg = Gravitational potential (force of gravity pulling the water)

ψm = Matric potential (force placed on the water by the soil matrix – soil water tension)

ψo = Osmotic potential (due to the difference in salt concentration across a semi–permeable membrane such as a plant root)

Matric potential, ψm, normally has the greatest effect on release of water from the soil to the plants.

Methods of expressing suctions

There are two units to express differences in energy levels of soil water.

1. pF scale - The free energy is measured in terms of the height of a column of water required to produce necessary suction or pressure difference at a particular soil moisture level. The pF, therefore, represents the logarithm of the height of water column (cm) to give the necessary suction.

2. Atmospheres or Bars - It is another common mean of expressing suction. Atmosphere is the average air pressure at sea level.

Table 56. pF values related to pressure and height of column of water

Height of column of water (cm)	Pressure (atmosphere)	pF value
1	0.001	0
10	0.01	1
10^2	0.1	2
10^3	1.0	3
10^4	10.0	4
10^5	100.0	5
10^6	1000.0	6
10^7	10,000.0	7

If the suction is very low as in the case of a wet soil containing the maximum amount of water that it can hold, the pressure difference is of the order of about 0.01 atmosphere or 1 pF equivalent to a column of water 10 cm in height. Similarly, if the pressure difference is 0.1 atmosphere, the pF will be 20. Soil moisture constants can be expressed in terms of pF values. A soil that is saturated with water has pF 0 while an oven–dry soil has pF 7.

18.3. MEASURING SOIL MOISTURE

Two general types of measurements relating to soil water are ordinarily used:

a. By some methods the moisture content is measured directly or indirectly

b. Techniques are used to determine the soil moisture potential (tension or suction)

18.3.1. Measuring Soil Moisture Content in Laboratory

1. Gravimetric method - In this method, a moist sample is dried in an oven at 105°C until there is no more weight loss. Then, the percentage of moisture is determined. The gravimetric method is time consuming and involves laborious processes of sampling, weighing and drying in laboratory.

2. Electrical conductivity method - This method is based upon the changes in electrical conductivity with changes in soil moisture. Gypsum blocks with two electrodes at a definite distance are used in this method. These blocks require previous calibration for uniformity. The blocks are buried in the soil at desired depths and the conductivity across the electrodes measured with a modified Wheatstone bridge. These electrical measurements are affected by salt concentration in the soil solution and are not very helpful in soils with high salt contents.

18.3.2. Measuring Soil Moisture Potential in situ (field)

Soil moisture potential is measured in situ by suction method or equilibrium tension method. In this method, field tensiometers measure the tension with which water is held in the soils. They are used in determining the need for irrigation. The tensiometer is a porous cup attached to a glass tube, which is connected to a mercury monometer. The tube and cup are filled with water and the cup is inserted in the soil. The water flows through the porous cup into the soil until equilibrium is established. These tension readings in monometer, expressed in terms of cm or atmosphere, measures the tension or suction of the soil.

If the soil is dry, water moves through the porous cup, setting up a negative tension (or greater is the suction). The tensiometers are more useful in sandy soils than in fine–textured soils. Once the air gets entrapped in the tensiometer, the reliability of readings is questionable.

18.4. CLASSIFICATION OF SOIL WATER

Soil water has been classified from a physical and biological point of view as physical classification of soil water, and biological classification of soil water.

18.4.1. Physical Classification of Soil Water

- Gravitational water
- Capillary water
- Hygroscopic water

1. Gravitational water

Gravitational water occupies the larger soil pores (macropores) and moves down readily under the force of gravity. Water in excess of the field capacity is termed gravitational water. Gravitational water is of no use to plants because it occupies the larger pores. It reduces aeration in the soil. Thus, its removal from soil is a requisite for optimum plant growth. Soil moisture tension at gravitational state is zero or less than 1/3 atmosphere.

Factors affecting gravitational water

a. Texture - It plays a great role in controlling the rate of movement of gravitational water. The flow of water is proportional to the size of particles. The bigger the particle, the more rapid is the flow or movement. Because of the larger size of pores, water percolates more easily and rapidly in sandy soils than in clay soils.

b. Structure - It also affects gravitational water. In platy structure, movement of gravitational water is slow and water stagnates in the soil. Granular and crumby structure helps to improve gravitational water movement. In clay soils having single grain structure, the gravitational water percolates more slowly. If clay soils form aggregates (granular structure), the movement of gravitational water improves.

2. Capillary water

Capillary water is held in the capillary pores (micropores). Capillary water is retained on the soil particles by surface forces. It is held so strongly that gravity cannot remove it from the soil particles. The molecules of capillary water are free and mobile, and are present in a liquid state. Due to this reason, capillary water evaporates easily at ordinary temperature though it is held firmly by the soil particle. Plant roots are able to absorb the capillary water. Therefore, it is known as available water. The capillary water is held between 1/3 and 31 atmosphere pressure.

Factors affecting capillary water

The amount of capillary water that a soil is able to hold varies considerably. The following factors are responsible for variation in the amount of capillary water:

a. **Surface tension -** An increase in surface tension increases the amount of capillary water.

b. **Soil texture -** The finer the texture of a soil, greater is the amount of capillary water. This is mainly due to the greater surface area and a greater number of micropores.

c. **Soil structure -** Platy structure contains more water than granular structure.

d. **Organic matter -** The presence of organic matter helps to increase the capillary capacity of a soil. Organic matter itself has a great capillary capacity. Undecomposed organic matter is generally porous having a large surface area, which helps to hold more capillary water. The humus that is formed on decomposition has a great capacity for absorbing and holding water. Hence, the presence of organic matter in soil increases the amount of capillary water in soil.

3. Hygroscopic water

The water that holds tightly on the surface of soil colloidal particle is known as hygroscopic water. It is essentially non–liquid and moves primarily in the vapour form.

Hygroscopic water is held so tenaciously (31–10,000 atmosphere) by soil particles that plants cannot absorb it. Some microorganism may utilize hygroscopic water. As hygroscopic water is held tenaciously by surface forces its removal from the soil requires a certain amount of energy. Unlike capillary water which evaporates easily at atmospheric temperature, hygroscopic water cannot be separated from the soil unless it is heated.

Factors affecting hygroscopic water

Hygroscopic water is held on the surface of colloidal particles by the dipole orientation of water molecules. The amount of hygroscopic water varies inversely with the size of soil particles. The smaller the particle, the greater is the amount of hygroscopic water it adsorbs. Fine–textured soils such as clay contain more hygroscopic water than coarse–textured soils. The amount of clay and also its nature influences the amount of hygroscopic water. Clay minerals of the montmorillonite type with their large surface area adsorb more water than those of the kaolinite type, while illite minerals are intermediate.

18.4.2. Biological Classification of Soil Water

There is a definite relationship between moisture retention and its utilization by plants. This classification is based on the availability of water to the plant. Soil moisture can be divided into three parts:

1. Available water - It is the water which lies between wilting coefficient and field capacity. It is obtained by subtracting wilting coefficient from moisture equivalent.

2. Unavailable water - This includes the whole of the hygroscopic water plus a part of the capillary water below the wilting point.

3. Super available or superfluous water - The water beyond the field capacity stage is said to be super available. It includes gravitational water plus a part of the capillary water removed from larger interstices. This water is unavailable for the use of plants. The presence of super available water in a soil for any extended period is harmful to plant growth because of the lack of air.

18.5. SOIL MOISTURE CONSTANTS

Earlier classification divided soil water into gravitational, capillary and hygroscopic water. The hygroscopic and capillary waters are in equilibrium with the soil under given conditions. The hygroscopic coefficient and the maximum capillary capacity are the two equilibrium points when the soil contains the maximum amount of hygroscopic and capillary waters, respectively. The amount of water that a soil contains at each of these equilibrium points is known as soil moisture constant.

Therefore, the soil moisture constant represents definite soil moisture relationship and retention of soil moisture in the field. The three classes of water (gravitational, capillary and hygroscopic) are, however, very broad and do not accurately represent the soil–water relationships that exist under field conditions.

- Though the maximum capillary capacity represents the maximum amount of capillary water that a soil can hold, the whole of capillary water is not available for the use of the plants.
- A part of it, at its lower limit approaching the hygroscopic coefficient, is not utilized by the plants.
- Similarly, a part of the capillary water at its upper limit is also not available for the use of plants.

Hence, two more soil constants viz., field capacity and wilting coefficient have been introduced to express the soil–plant–water relationships as it is found to exist under field conditions.

1. Field capacity

Assume that water is applied to the surface of a soil. With the downward movement of water all macro and micropores are filled up. The soil is said to be saturated with respect to water and is at maximum water–holding capacity or maximum retentive capacity. It is the amount of water held in the soil when all pores are filled.

Sometimes, after application of water in the soil all the gravitational water is drained away, and then the wet soil is almost uniformly moist. The amount of water held by the soil at this stage is known as the field capacity or normal moisture capacity of that soil. It is the capacity of the soil to retain water against the downward pull of the force of gravity. At this stage, only micropores or capillary pores are filled with water and plants absorb water for their use. At field capacity water is held with a force of 1/3 atmosphere. Water at field capacity is readily available to plants and microorganisms.

2. Wilting coefficient

As the moisture content falls, a point is reached when the water is so firmly held by the soil particles that plant roots are unable to draw it. The plant begins to wilt. At this stage even if the plant is kept in a saturated atmosphere it does not regain its turgidity and wilts unless water is applied to the soil. The stage at which this occurs is termed the wilting point and the percentage of water held by the soil at this stage is known as the wilting coefficient. It represents the point at which the soil is unable to supply water to the plant. Water at wilting coefficient is held with a force of 15 atmospheres.

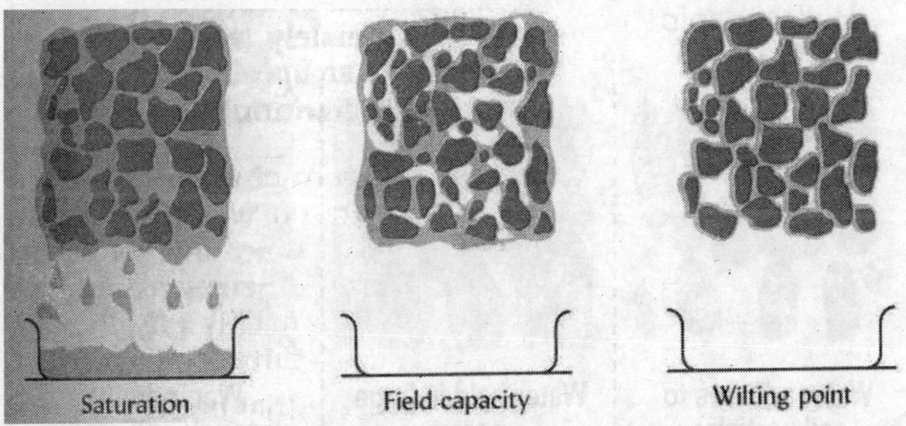

Fig. 45. Field Moisture Conditions

3. Hygroscopic coefficient

The hygroscopic coefficient is the maximum amount of hygroscopic water absorbed by 100 g of dry soil under standard conditions of humidity (50% relative humidity) and temperature (15°C). This tension is equal to a force of 31 atmospheres. Water at this tension is not available to plant but, may be available to certain bacteria.

4. Available water capacity

The amount of water required to apply to a soil at the wilting point to reach the field capacity is called the 'available water'. The water supplying power of soils is related to the amount of available water a soil can hold. The available water is the difference in the amount of water at field capacity (–0.3 bar) and the amount of water at the permanent wilting point (–15 bars).

5. Maximum water holding capacity

It is also known as maximum retentive capacity. It is the amount of moisture in a soil when its pore spaces both micro and macrocapillary are completely filled with water. It is a rough measure of total pore space of soil. Soil moisture tension is very low between $1/100^{th}$–$1/1000^{th}$ of an atmosphere or pF 1–0.

6. Sticky point moisture

It represents the moisture content of soil at which it no longer sticks to a foreign object. The sticky point represents the maximum moisture content at which a soil remains friable. Sticky point moisture values vary nearly approximate to the moisture equivalent of soils. Summary of the soil moisture constants, type of water and force with which it is held is given in Table 58.

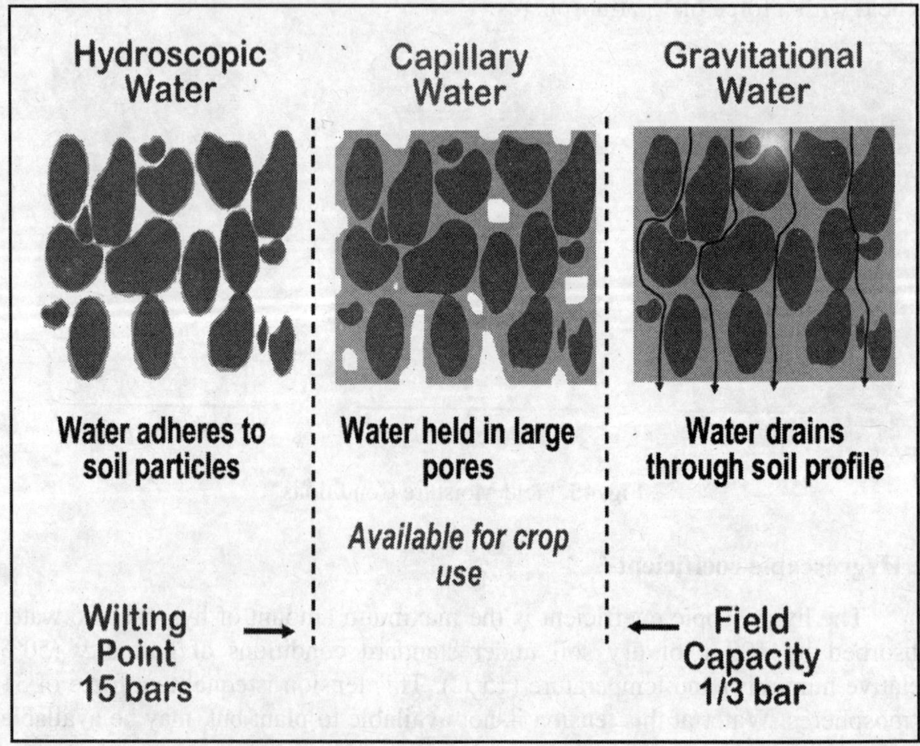

Fig. 46. Soil water held at different tensions

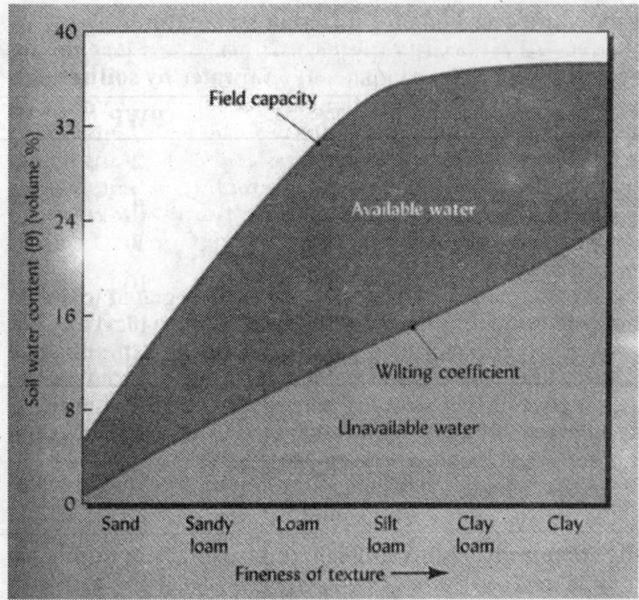

Fig. 47. The relationship between soil water content and texture

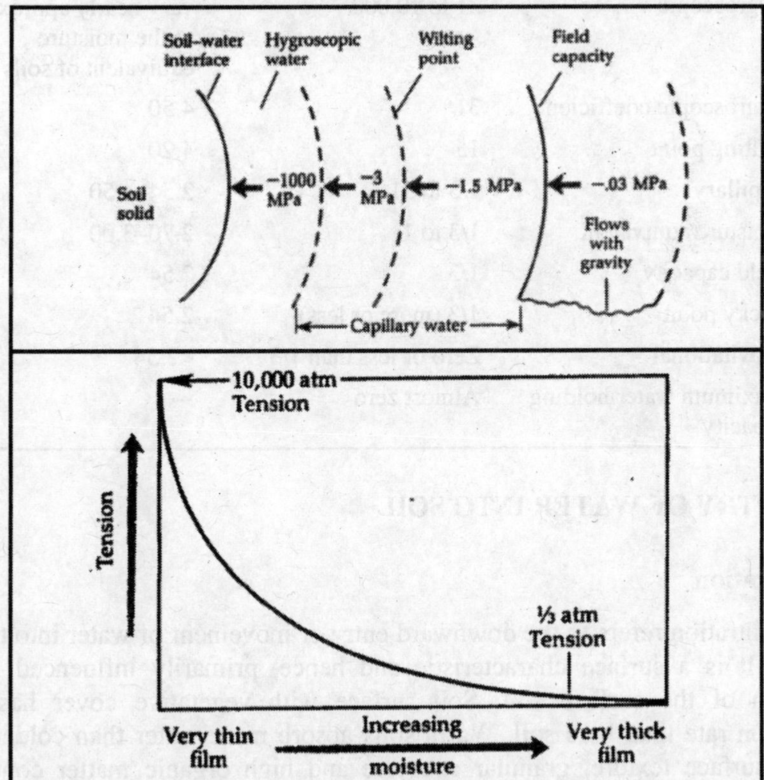

Fig. 48. Relationship between water films and soil particles and the tension with which water is held

Table 57. Soil moisture constants for different soil textures

Soil texture	% water by soil volume		
	FC	PWP	AW
Sand	9	2	7
Sandy loam	15	4	11
Silt loam	22	7	15
Clay loam	28	10	18
Clay	35	17	18
Fine clay	45	33	12

Table 58. Soil moisture constants and range of tension and pF

Sr. No.	Moisture class	Range of tension in atmosphere	Equivalent pF range
1.	Chemically combined	Very high	---
2.	Water vapour	Held at saturation point in the soil air	---
3.	Hygroscopic	31 to 10,000	vary nearly approximate to the moisture equivalent of soils
4.	Hygroscopic coefficient	31	4.50
5.	Wilting point	15	4.20
6.	Capillary	1/3 to 31	2.54–4.50
7.	Moisture equivalent	1/3 to 1	2.70–3.00
8.	Field capacity	1/3	2.54
9.	Sticky point	1/3 (more or less)	2.54
10.	Gravitational	Zero or less than 1/3	<2.54
11.	Maximum water holding capacity	Almost zero	---

18.6. ENTRY OF WATER INTO SOIL

1. Infiltration

Infiltration refers to the downward entry or movement of water into the soil surface. It is a surface characteristic and hence, primarily influenced by the condition of the surface soil. Soil surface with vegetative cover has more infiltration rate than bare soil. Warm soils absorb more water than colder ones. Coarse surface texture, granular structure and high organic matter content in surface soil, help to increase infiltration. Infiltration rate is comparatively lower in wet soils than in dry soils.

Factors affecting infiltration

- Clay minerals
- Soil texture
- Soil structure
- Moisture content
- Vegetative cover
- Topography

2. Percolation

The movement of water through a column of soil is called percolation. It is important for two reasons:

i. This is the only source of recharge of ground water, which can be used through wells for irrigation.

ii. Percolating waters carry plant nutrients down and often out of reach of plant roots (leaching).

Percolation is dependent on rainfall. In dry region, it is negligible and under high rainfall it is high. Sandy soils have greater percolation than clayey soil. Vegetation and high water table reduce the percolation loss.

3. Permeability

It indicates the relative ease of movement of water within the soil. The characteristic that determines how fast air and water move through the soil is known as permeability. The term hydraulic conductivity is also used, which refers to the readiness with which a soil transmits fluids through it.

Soil water movement

- Saturated flow
- Unsaturated flow
- Water vapour movement

Saturated flow - This occurs when the soil pores are completely filled with water. This water moves at water potentials larger than –33 kPa. Saturated flow is water flow caused by gravity's pull. It begins with infiltration, which is water movement into soil when rain or irrigation water is on the soil surface. When the soil profile is wetted, the movement of more water flowing through the wetted soil is termed as 'percolation'.

Hydraulic conductivity can be expressed mathematically as:

$$V = kf$$

Where,

V = Total volume of water moved per unit time

f = Water moving force

k = Hydraulic conductivity of soil

Factors affecting movement of water

- Texture
- Structure
- Amount of organic matter
- Depth of soil to hard pan
- Amount of water in the soil
- Temperature
- Pressure

Vertical water flow

The vertical water flow rate through soil is given by Darcy's law. The law states that the rate of flow of liquid or flux through a porous medium is proportional to the hydraulic gradient in the direction of flow of the liquid.

$$QW = -k\frac{(dw)\,At}{Ds}$$

Where,

QW = Quantity of water in cm^{-3}

k = Rate constant (cm/s)

dw = Water height (head) (cm)

A = Soil area (cm^2)

t = Time

ds = Soil depth (cm)

Unsaturated flow

At suctions less than one-third bar, water moves in all directions via unsaturated flow at a rate that is dependent on the square of the diameter of the water–filled pores. Water is pushed by pressure gradients from the point of its application where it is saturated locally, and pulled by capillary action due to adhesion force of water to the soil solids, producing a suction gradient from wet towards drier soil. Doubling the diameter of the pores increases the flow rate by a factor of four. Large pores, drained by gravity and not filled with water, do not greatly increase the flow rate for unsaturated flow. Water flow is primarily from coarse–textured soil into fine–textured soil and is slowest in fine–textured soils such as clay. Unsaturated flow is generally rapid through fine sand or well–

aggregated loams (medium–sized pores) and slower through very fine and poorly–aggregated clayey soil (very small pores).

Water vapour movement

Vapour flow occurs only in relatively dry soil where vapour pressure differences are significant. Water vapour movement can occur within the soil pores (internal vapour movement) and also at the soil surface where water vapour is lost by evaporation (external vapour movement). Water vapour moves in response to vapour pressure gradient. Water vapour pressure depends on the salt content. Zones of higher salt contents, such as around a fertilizer granule, have a lower vapour pressure and therefore, attract water vapour from surrounding zones. Vapour pressure also depends on the temperature, with warmer soil layers to cooler soil layers, if the soil is uniformly moist. This can be of significance for drought–resistant desert plants, when surface horizons cool at night causing vapour movement up from deeper layers. However, the amount of water vapours is very small, and hence it is of limited practical importance in the soils at the optimum water content for plant growth.

19 SOIL COLLOIDS

The colloidal state refers to a two–phase system in which one material (or materials) in a very finely divided state is dispersed through a second material. Few examples of the colloidal state are milk and cheese; clouds and fog; starch, gelatin, rubber, proteins, blood, and soil.

The rapidly–diffusing substances, mostly of crystalline nature in the solid state, are called 'crystalloids' while starch and proteins are called 'colloids'. But this distinction is not always true, because a crystalline substance like NaCl can be obtained in colloidal condition while albumin may be obtained in crystalline form. Therefore, it is more appropriate to discuss a colloidal state of a substance than a colloidal substance. The size of the particles in true solutions and in colloidal suspensions is 0.2–1 mμ and 1–200 mμ respectively. Colloidal particles float in a medium and do not tend to settle. Therefore, colloids are referred to as dispersed systems. The substance in solution is called the 'dispersed phase' while the medium in which the particles are dispersed is called the 'dispersion medium'. Soil particles less than 0.001 mm size possess colloidal properties and are known as 'soil colloids'.

19.1. NATURE AND IMPORTANT PROPERTIES OF SOIL COLLOIDAL PARTICLES

1. Adsorption - The surface adsorption is very large when large amount of colloidal materials are present in a medium, which provide large surface area. The adsorption of ions is governed by the type and nature of ion, and the type of colloidal particles. The higher the valence of cations, the more strongly it is adsorbed. Hydrogen (H^+) ions behave as polyvalent ions and hence, it is adsorbed on the surfaces of colloids more strongly than those divalent cations such as Ca^{2+}. Among the clay minerals, kaolinitic clay has a greater anion e.g., phosphate adsorbing capacity is higher than montmorillonitic or illitic clay.

2. Brownian movement - Colloidal particles are found to be in continual motion. The oscillation is due to the collision of colloidal particles or molecules of the liquid in the dispersion medium. This movement is primarily responsible for the coagulation or flocculation of soil colloidal materials (clay and humus).

3. Electrical charge - Colloidal particles usually have an electrical charge. Some colloidal particles are positively–charged and some are negatively-charged. When clay colloids are suspended in water, it carries a negative electric charge and thereby attracts positively charged ions (cations).

4. Flocculation - The colloidal particles are coagulated by adding an oppositely–charged ion. This process of formation of flocs is known as flocculation e.g., clay is coagulated by the use of aluminium (Al^{3+}). This condition is generally beneficial in relation to agriculture since it is the first step in the formation of stable aggregates or granules. The ability of common cations to flocculate soil colloids is in the order of:

$$Al > Ca \text{ and } H > Mg > K > Na$$

In soils of humid and sub–humid regions, the colloidal complexes are dominated by aluminium, hydrogen and calcium and these ions encourage flocculation and consequently help to form aggregates whereas, the soils of arid regions are dominated by Sodium on exchange complex. This results in dispersion of soil colloids and develops undesirable poor physical conditions of soil.

So, the phenomenon of flocculation plays an important role in the field of agriculture by improving the physical condition as well as other related condition of the soil, which is beneficial for the plant growth.

5. Plasticity - Soils containing more than 15 per cent of colloidal clay exhibit plasticity. Plasticity occurs only when soils are moist or wet. Due to this property clay colloids can be moulded to any shape. Plasticity phenomenon is extremely important because it encourages a change in soil structure, which is most related to the tillage operations.

6. Cohesion and adhesion - Colloidal clay particles exhibit the properties of cohesion and adhesion. Due to cohesion force clay particles are able to form aggregates and due to adhesive force clay particles envelope sand particles. Both the forces develop in the presence of water and these two forces help in the retention of water in the soil as well as absorption to the plants and microorganisms.

7. Swelling and shrinkage - Contraction and expansion in volume of soil with the variation in moisture are influenced by the presence of colloidal clay particles in soil. Swelling phenomenon occurs when colloidal clay particles are allowed to be placed in contact with moisture due to imbibition of water. Whereas shrinkage occurs in dry conditions of the soils containing sufficient amount of colloidal particles and causes decrease in volume with cracks in variable nature. Shrinkage and swelling are more pronounced in montmorillonite clays (expanding 2:1 type) than that of kaolinite (non–expanding 1:1 type) clays and these phenomena play a significant role in the formation of soil structure.

8. Non–permeability - Colloids are unable to pass through a semi permeable membrane, and this membrane allows passing water and other dissolved substances, but retains colloidal materials.

19.2. CONCEPT OF SILICATE CLAYS

1. Size and chemical composition - The chemical analysis of clay indicates the presence of silica, alumina, iron and chemically–combined water. These make up from 90-98 per cent of the colloidal clay. The soil colloidal matter contains plant nutrients like Ca, Mg and K. Clay is a mixture of hydrated alumino ferrosilicates of varying composition mixed in some cases with an excess of sesquioxides or silica.

The term 'clay' has three meanings in soil usage: (1) it is a particle fraction composed of any particles less than 2 microns (<2μ) in effective diameter; (2) it is a name for minerals of specific composition; and (3) it is a soil textural class. Many materials of the clay size fraction such as gypsum, carbonates or quartz are small enough to be classified as 'clay' on the basis of size but, are not 'clay minerals'. Sometimes some clay minerals have a size of 4 or 5μ (double the upper size limit of the 'clay size fraction').

2. Shape - Silicate clay minerals have been examined by electron microscope and it was found that the particles are laminated, made up of layers of plates or flakes or even rods. Each clay particle is made up of large number of plate–like structural units. The different units or flakes of clay minerals are held together with varying degrees of force depending upon the nature of the clay mineral.

The edges of some clay particles are clean cut and others are frayed or fluffy. In all cases, clay minerals are developed more in the horizontal axis than in the vertical axis.

3. Surface area - The surface area of a clay particle is usually defined as the area of the particle that is accessible to ions or molecules when the clay is in an aqueous solution. All clay particles (finer fraction of soil) must expose a large amount of external surface.

Some clays have extensive internal surfaces as well. This internal surface exists between the plate–like crystal units that make up each particle. So, the large surface area of clay colloids is not only due to its fineness but, also because of its plate–like structure.

Surface areas of clay particles can be measured by using cetyl pyridinium bromide (dissolving in water) for fully–dispersed clay suspensions.

Surface areas for different clays are:

- Sodium montmorillonite - 700–800 m^2/g
- Vermiculites and some mixed layer clays - 300–500 m^2/g
- Micaceous clays - 100–300 m^2/g
- Kaolinitic clays - 5–100 m^2/g
- Amorphous clays (oxygen and other atoms less regularly oriented) - 100–500 m^2/g

4. Electronegative charge - Clay micelles (microcells) carry negative charges and so a number of oppositely charged ions (cations) are attracted to each colloidal clay crystal. The colloidal clay particles have inner ionic layer (surfaces of highly negative charge) and the outer ionic layer (cations swarming layer).

5. Adsorbed cations - Clay micelles adsorb a number of cations. The common cations adsorbed are H^+, Al^+, Ca^{2+}, Mg^{2+}, Na^+ and K^+. The amount of these cations held by clay varies with its kind.

Cations adsorbed (if dominant) on the clay colloids very often determine the physical and chemical properties of the soil and thereby influence the plant growth.

Example - Alkali soil dominated by sodium on the surface of the clay micelle causes poor physical condition of the soil.

19.3. TYPE OF SOIL COLLOIDS

The individual crystalline particle is called 'micelle' or 'microcell'. Colloidal clay particles are considered as occupying an intermediate position being neither truely lyophilic (liquid loving) or hydrophilic (water loving) nor lyophobic (liquid hating) or hydrophobic (water repellent). Clay colloids possess the properties of hydrophilic because of its hydration. Its sensitivity to electrolytes is a hydrophobic nature.

Lyophobic colloids

The surface tension and viscosity of this colloidal system is similar to that of the dispersion medium. Small quantities of electrolytes cause precipitation. The particles are easily detected in the ultra–microscope and they migrate in one direction in an electric field.

Lyophilic colloids

The surface tension is usually lower and the viscosity is much higher than that of the dispersion medium. Large quantities of electrolytes cause precipitation. The particles cannot be easily detected in ultra–miscroscope and they migrate in different directions or not at all in an electric field.

Among all kinds of soil colloids, the silicate clays are the most dominant and important in the field of agriculture. Except sand fraction, all soils contain particles of colloidal size. Soil colloids are chemically–active portions of the soil and they largely determine the physical and chemical properties of a soil. They are broadly classified into inorganic colloids and organic colloids.

Inorganic colloids

Clay minerals, hydrous oxides usually make up the bulk of soil colloids. Colloids are particles less than 0.001 mm in size and the clay fraction includes particles less than 0.002 mm in size. Therefore, all clay minerals are not strictly colloidal.

Organic colloids

The organic colloids include highly decomposed organic matter generally called humus. Organic colloids are more reactive chemically and generally have a greater influence on soil properties per unit weight than the inorganic colloids. Humus is amorphous and its chemical and physical characteristics are not well–defined.

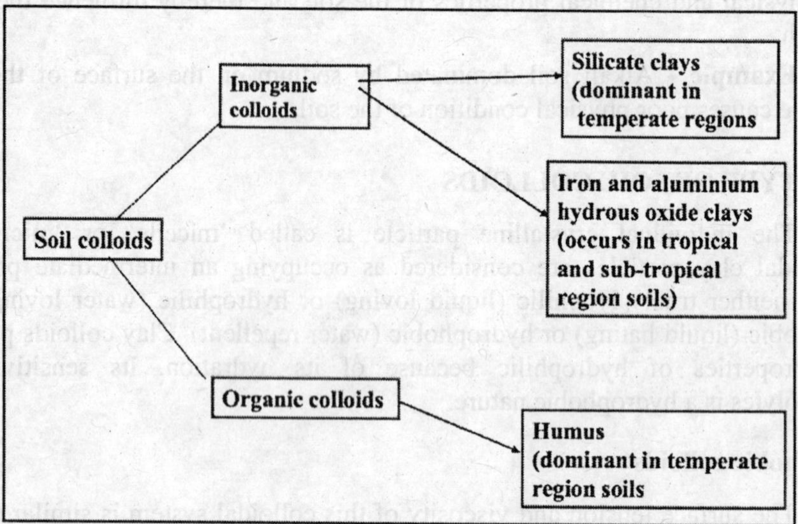

Fig. 49. Different types of colloids

Clay minerals are usually crystalline (although some are amorphous) and usually have a characteristic chemical and physical configuration. Both inorganic and organic colloids are intimately mixed with other soil solids. Thus, the bulk of the soil solids are essentially inert and the majority of the soil's physical and chemical characteristics are a result of the colloids present.

19.3.1. Types of Soil Colloids

- Crystalline Silicates

  1:1, 2:1 (expanding and non–expanding), 2:1:1
- Non–Crystalline Silicates

  Allophane and Immogolite (products of volcanic ash)
- Fe and Al oxides / Hydroxides / Oxyhydroxides

  Geothite, Gibbsite, Hematite, Ferryhydrite, Maghemite
- Organic Matter (humus)

  High and Low Molecular Weight Acids (Humic acid, Fulvic acid)

1. Structure of layer silicate clays

These are most important silicate clays, which are known as phyllosilicates (Phyllon - leaf) because of their leaf–like or plate–like structure. They are comprised of two kinds of horizontal sheets. One dominated by silicon and other by aluminum and/or magnesium.

Most of the silicate clays are made up of planes of oxygen atoms with silicon and aluminium atoms holding the oxygen together by ionic bonding, which is the attraction of positively and negatively–charged atoms. Three or four planes of oxygen atoms with intervening silicon and aluminium ions make up a layer. One clay particle is composed of many layers stacked like a deck of cards. Most silicate clays are aluminosilicates (aluminium and silicon components of the clay structure).

Silica tetrahedron

The basic building block for the silica–dominated sheet is a unit composed of one silicon atom surrounded by four oxygen atoms forming silica tetrahedron (because of its four–sided configuration) and the unit has tetrahedral coordination. The planes of oxygen held together by Si^{4+} are tetrahedral oriented and are referred to as a silica tetrahedral sheet.

Alumina octahedron

The aluminium octahedron is an eight–sided building block consisting of central aluminium atom surrounded by six hydroxyls or oxygens. Large numbers of aluminium octahedra, bound to each other by shared oxygen atoms in an octahedral layer, are arranged in a plane forming aluminium octahedral sheet.

Aluminium and/or magnesium ions are the key cations in the second type of sheet. An aluminium (or magnesium) ion surrounded by six oxygen atoms or hydroxyl group gives an eight–sided building block termed octahedron. An aluminium–dominated sheet is known as a dioctahedral sheet, whereas one dominated by magnesium is called a trioctahedral sheet. The distinction is due to the fact that two aluminium ions in a dioctahedral sheet satisfy the same negative charge from surrounding oxygen and hydroxyls as three magnesium ions in a trioctahedral sheet.

One silica sheet per one aluminium sheet is a 1:1 lattice. A 2:1 lattice has two silica sheets per (one) aluminium sheet. Different combinations of these two general structural units (tetrahedral and octahedral sheets) form the structures of the various layer silicates like mica, vermiculite, montmorillonite, chlorite, kaolinite and other interstratified and intergradient layer silicates.

On the basis of the number and arrangement of silica and alumina sheets, silicate clays may be classified into different groups:

a. 1:1 type minerals

b. 2:1 type minerals

i. Expanding minerals (expand between crystal units)

ii. Non–expanding minerals

c. 2:2 type minerals.

a. 1:1 type minerals

The layers of the 1:1 type minerals are made up of one tetrahedral (silica) sheet combined with one octahedral (alumina) sheet. In soils, kaolinite is the most prominent member of this group, which includes hallosite, nacrite and dickite.

The tetrahedral and octahedral sheets in a layer of a kaolinite crystal are held together tightly by oxygen atoms, which are mutually shared by the silicon and aluminium cations in their respective sheets. These layers, in turn, are held together by hydrogen bonding. Consequently, the structure is fixed and ordinarily no expansion occurs between layers when the clay is wetted.

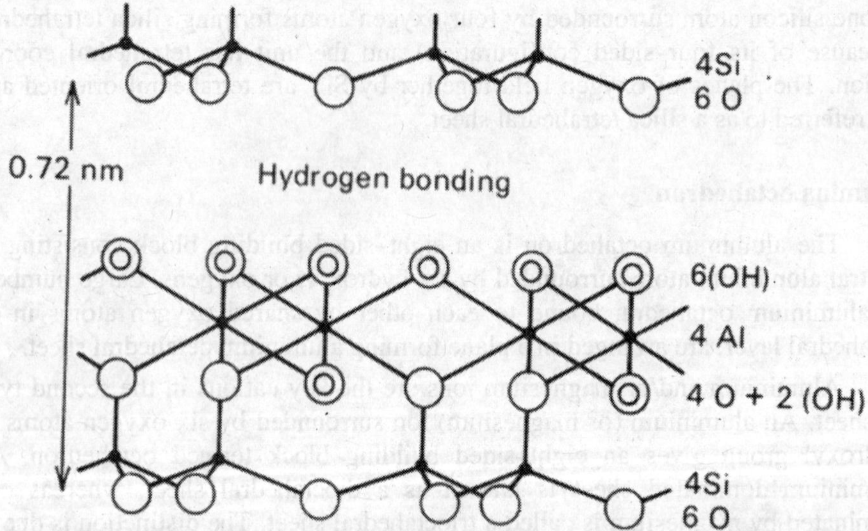

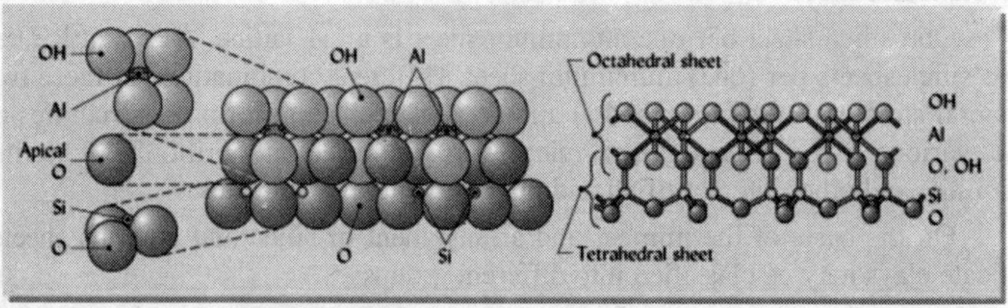

Fig. 50. Diagrams of the structure of Kaolinite

Cations and water do not enter between the structural layers of a 1:1 type mineral particle. The effective surface of kaolinite is thus, restricted to its outer faces or to its external surface area. Also, there is little isomorphous substitution in this 1:1 type mineral. Along with the relatively low surface area of kaolinite, this accounts for its low capacity to adsorb cations.

Kaolinite crystals are usually hexagonal in shape. In comparison with other clay particles, they are large in size, ranging from 0.10–5 nm across with the majority falling within the 0.2–2 nm range. Because of the strong binding forces between their structural layers, kaolinite particles are not readily broken down into extremely thin plates. Kaolinite exhibits very little plasticity (capability of being moulded), cohesion, shrinkage and swelling.

b. 2:1 type minerals

The crystal units (layers) of these minerals are characterized by an octahedral sheet sandwiched between two tetrahedral sheets. Three general groups of expanding and non–expanding type have this basic crystal structure.

i. Expanding type - smectites and vermiculite

ii. Non–expanding type - mica (illite)

i. Expanding Minerals - The smectite group is noted for interlayer expansion, which occurs by swelling due to wetting of minerals, when the water enters the interlayer space and forces the layers apart. Montmorillonite is the most prominent member of this group in soils, although beidellite, nontronite, and saponite are also found.

The flake–like crystals of smectites (e.g., Montmorillonite) are composed of an expanding lattice 2:1 type clay mineral. Each layer is made up of an octahedral sheet sandwiched between two tetrahedral (silica) sheets. There is little attraction between oxygen atoms in the bottom tetrahedral sheet of one unit and those in the top tetrahedral sheet of another. This permits a ready and variable space between layers, which is occupied by water and exchangeable cations. This internal surface far exceeds the external surface of the crystal. In montmorillonite, magnesium has replaced aluminium in some sites of the octahedral sheet. Similarly, some silicon atoms in the tetrahedral sheet may be replaced by aluminium. These substitutions give rise to a negative charge.

These minerals show high cation exchange capacity, marked swelling and shrinkage properties. Wide cracks commonly form as smectite–dominated soils (e.g., Vertisols) are dried. The dry aggregates or clods are very hard, making such soils difficult to till.

Vermiculites are also 2:1 type minerals in that an octahedral sheet occurs between two tetrahedral sheets. In most soil vermiculites, the octahedral sheet is aluminium–dominated (dioctahedral), although magnesium–dominated (trioctahedral) vermiculites are also common. In the tetrahedral sheet of most vermiculite, considerable substitution of aluminium for silicon has taken place. This accounts for most of the very high net negative charge associated with these minerals.

Water molecules, along with magnesium and other ions, are strongly adsorbed in the interlayer space of vermiculites. They act primarily as bridges holding the units together rather than as wedges driving them apart. The degree of swelling is, therefore, considerably less for vermiculites (partially expanding) than for smectites. For this reason, vermiculites are considered limited–expansion clay minerals, expanding more than kaolinite but, much less than the smectites.

The cation exchange capacity of vermiculites usually exceeds that of all other silicate clays including montmorillonite and other smectites, because of very high negative charge in the tetrahedral sheet. Vermiculite crystals are larger than those of the smectites but, much smaller than those of kaolinite.

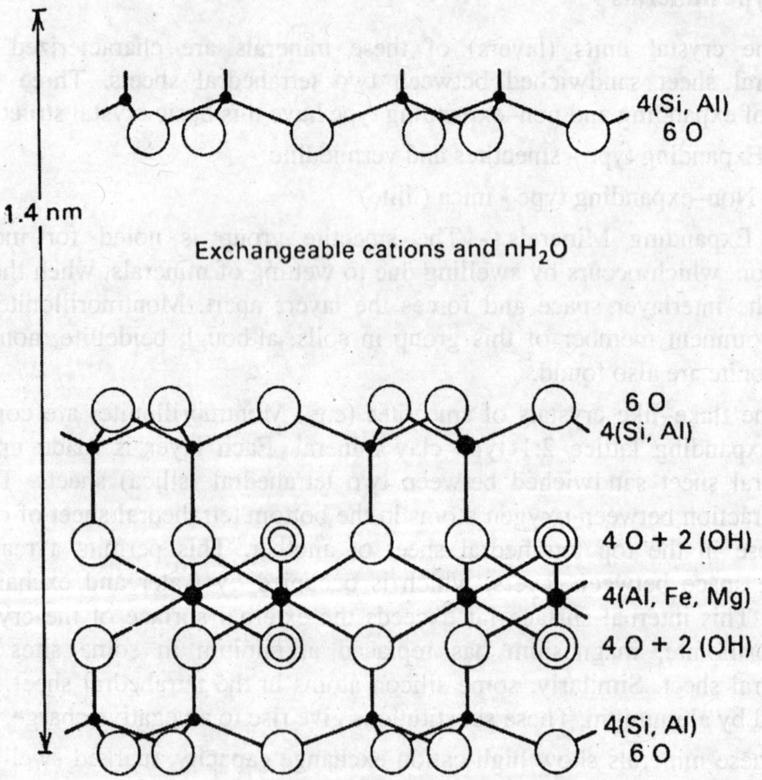

Fig. 51. Smectite Group (2:1)

ii. Non–expanding minerals - Micas are the type minerals in this group e.g., muscovite and biotite. Some weathered minerals similar in structure to these micas are found in the clay fraction of soils. They are called fine–grained micas. The clay formed from weathered biotite and muscovite is illite. Similar to smectites, fine–grained micas have 2:1 type crystal. However, the particles are much larger than those of the smectites. Also, the major source of charge is in the tetrahedral sheet where aluminium atoms occupy about 20 per cent of the silicon sites. These result in a high net negative charge in the tetrahedral sheet, even higher than that found in vermiculites. To satisfy this charge, potassium ions are strongly attracted in the interlayer space and are just the right size to fit into certain spaces in the adjoining tetrahedral sheets. The potassium, thereby, acts as a binding agent, preventing expansion of the crystal. Hence, fine–grained micas are quite non–expansive.

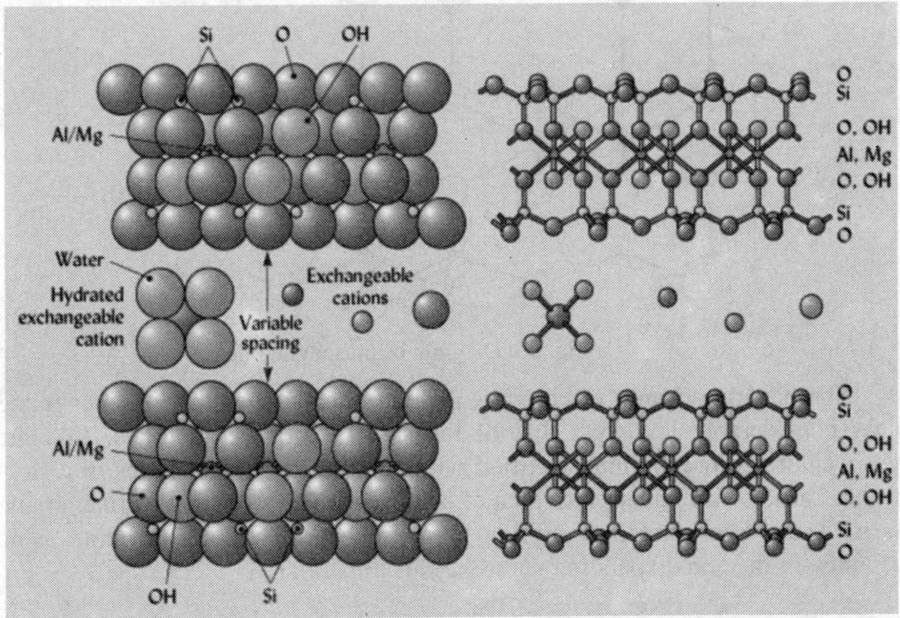

Fig. 52. Diagram of the structure of montmorillonite

The properties such as hydration, cation adsorption, swelling, shrinkage and plasticity are much less intense in fine–grained micas than in smectites. The fine–grained micas exceed kaolinite with respect to these characteristics, but this may be due in part to the presence of interstratified layers of smectite or vermiculite. In size, too, fine–grained mica crystals are intermediate between the smectities and kaolinites. Their specific surface area varies from 70–100 m^2/g, about one eighth of that for the smectites.

c. 2:2 or 2:1:1 type minerals

This silicate group is represented by chlorites, which are common in a variety of soils. **Chlorites** are basically iron magnesium silicates with some aluminium. A typical chlorite clay crystal has 2:1layers, similar to the layers in vermiculites, with alternate magnesium–dominated trioctahedral sheet, which gives 2:1:1 ratio. Magnesium also dominates the trioctahedral sheet in the 2:1 layer of chlorites. Thus, the crystal unit contains two silica tetrahedral sheets and two magnesium–dominated trioctahedral sheets giving rise to the term 2:1:1 or 2:2 type structure.

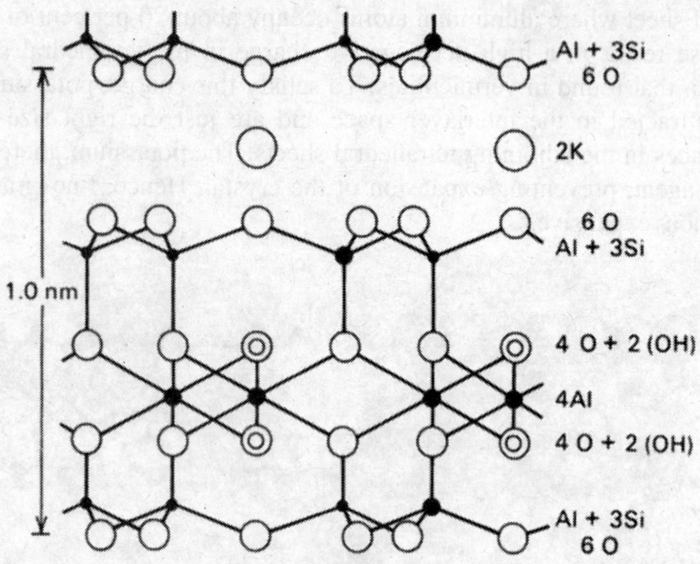

Fig. 53. Diagram of muscovite mica

The negative charge of chlorites is about the same as that of fine–grained micas considerably less than that of the smectites or vermiculites. Like fine micas, chlorites may be interstratified with vermiculites or smectites in a single crystal. Particle size and surface area of chlorites are same as that of fine–grained micas. There is no water adsorption between the chlorite crystal units, which accounts for the non–expanding nature of this mineral.

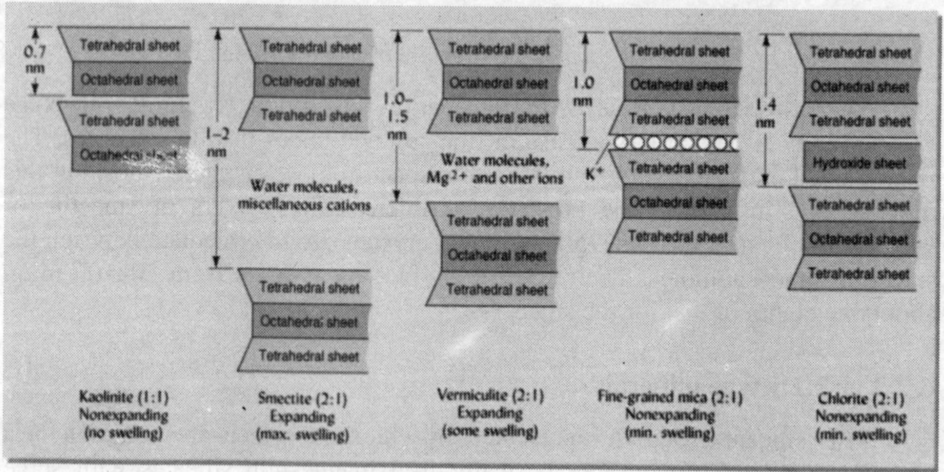

Fig. 54 Caly minerals comparison

Mixed and interstratified layers

Specific groups of clay minerals do not occur independently of one another. In a given soil, it is common to find several clay minerals in an intimate mixture.

Furthermore, some mineral colloids have properties and composition intermediate between those of any two of the well–defined minerals described. Such minerals are termed mixed layer or interstratified because the individual layers within a given crystal may be of more than one type. Terms such as 'chlorite-vermiculite' and 'fine–grained mica–smectite' are used to describe mixed–layer minerals. In some soils, they are more common than single–structured minerals such as montmorillonite.

Table 59. The common minerals (in descending order) as observed in the dominant soil groups of India

Sr. No.	Types of soils	Common Clay Minerals
1.	Alluvium derived soils	Illite, chlorite (intergrades) Montmorillonite (Flood plains and basins), Kaolinite
2.	Forest soils	Illite, chlorite (Vermiculite), Montmorillonite, Kaolinite
3.	Black (cotton) soils	Smectite, Illite, Chlorite / Chloritized Intergrades
4.	Desert soils	Illite, Attapulgite, Vermiculite
5.	Red and lateritic soils	Kaolinite (gibbsite), Chlorite, Illite

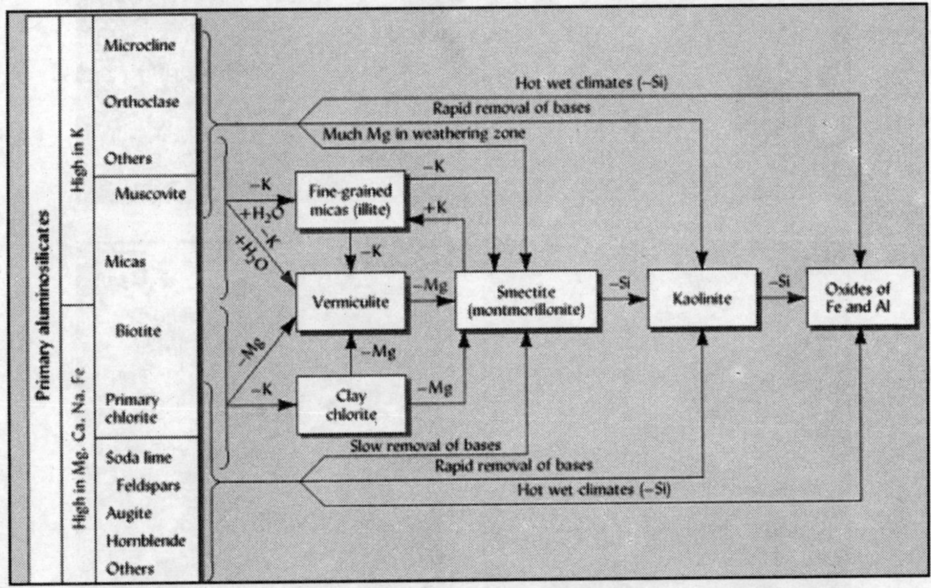

Fig. 55. Formation of various layer silicate clays and oxides of Fe and Al

19.3.2. Amorphous Clays

They are mixtures of silica and aluminium that have not formed well–oriented crystals. Technically, they are not minerals because they lack crystallinity. They are amorphous in nature. These clays occur where large amounts of weathered products existed but have not had the conditions or time for good crystal formation. Amorphous clays are common in soils formed from volcanic ash (e.g., Allophane). Their properties are inconsistent such as having high positive charges (high anion exchange capacities) or even high cation exchange capacity. Because almost all of their charge is from hydroxyl (OH^-) ions, which can gain a positive ion or lose the H^+ attached, these clays have a variable charge that depends on how much H^+ is in solution (the soil acidity).

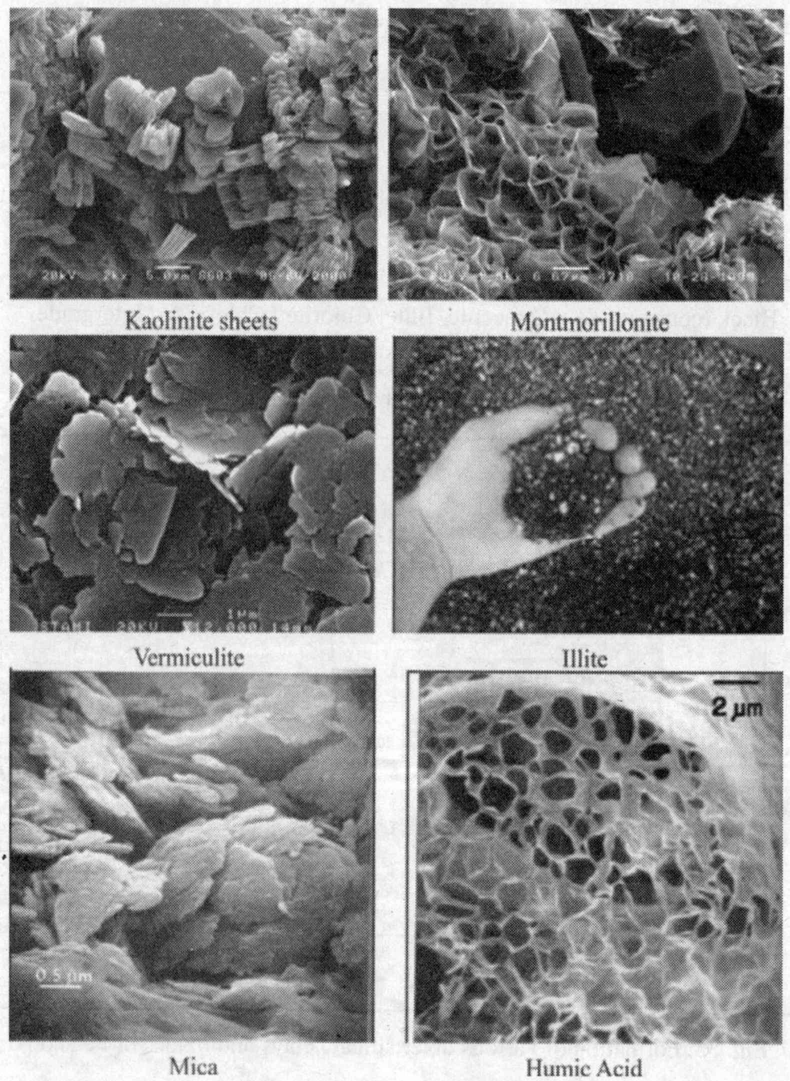

Fig. 56. Plates of layer silicates

19.3.3. Sesquioxide Clays (Metal Oxides and Hydrous Oxides)

Under conditions of extensive leaching by rainfall and long–time intensive weathering of minerals in humid warm climates, most of the silica and much of the aluminium are dissolved and slowly leached away. The remnant materials, which have lower solubility, are sesquioxides. Sesquioxides (metal oxides) are mixtures of aluminium hydroxide ($Al(OH)_3$) and iron oxide (FeO_3) or iron hydroxide ($Fe(OH)_3$). The latin word 'sesquioxides' refer to the clays of iron and aluminium because their formula can be written as Al_2O_3, xH_2O and Fe_3O_3, xH_2O - one and one–half times more oxygen than of Al or Fe.

These clays may be amorphous or crystalline and they do not swell. They are not sticky and they do not behave like silicate clays. Examples of iron and aluminium oxides common in soils are gibbsite ($Al_2O_3.3H_2O$) and geothite ($Fe_2O_3.H_2O$).

19.3.4. Humus (Organic Colloid)

Humus is amorphous, dark brown to black in colour, nearly insoluble in water, but mostly soluble in dilute alkali (NaOH or KOH) solutions. It is a temporary intermediate product left after considerable decomposition of plant and animal remains. This intermediate product is temporary because the organic substances continue to decompose slowly.

Table 60. Major properties of selected soil colloids

Colloid	Type	Size, μm	Shape	Surface area, m^2/g External	Surface area, m^2/g Internal	Interlayer spacing[a], nm	Net charge[b], cmol/kg
Smectite	2:1 silicate	0.01–1.0	Flakes	80–150	550-650	1.0–2.0	−80 to −150
Vermiculite	2:1 silicate	0.1–0.5	Plates, flakes	70–120	600–700	1.0–1.5	−100 to −200
Fine mica	2:1 silicate	0.2–2.0	Flakes	70–175	—	1.0	−10 to −40
Chlorite	2:1 silicate	0.1–2.0	Variable	70–100	—	1.41	−10 to −40
Kaolinite	1:1 silicate	0.1–5.0	Hexagonal crystals	5–30	—	0.72	−1 to −15
Gibbsite	Al-oxide	<0.1	Hexagonal crystals	80–200	—	0.48	+10 to −5
Goethite	Fe-oxide	<0.1	Variable	100–300	—	0.42	+20 to −5
Allophane & Imogolite	Noncrystalline silicates	<0.1	Hollow spheres or tubes	100–1000	—	—	+20 to −150
Humus	Organic	0.1–1.0	Amorphous	Variable[c]	—	—	−100 to −500

[a] From the top of one layer to the next similar layer, 1 nm = 10^{-9} m = 10 Å.
[b] Centimoles of charge per kilogram of colloid ($cmol_c/kg$), a measure of ion exchange capacity (see Section 8.9).
[c] It is very difficult to determine the surface area of organic matter. Different procedures give values ranging from 20 to 800 m^2/g.

The humus is often referred to as an organic colloid and consists of various chains and loops of linked carbon atoms. The humus colloids are not crystalline. They are composed basically of carbon, hydrogen and oxygen rather than of silicon, aluminium, iron, oxygen and hydroxyl groups.

The organic colloidal particles vary in size, but they may be at least as small as the silicate clay particles. The negative charges of humus are associated with partially–dissociated enolic (-OH), carboxyl (-COOH), and phenolic groups; these groups, in turn, are associated with central units of varying size and complexity.

20 ION EXCHANGE

As soils are formed during the weathering processes, some minerals and organic matter are broken down to extremely small particles. Chemical changes further reduce these particles until they cannot be seen by the naked eye. The very smallest particles are called colloids. The mineral clay colloids are plate–like in structure and crystalline in nature. In most soils, clay colloids exceed organic colloids in amount.

Colloids are primarily responsible for the chemical reactivity in soils. The kind of parent material and the degree of weathering determine the kinds of clays present in the soil. Since soil colloids are derived from these clays, their reactivity is also influenced by parent material and weathering. Each colloid (inorganic and organic) has a net negative (–) charge developed during the formation process. This means it can attract and hold positively (+) charged particles. An element with an electrical charge is called an ion. Potassium, sodium (Na), hydrogen (H), calcium (Ca) and magnesium (Mg) all have positive charges. They are called cations and ions with negative charges such as nitrate and sulfate, are called anions.

Table 61. Common cations and anions with their chemical symbols and ionic forms in soil

Cation	Chemical Symbol	Ionic form	Anions	Chemical symbol	Ionic form
Potassium	K	K^+	Chloride	Cl	Cl^-
Sodium	Na	Na^+	Nitrate	N	NO_3^-
Hydrogen	H	H^+	Sulphate	S	SO_4^{--}
Calcium	Ca	Ca^{++}	Phosphate	P	$H_2PO_4^-$
Magnesium	Mg	Mg^{++}			

Negatively–charged colloids attract cations and hold them like a magnet holds small pieces of metal. This characteristic explains why nitrate-N is more easily leached from the soil than ammonium-N. Nitrate has a negative charge like soil colloids. So, NO_3^- is not held by the soils, but remains as a free ion in soil water to be leached through the soil profile in some soils and under some rainfall conditions. The charges associated with soil particles attract simple and complex ions of opposite charge. Thus, a given colloidal mixture may exhibit not only a maze of positive and negative surface charges but also an equal complex complement of simple cations and anions such as Ca^{2+} and SO_4^{2-} that are attracted by the particle charges.

The adsorbed anions are commonly present in smaller quantities than the cations because the negative charges generally predominate the soil colloid.

20.1. SOURCES OF NEGATIVE CHARGES ON SILICATE CLAYS

There are generally two types of charges i.e., pH dependent and pH independent, that originate from exposed crystal surfaces (as a result of ionizable hydrogen ions) and isomorphous substitution (permanent) respectively.

Types of charges

- Permanent (due to isomorphous substitution)
- pH–dependent (variable, due to edges)

1. Exposed crystal edges - The negative charge on the silicate clays develops due to unsatisfied valances at the broken edges of silica and aluminium sheets. Besides, the flat external surfaces of silicate clay minerals also serve as the sources of negative charge. Ionizable hydrogen ions are hydrogens from hydroxyl ions on clay surfaces. The Al–OH or –Si–OH portion of the clay ionizes the H and leaves a temporary negative charge on the oxygen (–AI–O– or –Si–O–). The mechanism is shown in Fig. 57. Reactions are as follows:

$$-\text{SiOH} + \text{OH}^- \leftrightarrow -\text{SiO}^- + \text{H}_2\text{O}$$

$$-\text{AIOH}+\text{OH}^- \leftrightarrow -\text{AlO}^- + \text{H}_2\text{O}$$

Crystal edge

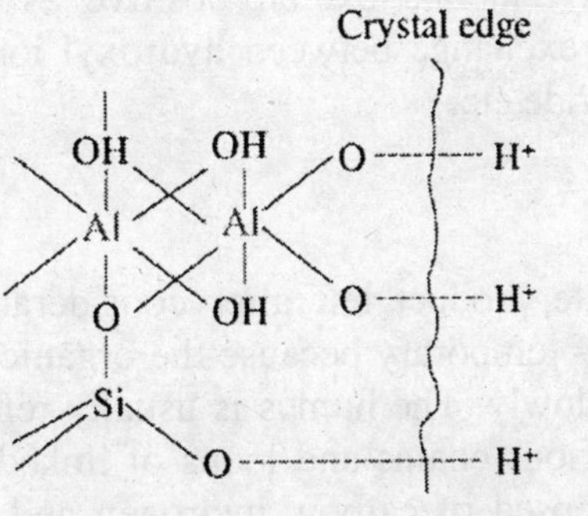

H^+ bound tightly, so the lower the pH, the less exchange there is (i.e., lower nutrient availability); Especially important in kaolinite, humus, where no internal charge imbalance.

Fig. 57. Exposed crystal edges

The extent of ionized hydrogen depends on pH of the solution; more ionization occurs in more alkaline solutions.

The magnitude of pH–dependent charge varies with the type of colloid. This type of charge is dominant for organic colloids.

2. Isomorphous substitution - Isomorphous substitution is the substitution of one ion for another of similar size but lower positive valence. Substitutions that are common are the Si^{4+} replaced by Al^{3+} and even more extensive replacement of Al^{3+} by one or more of these: Fe^{3+}, Fe^{2+}, Mg^{2+} or Zn^{2+}. The mechanism of isomorphous substitution is given in Fig. 58.

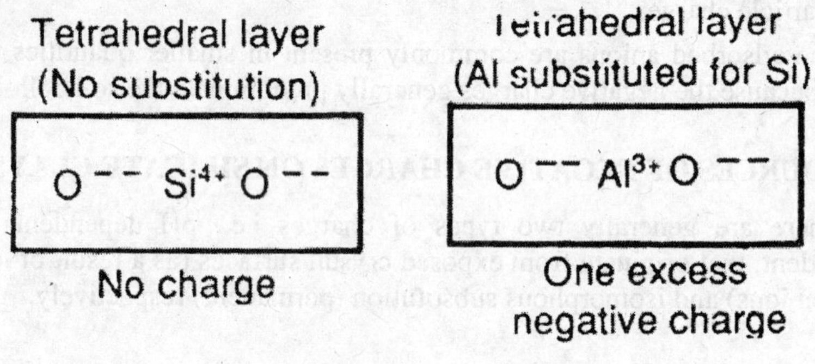

Octahedral sheet neutral Net negative charge

Fig. 58. Replacement of one ion for another of similar size within the crystalline structure of the clay

Without substitution, the positive and negative energy are in balance. Three positive charges of Al are fully satisfied by an equivalent of three negative charges from the surrounding oxygens or hydroxyls. There is no net negative or positive charge. However, when a magnesium (Mg^{2+}) ion of about same diameter as an aluminium ion (Al^{3+}) replaces one of the aluminiums by isomorphous substitution, an imbalance occurs. Consequently, the aluminium octahedral sheet assumes one negative charge for each Mg^{2+} for Al^{3+} substitution. In a similar way, the negative charge develops due to isomorphous substitution in the silica tetrahedral sheet.

3. Anion exchange - Some clay minerals exhibit positive as well as negative charges. This will cause anion exchange between hydroxyl ions (OH) and anions like phosphate, sulphate, chloride etc.

Organic Soil Colloids

Humus is a temporary intermediate product left after considerable decomposition of plant and animal residues - temporary because the organic substances remaining continue to decompose slowly. The humus is usually referred to as an organic colloid and consists of various chains and loops of linked carbon atoms. Humus micelle is basically composed of carbon, hydrogen and oxygen rather than of aluminium, silicon and oxygen in inorganic silicate clay minerals. The cation exchange capacity of humus micelle is much higher as compared to montmorillonite (2:1 type) minerals. Humus is amorphous and the size of the individual micelle is variable. Humus is not stable like clay and gets destroyed much more rapidly than clay. Humus has three components - fulvic acid, humic acid, and humin.

The charge on humus colloids is pH dependent. Under strongly acid conditions, hydrogen is rigidly bound and not easily replaceable by other cations. Humus colloid, therefore, exhibits a low negative charge.

20.2. CATION EXCHANGE (MECHANISM)

The exchange of cations has been explained on the basis of the electro–kinetic theory of ion exchange. According to this theory, the adsorbed cations forming the outer shell of the ionic double layer are supposed to be in a state of oscillation when suspended in water, forming a diffuse double layer. Due to these oscillations, some of the cations move away from the surface of the clay micelle. In the presence of the solution of an electrolyte, a cation of the added electrolyte slips in between the inner negative layer and the outer oscillating positive ion. The electrolyte cation is now adsorbed on the micelle and the surface cation remains in solution as an exchanged ion. Thus, the exchange of cations takes place. Cations are positively–charged nutrient ions and molecules while clay particles are negatively–charged constituents of soils. These negatively–charged particles (clay) attract, hold and release positively–charged nutrient ions (cations). Organic matter particles also have a negative charge to attract cations. Sand particles carry little or no charge and do not react. Cations held by soils can be replaced by other cations. This means they are exchangeable e.g., Ca^{++} can be exchanged for H^+ and /or K^+ and vice versa.

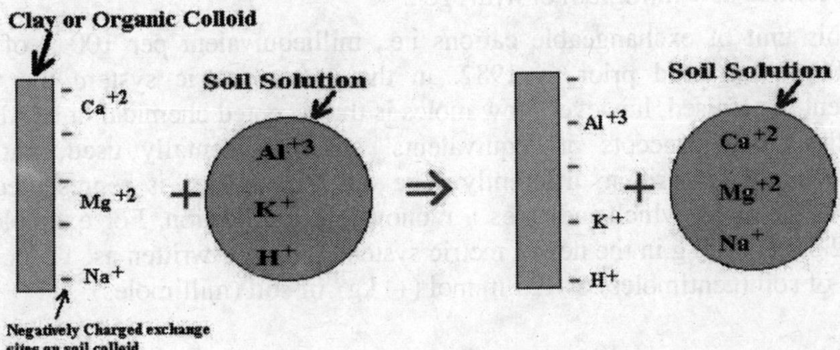

Fig. 59. Cation exchange phenomenon

Cation exchange capacity (CEC)

The CEC is the capacity of soil to hold and exchange cations. The cation exchange capacity is defined simply as the sum total of the exchangeable cations that a soil can adsorb. The higher the CEC of soil, the more cations it can retain. Soils differ in their capacities to hold exchangeable K^+ and other cations.

The CEC depends on amount and kinds of clay and organic matter present. A high–clay soil can hold more exchangeable cations than a low–clay soil. CEC also increases as organic matter increases. Clay minerals usually range from 10–150 meq/100 g in CEC values. Organic matter ranges from 200–400 meq/100 g. So, the kind and amount of clay and organic matter content greatly influence the CEC of soils. Clay soils with high CEC can retain large amounts of cations against potential loss by leaching. Sandy soils, with low CEC, retain smaller quantities.

This makes timing and application rates important in planning a fertilizer programme. For example, it may not be wise to apply K on very sandy soils in the middle of a monsoon, where rainfall can be high and intense. Fertilizer application should be split to prevent leaching and losses through erosion. Also, splitting N applications to meet peak crop demand are important to reduce the potential for nitrate leaching on sands as well as finer–textured soils.

Means of expression

The cation exchange capacity is expressed in terms of equivalents or more specifically, as milliequivalents per 100 gram and is written as meq /100g. The term equivalent is defined as one gram of atomic weight of hydrogen (or the amount of any other ion) that will combine with or displace this amount of hydrogen for monovalent ions such as Na^+, K^+, NH_4^+ and Cl^-, the equivalent weight and atomic weight are same, since they can replace one H ion. Divalent cations such as Ca^{++} and Mg^{++} can take the place of two H^+ ions. The milliequivalent weight of a substance is one thousandth of its atomic weight. Since the atomic weight of hydrogen is about 1 g, the term milliequivalent (meq) may be defined as 1 milligram of hydrogen.

This unit of exchangeable cations i.e., milliequivalent per 100 g of soil (meq/100g) was used prior to 1982. In the newer metric system the term equivalent is not used, however, now moles is the accepted chemical unit. All the calculations and concepts of 'equivalents' are still mentally used, but the notations must be written differently. The old 'equivalent' is represented by moles (+) or mole, which indicates a monovalent ion portion. For example, to write 12.5 meq/100 g in the newer metric system, it can be written as: 12.5 c mol (+) kg^{-1} of soil (centimoles) or 125 m mol (+) kg^{-1} of soil (millimoles).

Replacing power of cations

The replacing power of cations varies with the type of ion, its size and degree of hydration, valence and concentration, and the kind of clay mineral involved. As it is controlled by number of factors, no single order of replacement can be given. All other factors being equal, the replacing power of cations increases in the following order:

For monovalent cations, Li < Na < K < Rb < Cs < H

For divalent cations, Mg < Ca < Sr < Ba.

In case of mixture of monovalent and divalent cations as they exist in normal soils, the replacing power increases in the following order: Na < K < NH_4 < Mg < Ca < H.

This means Na is more easily replaced than K, and K more easily than NH_4 and so on.

Table 62. CEC of soil orders

Soil order	CEC	Key factor
Oxisols	Low	High Al/Fe oxides
Ultisols	3.5	1:1 Clays
Alfisols	9.0	Low pH
Mollisols	18.7	-
Vertisols	35.6	2:1 Clays
Histosols	128.0	OM

Percent base saturation

The per cent of total CEC occupied by the major cations has been used in the past to develop fertilizer programs. The idea is that certain nutrient ratios or 'balances' are needed to ensure proper uptake by the crop for optimum yields. However, research has shown that cation saturation ranges and ratios have little or no utility in a vast majority of agricultural soils. Under field conditions, ranges of nutrients can vary widely with no detrimental effects, though individual nutrients are present in sufficient levels in the soil to support optimum plant growth. It is a measure of the proportion of basic cations occupying the exchange sites. The base cations are those that do not form acids.

— $Ca_2+, Mg_2^+, K+, Na+, NH_4^+ \ldots$,

— ions other than H+ and Al^{3+}

$$\text{Base Saturation} = \Sigma \frac{Ca^{2+}, Mg^{2+}, K^+, NH_4^+, \ldots}{CEC} \times 100$$

Table 63. Base saturation of soil orders

Soil order	Base Sat
Oxisols	Low
Ultisols	< 35%
Alfisols	Medium–High
Mollisols	High > 50%

Cation exchange is an important reaction in soil fertility, in causing and correcting soil acidity and basicity. The changes in cation exchange reactions alter soil physical properties, and as a mechanism in purifying or altering percolating waters. The plant nutrients like calcium, magnesium, and potassium are supplied to plants in large measure from exchangeable forms.

Importance of cation exchange

- The exchangeable K is a major source of plant K.
- The exchangeable Mg is often a major source of plant Mg.
- The amount of lime required to raise the pH of an acidic soil is greater as the CEC is greater.
- Cation exchange sites hold Ca^{2+}, Mg^{2+}, K^+, Na^+, and NH_4^+ ions and slow down their losses by leaching.
- Cation exchange sites hold fertilizer K^+ and NH_4^+, and greatly reduce their mobility in soils.
- Cation exchange sites adsorb many metals (Cd^{2+}, Zn^{2+}, Ni^{2+}, and Pb^{2+}) that might be present in wastewater. Adsorption removes them from the percolating water, thereby cleansing the water that drains into groundwater.

20.3. ANION EXCHANGE

Anion adsorption - Adsorption of negative ion (anions) e.g., Cl^-, NO_3^-, $SO_4^=$, and $H_2PO_4^-$ on positively–charged sites of clay and organic matter is known as anion adsorption.

Sources of positive charge

Isomorphous substitution: Low valency cations replaced by high valency cations.

Surface and exposed broken bonds of clay lattice: OH group in certain acid soils.

Al–OH	+ H^+	----------	$Al–OH_2^+$
No charge	(Soil solution)		Positive charge
(Soil solids)			(Soil solids)

Complex aluminium and iron hydroxy ions in acid soils.

$$Al(OH)_3 + H^+ \rightarrow\rightarrow Al(OH)_2^+ + H_2O$$

No charge → Positive charge

pH dependent charges are important for anion exchange of organic matter.

Anion exchange (Mechanism)

The basic principles of cation exchange apply as well to anion exchange, except that the charges on the colloids are positive and the exchange is among negatively–charged anions.

A simple example of an anion exchange reaction is:

Micelle	NO_3^- + Cl^-	-------	Micelle	Cl^- + NO_3^-
(Soil solid)	(In soil solution)		(Soil solid)	(In soil solution)

Just as in cation exchange, equivalent quantities of NO_3^- and Cl^- are exchanged. The reaction can be reversed and plant nutrients can be released for plant absorption. Although simple reactions such as this are common; the adsorption and exchange of some anions including phosphates, molybdates, and sulphates are somewhat complex. The complexity is owing to specific reactions between the anions and soil constituents.

For example, the H_2PO_4 ion may react with the protonated hydroxyl group rather than remain as an easily exchanged anion.

$Al^{-}OH_2^+$ +	$H_2PO_4^-$	---------------	$Al^{-}H_2PO_4^-$	+ H_2O
(Soil solid)	(In soil solution)		(Soil solid)	(Soil solution)

This reaction actually reduces the net positive charge on the soil colloid. Also H_2PO_4 is held very tightly by the soil solids and is not readily available for plant uptake.

Despite these complexities anion exchange is an important mechanism for interactions in the soil and between the soil and plant. Together with cation exchange it largely determines the ability of soil to provide nutrients to plants promptly.

Anion exchange capacity

The sum total of exchangeable anions that a soil can adsorb is known as anion exchange capacity. It is expressed as C mol/kg or m.e./100 g soil. The capacity of adsorption and exchange of anions varies with the type of clay mineral, soil reaction, and the nature of anion. Kaolinitic minerals have a greater anion adsorbing and exchange capacity than montmorillonitic and illitic clays. The exchange capacity for anions is less in the case of montmorillonite and illite because the exchange is located at only a few broken bonds.

The capacity for holding anions increases with the increase in acidity. The lower the pH the greater is the adsorption. All anions are not adsorbed equally and readily. Some anions such as H_2PO_4 are adsorbed very readily at all pH values in the acid as well as alkaline range. Cl and SO_4 ions are adsorbed at slightly low pH but none at neutrality, while NO_3 ions are not adsorbed at all. The affinity for adsorption of some of the anions commonly present in soil is of the order: $NO_3 < Cl < SO_4 < PO_4$. Hence, at the pH commonly prevailing in cultivated soils, nitrate, chloride and sulphate ions are easily lost by leaching.

Importance of anion exchange

The phenomenon of anion exchange assumes importance in relation to phosphate ions and their fixation. The exchange is brought about mainly by the replacement of OH ions of the clay mineral. The reaction is very similar to cation exchange.

$$OH\text{ - Clay } + H_2PO_4 \text{ ----------- } H_2PO_4\text{ - Clay} + OH^-$$

The adsorption of phosphate ions by clay particles from soil solution reduces its availability to plants. This is known as phosphate fixation. As the reaction is reversible, the phosphate ions again become available when they are replaced by OH ions released by substances such as lime applied to soil to correct soil acidity. Hence, the fixation is temporary. The whole of the phosphate adsorbed by clay is, however, not exchangeable, as even at pH 7.0 and above. So, substantial quantities of phosphate ions are still retained by clay particles. The OH ions originate not only from silicate clay minerals but, also from hydrous oxides of iron and aluminium present in the soil.

The phosphate ions, therefore, react with the hydrous oxides also and get fixed as in the case of silicate clay, forming insoluble hydroxy-phosphates of iron and aluminium.

$$\underset{\text{(Soluble)}}{Al(OH)_3} + H_2PO_4 \text{ ---------------- } \underset{\text{(Insoluble)}}{Al(OH)_2.H_2PO_4} + OH^-$$

If the reaction takes place under conditions of slight acidity, it is reversible; and soluble phosphate is again liberated when hydroxy phosphate comes in contact with ions. If the reaction takes place at a low pH under strongly acid conditions, the phosphate (ions) are irreversibly fixed and are totally unavailable for the use of plants.

Table 64. pH scale and ranges in soil pH

Acidity						Neutral	Alkalinity			
Ultra acid	Extreme acid	Very Strong acid	Strong acid	Moderate acid	Slight acid	Neutral	Slightly alkaline	Mode-rately alkaline	Strongly alkaline	Very Strongly alkaline
< 3.5	3.5–4.4	4.5–5.0	5.1–5.5	5.6–6.0	6.1–6.5	6.6–7.3	7.4–7.8	7.9–8.4	8.5–9.0	> 9.0

20.4. SOIL REACTION

The acidity or alkalinity of a solution can be expressed on the scale of acidity and alkalinity. The scale of acidity or alkalinity is called pH scale. The unit of this scale is called pH value. This scale runs from 0–14. Negative log of hydrogen ion concentration is called as pH.

$$pH = -\log [H^+]$$

The neutral point in this scale is at pH 7. All the values above pH 7 represent alkalinity and below 7 denote acidity. The degree of alkalinity increases as values go above pH 7 and the degree of acidity increases as the pH decreases below 7.

According to the theory of electrolytic dissociation, when liquids have number of H ions just equal to the number of OH ions, the solution is neutral. While H ion exceeds OH ion, the solution is acidic. Conversely if OH ions are in excess, the solution is alkaline. Pure water ionizes to a very small degree though to a measurable extent according to the following equation:

$$H_2 \longrightarrow OH^+ + OH^-$$

According to the law of mass action

$$\frac{\text{Concentration of H ion} \times \text{Concentration of OH ion}}{\text{Concentration of undissociated } H_2O} = \text{kW (A constant)}$$

or

$$[\overset{+}{H}]\,[\overset{-}{OH}] / H_2O = KW$$

Since the amount of undissociated water is extremely large, it can be taken as unity and the above equation, therefore becomes:

$$[H^+]\,[\overset{-}{OH}] = KW$$

This ionic product of water has been measured by electrical conductivity measurements and is found to be of the order of 1×10^{-14} ions per litre at 22°C.

Since in pure water, the number of H ions is equal to the number of OH ions, the concentration of each ion type must be 1×10^{-7} g ions per litre. If the concentration of H ions is more than 10^{-7}, the solution is acidic; if it is less, solution is alkaline.

This method of expressing the H ions concentration is very inconvenient and therefore, Sorenson (1969) suggested that H ions concentration be generally expressed as the numerical value of the negative power to which 10 must be raised in order to express the required concentration and this value be designated by the symbol pH.

Thus, technically, pH is the negative logarithm of the H ion concentration or the logarithm to the base ten of the reciprocal of hydrogen ion concentration.

$$H^+ = 10^{-pH}$$

$$\log [H^+] = -pH \log 10$$

$$-pH = \log [H^+] / \log 10$$

$$pH = \log [H^+] / \log 10$$

Since, $\log 10 = 1$,

Therefore, $pH = -\log [\overset{+}{H}]$ OR $pH = \log 1 / [\overset{+}{H}]$

Soil reaction is one of the most important physiological characteristics of the soil solution. The presence and development of microorganisms and higher plants depend upon the chemical environment of soil. Therefore, study of soil reaction is important in soil science.

20.4.1. Types of Soil Reactions

There are three types of soil reactions:

1. Acidic

It is common in region where precipitation is high. The high precipitation leaches appreciable amounts of exchangeable bases from the surface layers of the soils so that the exchange complex is dominated by H ions. Acid soils, therefore, occur widely in humid regions and affect the growth of plants in a marked way.

2. Alkaline

Salts like carbonates of calcium, magnesium and sodium also give a preponderance of OH ions over H ions in the soil solution. Alkali soils occur when there is comparatively high degree of base saturation. When salts of strong base such as sodium carbonate go into soil solution and hydrolyze, consequently they give rise to alkalinity.

The reaction is as follows:

$$Na_2CO_3 \longrightarrow 2Na^+ + CO_3^=$$

$$2Na + CO_3^= + 2HOH \longrightarrow 2Na^+ + 2OH^- + H_2CO_3$$

Since sodium hydroxide dissociates to a greater degree than the carbonic acid, OH ions dominate and give rise to alkalinity. This may be as high as 9 or 10. These soils most commonly occur in arid and semi–arid regions.

3. Neutral

Neutral soils occur in regions where H ions just balance OH ions.

20.5. SOIL pH

The reaction of a solution represents the degree of acidity or basicity caused by the relative concentration of H ions (acidity) or OH ions present in it. Acidity is due to the excess of H ions over OH ions, and alkalinity is due to the excess of OH ions over H ions. A neutral reaction is produced by an equal activity of H and OH ions. According to the theory of dissociation, the activity is due to the dissociation or ionization of compounds into ions.

Method of expressing acidity or alkalinity

Equivalent quantities of all acids or alkalis contain the same number of total H or OH ions respectively. But, when they are dissolved in water they do not ionize to the same extent. The amount of acid or alkali ionized depends upon the content of free H and OH ions.

Ex., when the dissociation is high, Hydrochloric acid dissociates to a larger extent than the weak acetic acid. Acetic acid dissociates only to about 10% as compared to hydrochloric acid. In a 1 N solution of hydrochloric acid, there will be 1 gram of H ions per litre, while in a normal solution of acetic acid there will be 1/10 gram of H ions per litre. But in titration 1 ml of 1 N hydrochloric acid and 1 ml of 1 N acetic acid will require 1 ml of 1 N sodium hydroxide for neutralization separately, because the total acidity is the same and titration determines both the ionized and unionized H or OH ions.

Table 65. The categories of soils with respect to soil pH

Soil pH	Categories
Below 5.5	Acid
5.5-6.5	Slightly acid
6.5-7.5	Neutral
7.5-8.5	Tending to become alkali
Above 8.5	Alkali

21 PROBLEM SOILS

Problem soils are classified into:

- Acid soils
- Acid sulphate soils
- Saline soil
- Alkali soil (Sodic soil)
- Saline alkali soil
- Degraded alkali soil
- Calcareous soils
- Waterlogged soil

21.1. SOIL ACIDITY

There are three kinds of acidity:

1. Active acidity - It is due to the H+ ion in the soil solution.
2. Salt replaceable acidity - It is represented by the hydrogen and aluminium that are easily exchangeable by other cations in a simple unbuffered salt solution such as KCl.
3. Residual acidity - It is the acidity, which can be neutralized by limestone or other alkaline materials, but cannot be detected by the salt–replaceable technique. Obviously, all these types of acidity add up to the total acidity of a soil.

1. Active acidity

The active acidity is a measure of the H^+ ion activity in the soil solution at any given time. However, the quantity of H^+ ions owing to active acidity is very small compared to the quantity in the exchange and residual acidity forms.

For example, only about 2 kg of calcium carbonate would be required to neutralize the active acidity in a hectare-furrow slice of an average mineral soil at pH 4. Even though the concentration of hydrogen ions owing to active acidity is extremely small, it is important because it is the environment to which plants and microbes are exposed to.

2. Salt replaceable (exchangeable) acidity

This type of acidity is primarily associated with the exchangeable aluminium and hydrogen ions that are present in largest quantities in very acid soils. These ions can be released into the soil solution by an unbuffered salt such as KCl.

$$1.)\ \text{Al-soil} + \text{KCl} \rightleftharpoons \text{K-soil} + Al^{3+}$$

$$2.)\ Al^{3+} + H_2O \rightleftharpoons Al(OH)_2^{+} + H^{+}$$

In moderately acid soils, the quantity of easily exchangeable aluminium and hydrogen is quite limited. Even in these soils, however, the limestone needed to neutralize this type of acidity is commonly more than 100 times that needed for the soil solution (active acidity).

At a given pH value, exchangeable acidity is generally highest for smectites, intermediate for vermiculites, and lowest for kaolinite. However, in any case, it accounts for only a small portion of the total soil acidity.

3. Residual acidity

Residual acidity is that which remains in the soil after active and exchange acidity has been neutralized. Residual acidity is generally associated with aluminium hydroxy ions and with hydrogen and aluminium atoms that are bound in non–exchangeable forms by organic matter and silicate clays.

If lime is added to a soil, the pH increases and the aluminium hydroxy ions are changed to uncharged gibbsite as follows:

$$Al(OH)_2^{+} + H_2O \leftrightarrow Al(OH)_3\text{,gibbsite} + H^{+}$$

In addition, as the pH increases bound hydrogen and aluminium can be released by calcium and magnesium in the lime materials ($Ca(OH)_2$ is used as an example of the reactive calcium liming material). The residual acidity is commonly far greater than either the active acidity or salt replaceable acidity. Conservative estimates suggest that the residual acidity may be 1,000 times greater than the soil solution or active acidity in a sandy soil, and 50,000 or even 1,00,000 times greater in a clayey soil high in organic matter. The amount of ground limestone recommended to at least partly neutralize residual acidity is commonly 4–8 metric tons (Mg) per hectare furrow slice (1.8–3.6 tons/AFS). It is obvious that the pH of the soil solution is only 'the tip of the iceberg' in determining how much lime is needed.

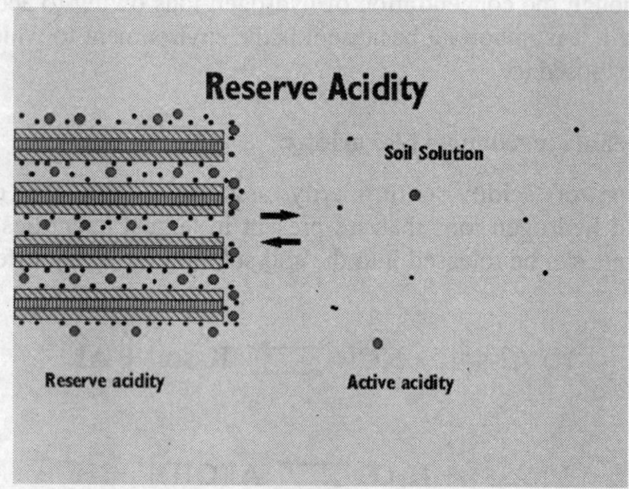

Fig. 60. Reserve and active acidity

Buffering and soil reaction (buffer action)

Buffering refers to resistance to a change in pH. If 1 ml HCI (0.1 N) is added to one litre of pure distilled water of pH 7, the resulting solution would have a pH of about 5. On the other hand, if the same amount of acid is added to a litre of soil suspension, the resulting change in pH would be very small. There is a distinct resistance to change in pH. This power to resist a change in pH is called buffer action. A buffer solution is one, which contains reserve acidity and alkalinity, and does not change pH with small additions of acids or alkalies.

Buffer capacity

The colloidal complex acts as a powerful buffer in the soil, and does not allow rapid and sudden changes in soil reaction. Buffering depends upon the amount of colloidal material present in the soil. Clay soils rich in organic matter are highly buffered than sandy soils. Buffer capacity of the soil varies with its Cation Exchange Capacity (CEC). The greater the CEC, the greater will be its buffer capacity. Thus, heavier the texture and greater the organic matter content of a soil, the greater is the amount of acid or alkaline material required to change its pH.

Importance of buffering to agriculture

Changes in soil reaction (pH) have a direct influence on the plants and it also affects the availability of plant nutrients. Deficiency of certain plant nutrients and excess availability of others in toxic amounts would seriously upset the nutritional balance in the soil. Buffering prevents sudden changes and fluctuation in soil pH, so it regulates the availability of nutrients and also checks direct toxic effect to plants.

21.2. FACTORS CONTROLLING SOIL REACTIONS

Soil reaction varies due to following factors:

1. Nature of soil colloids - The colloidal particles of the soil influence soil reaction to a very greatest extent. When hydrogen (H^+) ion forms the predominant adsorbed cations on clay colloids, the soil reaction becomes acid.

2. Soil solution - The soil solution carries a number of salts dissolved in capillary water. The cations of the salts intermingled with those of the diffuse double layer of the clay particle and increase the concentration. The concentration of cations in bulk of the solution is more or less (or nearly) the same as that near the particle surfaces. For an unsaturated soil (Clay), the more compact the layer, the greater is the number of hydrogen ions dissociating into the solution. This increases the acidity of the soil solution or lowers its pH. Under field conditions, the concentration of salts varies with the moisture content of the soil. The more dilute the solution, the higher the pH value. Hence, the pH tends to drop as the soil gets progressively dry. Soil reaction is also influenced by the presence of CO_2 in soil air. As the CO_2 concentration increases, the soil pH falls and increases the availability of the nutrients. Under field conditions, plant roots and microorganism liberate enough CO_2, which results in lowering the pH appreciably. This principle of increasing the concentration of CO_2 in soil air is also used in the reclamation of alkali soils.

3. Climate - Rainfall plays an important role in determining the reaction of soil. In general, soils formed in regions of high rainfall are acidic (low pH value), while those formed in regions of low rainfall are alkaline (high pH value).

4. Soil management - Cultural operations, in general, tend to increase soil acidity. They make the acid soil more acidic, and an alkaline soil less alkaline. As a result of constant cultivation, basic elements are lost from the soil through leaching and crop removal. This leads to change the soil reaction to the acid side.

5. Parent materials - Soils developed from parent material of basic rocks, generally, have higher pH than those formed from acid rocks (e.g., granite). The influence of parent material is not very important as it is completely masked by the climatic conditions under which the soil is developed.

6. Precipitation - As water from rainfall passes through the soil, basic nutrients such as calcium (Ca) and magnesium (Mg) are leached. They are replaced by acidic elements including Al, H and manganese (Mn). Therefore, soils formed under high rainfall conditions are more acidic than those formed under arid conditions.

7. Decomposition of organic matter - Soil organic matter is continuously being decomposed by microorganisms into organic acids, carbon dioxide (CO_2) and water, forming carbonic acid. Carbonic acid, in turn, reacts with the Ca and

Mg carbonates in the soil to form more soluble bicarbonates, which are leached away, leaving the soil more acidic.

8. Native vegetation - Soils often become more acidic when crops are harvested because of removal of bases. Type of crop determines the relative amounts of removal. For example, legumes generally contain higher levels of bases than grasses. Calcium and Mg contents also vary according to the portion (s) of the plant harvested. Many legumes release H ions into their rhizosphere when actively fixing atmospheric N_2. The acidity generated can vary from 0.2–0.7 pH units per mole of fixed N.

9. Soil depth - Except in low rainfall areas, acidity generally increases with depth, so the loss of topsoil by erosion can lead to more acidic pH in the plough layer. The reason is that more subsoil is included in the plough layer as topsoil is lost. However, there are areas where subsoil pH is higher than that of the topsoil.

10. Nitrogen fertilization - Nitrogen from fertilizer, organic matter, manure and legume N fixation produces acidity. Nitrogen fertilization speeds up the rate at which acidity develops. At lower N fertilizer rates, acidification rate is slow, but is accelerated as N fertilizer rates increase.

11. Flooding - The overall effect of submergence is an increase of pH in acid soils and a decrease of pH in basic soils. Regardless of their original pH values, most soils reach to pH of 6.5–7.2 within one month after flooding and remain at the level until dried. Consequently, liming is of little value in flooded rice production. Further, it can induce deficiencies of micronutrients such as zinc (Zn).

Influence of soil reaction on availability of nutrients

The unproductiveness of acid and alkali soils is very often due to the lack of available plant nutrients. In highly acid soils (low pH), the availability of some of the nutrients such as aluminium, iron, manganese etc., is increased to a point to become toxic to the plant. At the same time the supplies of available calcium, nitrogen, phosphorus etc., are reduced to starvation level (become unavailable). The same is in the case at high pH (alkaline conditions). Plant growth suffers due to the unavailability of nutrients like nitrogen, phosphorus and some minor elements (e.g., iron, manganese, boron etc.). Another indirect effect occurs through the activity of microorganisms. Most microorganisms function at their best within a pH range of 6.0–7.5. If soil reaction is changed beyond this range, the microorganisms become functionless. Consequently, the supply of some of the essential plant nutrients such as nitrogen is considerably reduced.

1. Nitrogen - Plant absorbs most of their nitrogen in the form of nitrate. The availability of nitrate depends on the activity of nitrifying bacteria. The microorganisms responsible for nitrification are most active when the pH is between 6.5 and 7.5. They are adversely affected if the pH falls below 5.5 and rises above 9.0. Nitrogen fixing bacteria (like Azotobacter) also fail to function below pH 6.0. The decomposition of organic matter, which is the primary source of nitrogen, is also slowed down under acidic condition.

2. Phosphorus - The availability of phosphorus is at its highest when the reaction is between 6.5 and 7.5. When the reaction is above or below this range, availability is reduced. In the strongly acidic soil (pH 5.0 or less), iron, aluminium, manganese and other bases are present in a soluble state and in more quantity. The phosphates of these elements are formed and become unavailable.

3. Potassium - The availability of potassium does not influence soil reaction to any great extent. In acid soil, potassium is lost through leaching. The unavailability of K is due to the conversion of exchangeable potassium to non-exchangeable potassium in alkaline soil. Particularly if the alkalinity is due to $CaCO_3$ (brought about by over liming in acid soil), the solubility of soil potassium is depressed.

4. Calcium and magnesium - Acid soils (base unsaturated) are poor in available calcium and magnesium. In alkaline soil (pH not exceeding 8.5) the availability of Ca and Mg nutrients are always high. When the pH is above 8.5, the availability of these nutrients again decreases.

5. Iron, aluminium and manganese - When the pH is low the solubility of iron, aluminium and manganese compounds are increased and hence, they are readily available in acid soils. At the pH range of 5.5–7.0, iron and manganese are present in the soluble ferrous (Fe^{++}) and manganous (Mn^{++}) forms. At pH below 5.5 the solubility of these compounds is considerably increased with the result that they have a toxic influence on plant growth. Under neutral and alkaline conditions, iron and manganese are usually present in ferric (Fe^{3+}) and manganese (Mn^{++++}) states. Hence, in soils with pH 7.5 and above, they become unavailable and sometimes produce deficiency diseases like chlorosis in plants.

6. Sulphur - The availability of sulphur is not affected by soil reaction as sulphur compounds are soluble in low pH range. However, it is more soluble in acid soil and lost in leaching. Acid conditions, which retard the decomposition of organic matter, therefore, retard the release of available sulphur. The availability of sulphur present in organic matter depends upon the decomposition of organic matter.

7. Micronutrients - In general, the availability of boron, copper and zinc is reduced in alkaline soils and that of molybdenum in acid soils. The availability of these nutrients progressively decreases as the soil pH increases. Their availability also decreases under highly acidic conditions when the pH is below 5.0. Zinc availability in alkaline soils from insoluble zinc salts (calcium zincate) is reduced. Zinc and copper are adsorbed on the clay colloids and not easily displaced and hence, not available for plant growth. The availability of molybdenum is reduced under acid soils. It is more available in neutral and alkaline soils.

ACID SULPHATE SOILS

These soils are extremely acid soils (pH below 4.0). The acidity is due to the presence of sulphuric acid, and iron and aluminium sulphates. They are derived from marine sediments high in pyrites and poor in bases.

21.3. SOIL SALINITY AND ALKALINITY

Electrical conductivity (EC)

Electrical conductivity is a measure of the ability of a solution to carry a current. It is defined as the reciprocal of resistance, in ohms, of a 1 cm^3 of liquid at a specified temperature which is expressed as siemens/cm. As the range of the conductivity is very low it is frequently expressed as millionths of a siemens or microsiemens/cm (μS/cm). Current in liquid is carried by ions. Any change in concentration of ions changes the conductivity of a liquid. Ions are formed in a liquid when a solid such as a salt is dissolved. NaCl (table salt) is dissolved to form Na^+ and Cl^- and contributes to the overall conductivity of the solution.

Salinity is described in terms of electrical conductivity (EC). The higher the EC value, the more is the salinity of soil or water.

EC of soil or water can be easily measured with a handheld meter. To read the EC of a soil, a mixture of 1 part soil to 5 parts water is made and the reading from the meter is expressed as EC 1:5.

The common unit for Electrical conductivity is deciSiemens/metre (dS/m), which is equal to another unit milimhos/cm (mmhos/cm).

As a guide for soils

- Sandy or loamy soils are saline if $EC_{1:5}$ is above 0.4 dS/m
- Clay soils are saline if $EC_{1:5}$ is above 0.7 dS/m
- Measuring salinity in this way is not very accurate because it does not take into account soil texture. Hence, in the laboratory EC is measured by making a saturated extract of soil and water.
- The EC value is then expressed as EC_e or EC_{se}.

Table 66. EC Values of soil salinity classes

Class	ECe (dS/m)	Comments
Non–saline	<2	Salinity has no significant effect
Slightly saline	2–4	Yields of very sensitive crops may be affected
Moderately saline	4–8	Yields of many crops affected
Very saline	8–16	Only tolerant crops yield satisfactorily
Highly saline	>16	Only a few very tolerant crops yield satisfactorily

Table 67. Unit conversion

To convert ...		
from this unit	**to this unit**	**do this**
EC (μS/cm)	dS/m	divide by 1,000
ppm (mg/l)	dS/m	divide by 640
dS/m	EC	multiply by 1,000
dS/m	ppm	multiply by 640

Some EC values for water

- A maximum desirable figure for domestic drinking water is approximately 1.6 dS/m.
- For sheep, the most tolerant of farm stock, the desirable maximum for healthy growth is 10 dS/m.
- When describing a soil as saline, you need to take into account what crops or plant species are being grown. Some are much more tolerant than others.

What is a salinity problem?

- Salinity is usually linked with waterlogging.
- One of the first signs of dryland salinity is boggy ground - often referred to as a seep.
- But the water is not the problem. The problem lies with the salt carried by the groundwater, and its effects on plant growth. Plants can be affected in a number of ways:
  - The plant is 'starved' of moisture. In simple words, water that is full of salt cannot move into the plant cells. Water in the soil becomes unavailable to the plant and the plant becomes water stressed.
  - Some ions (particularly chloride) are toxic to plants. If the concentration of these ions increases, the plant is poisoned.
  - Increasing the salt content can cause an imbalance of plant nutrients. In particular, nitrate uptake in saline soils may be limited, and plants may be starved of nitrogen.

Causes of salinity

- Farming practices - The land is the mean of earning wealth, but nowadays for extracting this wealth the natural vegetation is getting replaced with improved crop varities and pasture species.

- The process of land management had disastrous long-term consequences when combined with climate and certain catchments shapes.
- Groundwater recharge occurs when water passes below the root zone of plants. Once it passes through the root zone, it cannot be used, and it moves down to the water table.

Farming practices that can increase the rate of recharge include:

- Growing annual crops and pastures that are shallow rooted and only grow for part of the year.
- Overgrazing - Plants whose leaves have been removed do not use much water.
- Poor crop and pasture growth - If plant growth is limited by pests, disease or lack of nutrients, less water is taken up.
- Long fallows - This is the period between cultivation and sowing, when there is no plant growth in a field. The soil can be filled with water, which passes through to the groundwater.

Sodicity

- Sodic soils contain a higher than desirable amount of sodium attached to the clay particles. But in saline soils, the salts causing the problems are in the soil water. When in contact with water, sodic soils disperse into tiny fragments. On drying, these tiny fragments block the soil pores.
- Sodic soils are difficult to manage, often hard–setting, and are susceptible to waterlogging, poor aeration and erosion.
- The sodicity of soil can be tested quite simply using a turbidity test (testing the cloudiness of a soil/water mixture)
- Sodicity may be a problem in the topsoil and/or the subsoil, so both should be checked.
- Soil acidity and soil salinity may be linked to soil sodicity, and each condition should be treated to get the soil back into a productive state.
- Gypsum helps in controlling the problem of sodicity by providing calcium to replace sodium.

Degraded alkali soil

The soils do not contain free calcium carbonate as a result of prolong leaching. Na–clay hydrolyzes NaOH, which combines with CO_2 to form Na_2CO_3 and creates alkaline condition. The process of breakdown of H–clay under alkaline condition is known as solodization and the soil so formed is solod or soloth or degraded alkali. The process of leaching of soluble salts as a result of the increase in rainfall is called Alkalization.

pH = > 8.5

ESP = >15 %

EC = < 4.0

Table 68. Comparisons of saline and alkali soils

	pH	EC (dS/m)	ESP
Normal soils	6.5-7.2	<4	<15
Acid soils	<6.5	<4	<15
Saline soils	<8.5	>4	<15
Sodic soils	>8.5	<4	>15
Saline-sodic	<8.5	>4	>15

Calcareous soils

Calcareous soils are relatively alkaline. In other words, they have a high pH. This is because of the very weak acidity of carbonic acid. Note that this is not the only reason for a high soil pH. They are characterized by the presence of calcium carbonate in the parent material and by a Calcic horizon, a layer of secondary accumulation of carbonates (usually Ca or Mg) in excess of 15% calcium carbonate equivalent and at least 5 per cent more carbonate than an underlying layer.

Waterlogged soils or wetland soils

These are the soils whose development and properties are strongly influenced by temporary or permanent saturation in the upper part of the soils.

Types of waterlogging

- **Riverine flood waterlogging -** Rainy season, flood may occur to the nearby land from the river having excess flood water.
- **Oceanic flood waterlogging -** Ocean water spreads in the nearby land causing waterlogging.
- **Seasonal waterlogging -** Runoff water accumulates in the low land and depressions in the rainy season.
- **Perennial waterlogging -** Deep water, swamp get rain water runoff and seepage water from canals.
- **Subsoil waterlogging -** High water table in the rainy season.

22 PHYSICAL CONSTRAINTS OF SOIL

An optimum physical environment of soils is essential for better growth of plants, consequently for better yields. Based on the physical properties of soil viz., infiltration, bulk density, hydraulic conductivity, porosity (capillary and non –capillary), aggregates etc., physical constraints of soil are identified as below:

- Slow–permeable soils
- Excessively–permeable soils
- Subsoil hardening
- Surface crusting
- Fluffy paddy soils
- Shallow soils

1. Slow–permeable soils

Slow–preamble soils are those having infiltration rates less than 6 cm/day due to high clay content of the soil. Due to low infiltration rates, the amount of water entering the soil profile is reduced thus, increasing the run-off. Further, it encourages erosion of surface soil leading to nutrient removal in the running water. Moreover, due to heavy clay content, the capillary porosity is relatively high resulting in impeded drainage and reduced soil conditions. This results in increase of some soil elements to the level of toxicity to the plants. It also induces nutrient fixation in the clay complex, thereby making the nutrient becoming unavailable to the crop; eventually causing deficiency of nutrients. Such soils are spread over Tamil Nadu in an area of 7,54,631 ha, which is 7.5 per cent of total geographical area.

Management

The constraints in such soils can be managed by adopting suitable practices such as:

- Provision of drainage facilities either through open or closed sub surface drains.
- Forming contour and compartmental bunding to increase the infiltration rates of soils.
- Application of huge quantities of river sand or red soils of coarser texture to dilute heaviness of the soil.
- Application of liberal doses of organic manures like farm yard manure, compost, green manure, composted coir pith, sewage waste, press mud etc.
- Adopting ridges and furrows, raised beds, broad bed and furrow systems.
- Application of soil conditioners such as H–concentrate, Vermiculite, Jalasakti etc. to reduce run–off and soil erosion.

2. Excessively–permeable Soils

Excessively–permeable soils are those having high amount of sand exceeding 70 per cent. Due to this, the soils are inert and unable to retain nutrient and water. These soils being devoid of finer particles and organic matter, the aggregates are weakly formed, and the non–capillary pores dominating with very poor soil structure. Due to low retaining capacity of the soils, the fertilizer nutrients are also lost in the drainage water. These soils are spread over 24,12,086 ha in Tamil Nadu (23.97 per cent of total geographic area).

Management

The excessively–permeable soils can be managed by adopting the techniques given below:

- Compacting the field with 400 kg stone roller (tar drum filled with 400 kg of sand or stones can also be used) for 8–10 times at optimum moisture conditions.
- Application of clay soil up to a level 100 t ha^{-1} based on the severity of the problem and availability of clay materials.
- Application of organic materials like farm yard manure, compost, press mud, sugar factory slurry, composted coir pith, sewage sludge etc.
- Providing asphalt sheet, polythene sheets etc. below the soil surface to reduce the infiltration rate.
- Crop rotation with green manure crops like sunhemp, sesbania, daincha, kolinchi etc.

3. Sub–soil hardening / hard pan

The sub–soil hard pan in red soils is due to illuviation of clay to the sub–soil horizon coupled with cementing action of oxides of Fe, Al and Calcium carbonate, which increases the bulk density of soil to more than 1.8 Mg m^{-3}. Further, the hard pan can also develop due to continuous cultivation of crops using heavy implements up to certain depth constantly. Besides, the higher exchangeable sodium content in black soils areas also results in compactness. All put together lowered the infiltration and percolation rates, nutrient movement and free air transport within the soils profile. It prevents root proliferation and limits the volume of soils available for nutrients uptake resulting in depleted, less fertile surface soil. Due to this, the contribution of sub soil fertility to crop growth is hampered.

Management

These soils are managed by adopting following practices:

- Ploughing the soil with chisel plough at 0.5 m interval criss–cross at 0.5 m depth once in 2–3 years.
- Application of organics to improve the aggregation and soil structure so as to prevent further movement of clay to the lower layers.
- Deep ploughing of the field during summer season to open up the sub soils.
- Cultivating deep–rooted crops like tapioca, cotton so as to encourage natural breaking of the hard pan.
- Raising deep–rooted semi–perennial crops like Mulberry, Jasmine, Match wood tree etc. can also help in opening up the sub–surface hard pan.

4. Surface crusting

Surface crusting is due to the presence of colloidal oxides of iron and aluminium in Alfisols, which binds the soil particles under wet regimes. On drying, it forms a hard mass on the surface.

The ill effects of surface crusting are:

- Prevents germination of seeds
- Retards/inhibits root growth
- Results in poor infiltration
- Accelerates surface runoff
- Creates poor aeration in the rhizosphere
- Affects nodules formation in leguminous crops area: 4,51,584 ha (4.49 per cent TGA) in Tamil Nadu.

Management

Surface crusting can be managed as below:

- Ploughing should be done when the soil is at optimum moisture regime.
- Lime at 2 t ha^{-1} may be uniformly spread and another ploughing is to be given for blending of amendment with the surface soil.
- Farm yard manure at 10 t ha^{-1} or composted coir pith at 12.5 t ha^{-1} or other organics may be applied to improve the physical properties of the soils, after preparation of land to optimum tilth.
- Scraping of surface soil by tooth harrow can be useful.
- Bold–grained seeds may be used for sowing on the crusted soils.
- More number of seeds/till may be adopted for small–seeded crops.
- Sprinkling water at periodical intervals may be done whenever possible.
- Resistant crops such as cowpea can be grown.

5. Fluffy paddy soils

The traditional method of preparing the soil for transplanting rice consists of puddling, which substantially breaks soil aggregates into a uniform structureless mass. Under continuous flooding and submergence of soil for rice cultivation in a cropping sequence of rice-rice-rice, the soil particles are always in a state of flux and the mechanical strength is lost leading to the fluffiness of the soil. Impact of fluffiness is sinking of drought animals and labourers during puddling. Thus, this has been an invisible drain of finance for the farmers due to high pulling power needed for the bullocks and slow movement of labourers during the puddling operations. Further fluffiness of the soil leads to very low bulk density and thereby leading to very rapid hydraulic conductivity, and in turn, the soil does not provide a good anchorage to the roots. The potential yield of crops is adversely affected. About 25,919 ha (0.26% TGA) in Tamil Nadu have this problem.

Management

Following practices are needed to be adopted to overcome this problem:

- The irrigation should be stopped 10 days before the harvest of rice crop.
- After the harvest of rice, when the soil is under semi–dry condition, compact the field by passing 400 kg stone roller or a tar drum filled with 400 kg of sand for eight times.
- The usual preparatory cultivation is carried out after compaction.

6. Shallow soils

The shallow soils are characterized by the presence of the parent root immediately below the soil surface at about 15–20 cm depth. This restricts the root elongation and spreading. Hence, the crops grown in these soils necessarily are shallow–rooted crops, which can exhaust the soil within 2–3 seasons. Therefore, frequent renewal of soil fertility is must in these soils. These soils can be managed by growing crops, which can withstand the hard rocky sub–soils like mango, ber, fig, country gooseberry, West Indian cherry, anona, cashew, and tamarind etc.

23 QUALITY OF IRRIGATION WATER

Water quality is determined according to the purpose for which it is used. The continuous use of poor quality irrigation water affects the physical and chemical properties of soil over a long period. Ultimately the soils become unfit for any agricultural operation. Hence, it is essential to assess the quality of irrigation water before using them.

The important characteristics of irrigation water used in determining its quality are:

- Salinity hazard
- Sodicity hazard
- Carbonate hazard
- Permeability hazard
- Specific ion toxicity

1. Salinity hazard

Accumulation of soluble salts in the soil is directly related to the neutral soluble salt content of the irrigation water. Salts accumulating in the root zone affect crop growth largely by osmotic suction and restrict the availability of soil water to plants. The concentration of soluble salts in irrigation water is determined in terms of electrical conductivity and expressed as dS m^{-1} at 25°C.

Table 69. Water salinity levels

EC (dS m^{-1})	Salinity Class
< 0.25	Low
0.25-0.75	Medium
0.75-2.25	High
2.25-5.0	Very High

2. Sodicity hazard

Irrigation water containing higher proportion and sodium to other cations leads to the problem of sodicity. The sodicity hazard of irrigation water is usually evaluated by Sodium Adsorption Ratio (SAR). SAR is expressed as the relative proportion of sodium to other cations viz., Ca and Mg in water, and is calculated by using the following formula, where the concentration of ions is expressed in me L^{-1}:

$$SAR = Na^+ / \left((Ca^{++} + Mg^{++}) / 2 \right)^{1/2}$$

SAR	Quality Class
< 10	Safe
10 - 20	Moderate
>20	Unsafe

3. Carbonate hazard

Carbonates and bicarbonate ions are important because they have the tendency to precipitate Ca and Mg in soil solution. The effect of bicarbonate and carbonate ions on water quality is expressed in terms of Residual Sodium Carbonate (RSC) concept, where the concentration of ions is expressed in me L^{-1}.

$$RSC = (CO_3^{2-} + HCO_3^-) - (Ca^{2+} + Mg^{2+})$$

Irrigation water, containing high concentration of carbonate and bicarbonate ions, has a tendency to precipitate Ca and Mg, in the form of carbonates. As solution becomes more concentrated with soluble Na^+, SAR of soil solution increases, and consequently an increase in ESP occurs.

Table 70. RSC levels of irrigation waters

RSC	Quality Class
< 1.25	Safe
1.25–2.50	Moderate
>2.50	Unsafe

4. Permeability hazard

Permeability hazard occurs when the rate of water infiltration into the soil is reduced by the effect of specific salts. High sodium in irrigation water results in severe soil permeability problem. Permeability is also affected by carbonate and bicarbonate content in irrigation water. Permeability Index can be calculated by using the following formula, where the ionic concentrations are expressed in me L^{-1}:

$$\text{Permeability Index (PI)} = (Na + HCO_3^-) / (Ca + Mg + Na) \times 100$$

Permeability problems due to excess of sodium or shortage of calcium are evaluated by a relatively new concept, the 'adjusted Sodium Adsorption Ratio' (adj. SAR), formerly evaluated through the Sodium Adsorption Ratio (SAR) and the Residual Sodium Carbonate (RSC). This new concept adds to the older SAR, the effect of carbonate/bicarbonate through a theoretical and calculated pHc value added to the SAR. The pHc evaluates the tendency of the irrigation water to dissolve lime from the soil: adding to soluble calcium, or the tendency to precipitate lime: reducing the soluble calcium. The presence of appreciable carbonate/bicarbonate may markedly influence the calcium availability. This effect on SAR is evaluated by the new adjusted Sodium Adsorption Ratio equation (adj.SAR) as recently developed by the U.S. Salinity Laboratory as follows:

$$\text{adj.SAR} = \frac{\text{Na}}{\sqrt{\frac{\text{Ca} + \text{Mg}}{2}}}[1 + (8.4 - \text{pHc})]$$

The values for sodium, calcium and magnesium (reported in milliequivalents per litre) are taken directly from the water analysis. The pHc is a calculated, theoretical value related to 1) the total salinity as measured by sodium + calcium + magnesium, 2) the calcium + magnesium supply in the water, and 3) the carbonate + bicarbonate present.

This particular adj.SAR procedure is most applicable at about 70 per cent field efficiency of irrigation. At greater efficiency, these values may be a little low; at lower efficiency they may be a little high.

Still there are other permeability problems in agriculture that are unrelated to adj.SAR or to low salts, which must be evaluated separately. These may include problems that are more closely related to soil texture, to compaction and plough pans, or to other physical as well as chemical properties of soil.

$$\text{pHc} = (\text{pK}_2' - \text{pK}_c') + \text{p}\ (\text{Ca} + \text{Mg}) + \text{pAlK}$$

pHc can be calculated using Table 71.

Where $\text{pK}_2' - \text{pK}_c'$ is obtained from Ca + Mg + Na

p (Ca+Mg) is obtained from Ca + Mg

p (Alk) is obtained from $CO_3 + HCO_3$

5. Specific ion toxicity

Ground waters having toxic ions such as B, Cl, F, NO_3, Se etc. become problematic for irrigating crops and have consequence of entering human food chain. Apart from boron and fluorine, some trace elements present in irrigation waters may also cause toxic effect. The permissible maximum concentrations of these toxicants in irrigation waters are based on the limitations of soils for crop production.

Table 71. Degree of problem of common ions in irrigation water

Specification	No problem	Increasing problem	Severe problem
Sodium (adj.SAR)	<3	3–9	>9
Chloride (mg L^{-1})	< 4	4–10	>10
Boron (mg L^{-1})	< 0.75	0.75–2	>2
$NO_3 - N$ (mg L^{-1})	< 5	5–30	>30
HCO_3-N (mg L^{-1})	< 1.5	1.5–8.5	>8.5
Fluoride (mg L^{-1})	< 1	1–15	>15

Table 72. Permissible concentration of trace elements in irrigation waters (ppm)

Element	For using waters continuously on all normal soils (mg L^{-1})	For use up to 20 years on fine–textured soils of pH 6.0–8.5 (mg L^{-1})
Aluminium	5	20
Arsenic	0.1	2
Beryllium	0.1	0.5
Boron	0.75	2
Cadmium	0.1	0.05
Chromium	0.10	1
Cobalt	0.25	5
Copper	0.2	5
Fluorine	1	15
Iron	5	20
Lead	2.5	10
Lithium	5	2.5
Manganese	0.2	10
Molybdenum	0.01	0.05
Nickel	0.2	2.0
Selenium	0.02	0.02
Vanadium	0.1	1.0
Zinc	2	10

24 ESSENTIAL NUTRIENTS

Plants absorb more than 90 elements from soil, water and air.

Essential Nutrients

Sixteen elements are considered essential for plants. They are classified into groups based on their relative abundance in plants.

Many non–essential elements are also found in plants that account for over 60 elements. Al is absorbed when soil solution is rich in Al ions. When plant material is burned, the remaining plant ash contains all the essential and non–essential elements except C, H, O, N, S. Macronutrients are absorbed 30–100 times in proportion to that of micronutrients.

Table 73. Essential elements

Non–mineral Nutrients	Macronutrients		Micronutrients		
	Primary	Secondary	By all plants		Some
C	N	Ca	Fe	B	Na
H	P	Mg	Mn	Cl	Co
O	K	S	Zn	Mo	Va
			Cu		Ni
					Si

Chemical nature

Metals	K, Ca, Mg, Fe, Mn, Zn, Cu,
Non-metals	C, H, O, N, P, S, Cl

The content of mineral elements in plant is affected by many factors and their concentration in crops varies considerably.

Table 74. Relative nutrient concentrations in plants

Nutrient		Relative concentration (Times)	Concentration (%)
Primary	N, K	400–1,000	1–1.5
	P	30	0.1–0.3
Secondary	Ca, Mg	100–200	0.2–0.5
	S	30	0.1
Micro	Fe, Mn, B	1–2	0.002–0.01
	Others	<1	<0.002

The structural elements viz., C, H, O are mostly comes from air and water in the forms of CO_2 and H_2O. Other 13 elements are absorbed from soil. Macronutrients like N, P, K, Ca, Mg, S are needed in relatively larger amounts (>1 ppm). Micronutrients viz., Zn, Cu, Fe, Mn, B, Cl, Mo are needed in relatively small amounts (< 1 ppm). Ultra – micronutrients or nanonutrients such as Mo, Co are needed in smallest amount (<1 ppb). Beneficial elements viz., Co, Ni, Si, Na, V are established as essential elements in some plants

Discovery of elements

An element is essential for plant growth when the plant is unable to complete its life cycle without that element, when no other element can take its place and when the element must be directly involved in the plant nutrition. The essentiality of most of the elements was found only during the 1st half of the 20th century as shown in Table 75.

Table 75. Essentiality of nutrients

Essential element	Essentiality 1st established by	Year
H	Essentiality in plant is established since time immemorial	-
O	Since time immemorial jointly by Priestly senctier and Theodre De saussure	-
C	Theodre de Saussure	1804
P, K, Mg, S, Ca	C. Sprengel	1839
Fe	I. Gris	1844
Mn	J.M. Mc. Hargue	1922
Zn	A.I. Sommer and C.P. Lipman	1926
Cu	AI. Sommer, C.P. Lipman and G.Mc. Kinney	1931
Mo	D.I. Arnon and P.R. Stout	1931
Na	P.P. Brownell and I.G. Wood	1959
Co	S. Ahmed and H.T. Evans	1959
B	K. Warington	1923
Cl	T.C. Broyer, A.B. Carlton C.M. Johnson, P.R. Stout	1954
V	D.I. Aronon and G.Wessel	1953
Si	T.C. Lewin	1962

Criteria of essentiality of plant nutrients

To consider a nutrient as an essential nutrient for plant growth, Arnon (1954) defined the following criteria for essentiality:

- The plant must be unable to grow normally or complete its life cycle in the absence of the element.
- The element is specific and cannot be replaced by other element.
- The element plays a direct role in metabolism.
- The deficiency symptoms of the element can be corrected or prevented by the application of that element only.

Functional nutrients

In view of the limitations of the experimental techniques and in the absence of complete knowledge, many situations arise in which an element is found to meet 1 or 2 instead of all 3 criteria of essentiality laid down by Arnon. Many instances are also known in which Rubidium (Ru) can replace K and strontium (St) can replace Ca, and Na can replace K. Nicholas (1963) suggested the use of the term 'functional nutrients' for all such nutrients, which play a specific role in metabolism but, the requirement of which can be bypassed by supplying the product of the reaction mediated by the elements.

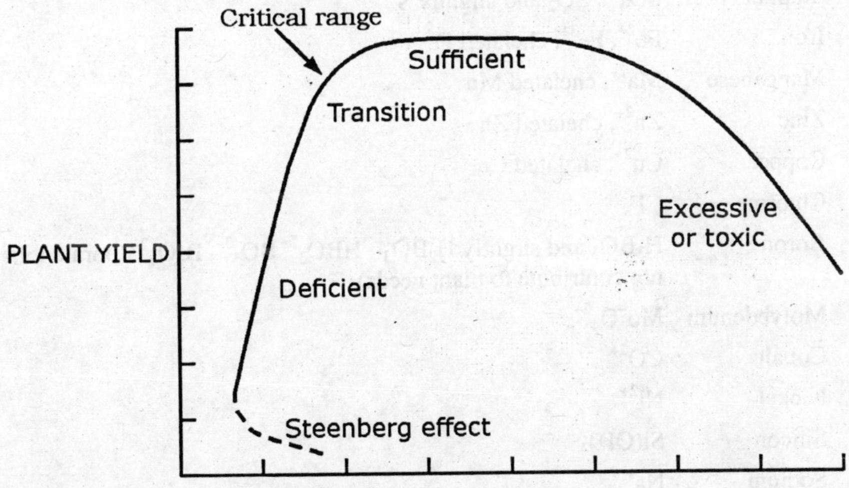

Fig. 61. Critical ranges

Level of nutrients in plants

A mineral element is considered essential to plant growth and development if the element is involved in plant metabolic functions and the plant cannot complete its life cycle without the element. Terms commonly used to describe level of nutrients in plants are given below:

Deficient

When the concentration of an essential element is low enough to limit yield severely, and when distinct deficiency symptoms are visible then that element is said to be deficient. Extreme deficiencies can result in plant death. With moderate or slight deficiencies, symptoms may not be visible, but yields will still be reduced.

Table 76. Forms of nutrient elements absorbed by plants

S.N.	Elements	Forms
1.	Carbon	CO_2
2.	Hydrogen	H_2O, H^+
3.	Oxygen	CO_2, O_2, H_2O
4.	Nitrogen	NH_4^+, NO_3^-, $CO(NH_2)_2$, slightly NO^-_2, amino acid and Nucleic acid
5.	Phosphorus	$H_2PO^-_4$, HPO_4^{2-} and slightly pyrophosphate and meta phosphate
6.	Potassium	K^+
7.	Calcium	Ca^{2+}
8.	Magnesium	Mg^{2+}
9.	Sulphur	SO_4^{2-}, SO_2 and slightly S
10.	Iron	Fe^{2+}, Fe^{3+}, chelated Fe
11.	Manganese	Ma^{2+}, chelated Mn
12.	Zinc	Zn^{2+}, chelated Zn
13.	Copper	Cu^{2+}, chelated Cu
14.	Chlorine	Cl^-,
15.	Boron	H_3BO_3 and slightly $H_2BO_3^-$, HBO_3^{2-} BO_3^{3-}, $B_4O_7^{2-}$ (borate ions do not contribute to plant needs)
16.	Molybdenum	$Mo^-O_4^{2-}$
17.	Cobalt	CO^{2+}
18.	Nickel	Ni^{2+}
19.	Silicon	$Si(OH)_4$
20.	Sodium	Na^+
21.	Vanadium	VO_3^-

Critical range

It is the level of nutrient concentration in the plant below which a yield response to added nutrient occurs. Critical level or ranges vary among plants and nutrients but occur somewhere in the transition between nutrient deficiency and sufficiency.

Sufficient

Concentration range in which added nutrients will not increase yield but, can increase nutrient concentration. The term luxury consumption is used to describe nutrient absorption by the plant that does not influence yield.

Excessive or toxic

When the concentration of essential or other elements is high enough to reduce plant growth and yield, it is said to be toxic. Excessive nutrient concentration can cause an imbalance in other essential nutrients, which also can reduce yield.

Concentration and plants yield

Yield is severely affected when a nutrient is deficient. When deficiency is corrected, growth increases more rapidly. Under severe deficiency, rapid increase in growth with added nutrient can cause a small decrease in nutrient concentration due to dilution effect. This is called the 'Steenberg effect'.

Luxury consumption

Nutrient sufficiency occurs over a wide concentration range, wherein yield is unaffected. Increase in nutrient concentration above the critical range indicates that the plant is absorbing nutrient above that needed for maximum yield. This 'luxury consumption' is common in most plants. Elements absorbed in excessive quantities can reduce plant yield directly through toxicity or indirectly by reducing concentration of other nutrients below critical ranges.

25 SOIL FERTILITY AND ITS EVALUATION

Soil fertility evaluation

Soils are very heterogeneous in respect to forms of nutrients they contain, which greatly complicates the interpretation of soil test for assessing fertilizer needs. The total amount of a nutrients present in a soil offers little information relative to the amount of that nutrient which is available. Numerous extracting solutions and procedures have been used to remove nutrient elements from the soil, but none remove exactly the amount that plant roots obtain.

This means that to interpret the data, the results from each analytical procedure must be correlated with the plant response obtained in field experiments by applying that fertilizer nutrient.

By this calibration, the requirement of fertilizer is calculated for achieving specific yield target. Soil fertility evaluation preferably employs a particular method of calibration as per the utility of the outcome.

The techniques that are commonly employed for the evaluation of soil fertility are:

- Nutrient deficiency symptoms of plants
- Analysis of tissue (Tissue test) or plant parts from plants growing on the soil
- Biological tests in which the growth of either higher plants or certain microorganisms is used as a measure of soil fertility
- Chemical Soil tests

25.1. SOIL TESTING RESEARCH IN INDIA

Farming is a business and like all business persons, the farmer operates farming business for profit. Obviously, to reach a decision before planting a crop, a farmer needs all the reliable information from the country's soil evaluation programme. In our country, system of fertility evaluation is being modified from time to time in various stages, incorporating crop response data.

Table 77. Some important landmarks in soil testing research in India

Year	Landmark
1953	Soil Fertility and Fertilizer Use Project
1955	Establishment of Soil Testing Laboratories
1956	All India Coordinated Agronomic Research Project
1960s	Establishment of Agricultural Universities
1967	All India Coordinated Research Project on Soil Test Crop Response (STCR) Correlation
1967	All India Coordinated Research Project on Micronutrients in Soils and Plants
1970	All India Coordinated Research Project on Long–term Fertilizer Experiments
1980s	Emphasis on fertilizer prescription for whole cropping system based on initial soil tests.

Since Liebig's time i.e., around 1840 many methods and approaches have been tried to get a precise or workable basis for predicting the fertilizer requirements of crops. The Fertilizer–use Project initiated in 1953 following a study by Stewart (1947) was the first systematic attempt in the whole of the country to relate the knowledge of soils to the judicious use of fertilizers. The establishment of soil–testing laboratories was initiated in 1955–56.

The project on Model Agronomic Experiments on Experimental Farms and Simple fertilizer trials on cultivators' fields was started in 1957. The Soil Test Crop Response (STCR) correlation work carried out at Indian Agricultural Research Institute, New Delhi had resulted in the selection of soil test methods and categorizing the tests into low, medium, and high soil fertility classes.

Crop response data

In nutrient–rating experiments, soil test data is correlated with the response of crop. The response is measured in terms of 'per cent yield' or 'per cent yield increase'. Both represent the ratio of the yield obtained in unfertilized soil (nutrient limiting or deficient soil) to the yield in nutrient–fertilized soil (non–limiting or nutrient sufficient soil). The yield in non–limiting soil is otherwise known as 'maximum attainable yield' or 'yield of standard treatment with all nutrients applied'.

While comparing native fertility, per cent yield is used in experiments that are conducted at different locations.

$$\text{Percent yield} = \frac{\text{Yield without fertilizer nutrient}}{\text{Yield in standard treatment}} \times 100$$

While conducting multi–location trials or a pot–experiment in a single location with soils brought from different locations, per cent yield increase is worked out for every level of fertilizer application. The highest yield obtained in the experiment with all fertilizer nutrient applied, is taken as maximum attainable yield.

$$\text{Percent yield increase} = \frac{\text{Yield at fertilizer nutrient level} - \text{Yield without fertilizer nutrient}}{\text{Maximum attainable yield}} \times 100$$

The soils analyzed by standard nutrient extraction method are grouped together. The soil test values are calibrated into different fertility groups by various procedures.

Soil Analysis and correlation

This approach is extensively used in India for formulating fertilizer recommendations.

Soil testing programme was initially started adopting the International Soil Fertility Evaluation and Improvement Programme (Fitts 1956). In this approach much stress is laid on laboratory characterization, followed by potted–plant studies ultimately leading to field–verification trials. These two initial steps help in eliminating arbitrariness of field trials. However, this approach had low adoption though it gives one critical value below which economic responses are possible. Later, many improvements were suggested for fertility grouping of soils.

The composite samples obtained from fields are used in pot–culture plant studies to assess the crop responses (in terms of dry matter) to added fertilizers. Calibration of the responses to soil–test values was obtained by using different methods. Fixation studies are conducted to ascertain the capacity of soil to fix the applied nutrient. Finally, the results are interpreted to confirm nutrient deficiencies, index soil fertility, and establish 'critical level' for each nutrient. The results are verified in the next stage, in field experiments.

Fertility grouping

The International Soil Fertility Evaluation and Improvement Programme (Waugh and Fitts 1965) advocated grouping of soils into low, medium and high categories.

Method of fertility grouping

- Collect representative soil samples and analyze by using different extractants for the 'available' nutrient.
- Conduct potted–plant studies with graded doses of all the nutrients to meet the nutrient requirement of the crop.
- Compute the percentage yield responses.
- Plot the soil–test data obtained with different methods and the per cent yield response, and find the scatter distribution for the best correlating

soils test methods. The scatter diagram will be generally curvilinear. The response is less, as the soil test value increases.

- The general groups are low, medium and high.

Table 78. The fertility groups followed in Tamil Nadu

Fertility group	Available nutrients (kg/ ha)		
	$KMnO_4$-N	$NaHCO_3$-P	NH_4OAc-K
Low	<280	< 11	< 118
Medium	280-450	11- 22	118-280
High	>450	> 22	>280

This classification indicates that the soils of low group would markedly respond to added fertilizers and the soils of high group do not respond to them. In the medium group, nothing could be predicted. In addition, by this grouping, it was not possible to indicate how much fertilizer was required to be added to get economic yields. Thus, this grouping is qualitative. Arbitrarily, the recommended dose of fertilizer for a crop is increased by 25 per cent in low group and reduced by 25 per cent in high group (Fig. 62).

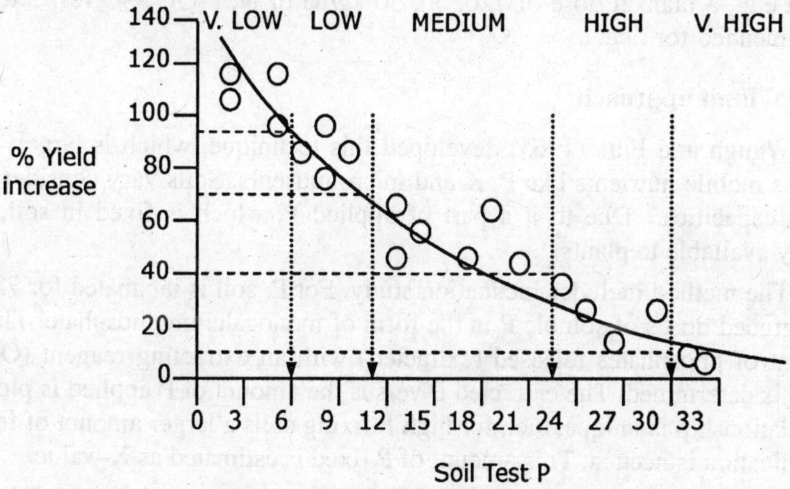

Fig. 62. Percent yield increase with P

Fertility index

Fertility index expresses the relative sufficiency as a percentage of soil nutrient amounts adequate for optimum yields. The probability of a response to fertilizer application increases with decreasing soil test level.

More than 85 per cent of soils, having very low soil–test value may give greater response and profit. About 60–85 per cent of soils, having medium soil test value may give little profit. Around 15 per cent soils with very high soil test value may give little response.

In general, the very low to very high classification is easily understood by the grower. However, separate group classification needs to be done accounting for greater variability associated with crops and soils.

Table 79. Fertility Index

Soil test rating	Probability of crop response (% Yield increase)	Fertility index
Very low	95–100	0–10
Low	70–95	10–25
Medium	40–70	25–50
High	10–40	50–100
Very High	0–10	100+

Agronomic approach

This is based on fertilizer rate experiments conducted at many locations. The level at which yields are high are recorded and averaged. From these results, the optimum dose of fertilizer is recommended for a crop at given agro–climatic region e.g., a blanket dose of 120, 50, 50 kg/ha of N, P_2O_5, K_2O respectively is recommended for rice.

Critical limit approach

Waugh and Fitts (1965) developed this technique, which is largely meant for less mobile nutrients like P, K and micronutrients. Soils vary considerably in fixing capacities. Due to it a part of applied P, which is fixed in soil, is not readily available to plants.

The method includes incubation study. For P, soil is incubated for 72 hours with graded doses of soluble P in the form of monocalcium phosphate. Then, the amount of phosphates released (extracted) with an extracting reagent (Olsen or Bray) is determined. The extracted P versus the amount of P applied is plotted. If the relationship is unique, then for high P fixing soils a larger amount of fertilizer P application is needed. This amount of P fixed is estimated as X–value.

Then, in potted–plant experiment, fertilizer P is added at 0, 0.5X, X, 2X levels, and the test crop is grown to find out the soil critical value. The percentage yield obtained is plotted against the soil test value for different soils. By using a plastic overlay, these data are grouped into two populations as described by Cate and Nelson (1965). The soil test value where the vertical line crosses the x–axis is designated as the soil critical limit.

Critical limit for the soil test value is the limit below which a positive or economic response to added fertilizer is possible and above which the response diminishes at a faster rate or vanishes. The validity of critical value is verified by conducting verification trials in the field locations from where the soil samples have been collected for pot studies.

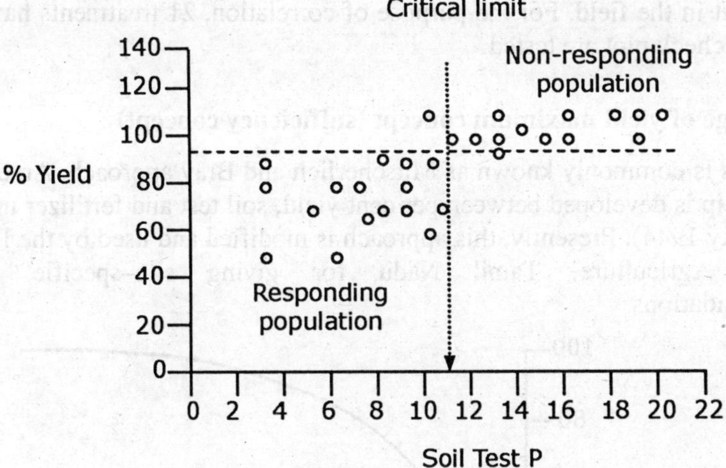

Fig. 63. Critical limit approach

Predicting yields using nutrient functions

Nutrient function

Crop yield is a function of four major factors:

Yield = f (crop, soil, climate, management)

Deductive approach

Deductive approach utilizes the natural variation in soil fertility for calibrating soil test values. Multi–location trials are carried out with same set of treatments. The responses are then fitted in the fertilizer prescription models.

Colwell (1967) developed this approach on the basis that if some variables, which affect the response of crop to fertilizers, are omitted, they lead to poor correlations between soil test and crop response. To prevent this difficulty, a calibration model was suggested based on the generalization of coefficients of an 'orthogonal polynomial yield–response model', which could include all the variables affecting the responses to fertilizers.

Inductive approach

This approach is by creating fertility gradient artificially, in a particular experimental location, by addition of fertilizers. The approach of inducing fertility gradient (Ramamoorthy 1970) aims at eliminating influence of the 3 out of 4 factors in the yield function, viz., crop, climate and management.

A large field of a certain amount of natural variation in fertility is chosen in a location. It is divided into four strips, which are treated with four different doses of N, P and K fertilizers. These strips are named as control ($N_0P_0K_0$), ½ normal dose ($N_{½}$ $P_{½}$ $K_{½}$), normal dose ($N_1P_1K_1$) and double dose ($N_2P_2K_2$). Exhaustive crops like maize or rice are grown. The calibration experiment is then carried out in the field. For the purpose of correlation, 21 treatments having one untreated check plot are tested.

Percentage of yield maximum concept (sufficiency concept)

This is commonly known as Mitscherlich and Bray approach. An empirical relationship is developed between per cent yield, soil test and fertilizer maximum yield (Bray 1944). Presently, this approach is modified and used by the Department of Agriculture, Tamil Nadu for giving site–specific fertilizer recommendations.

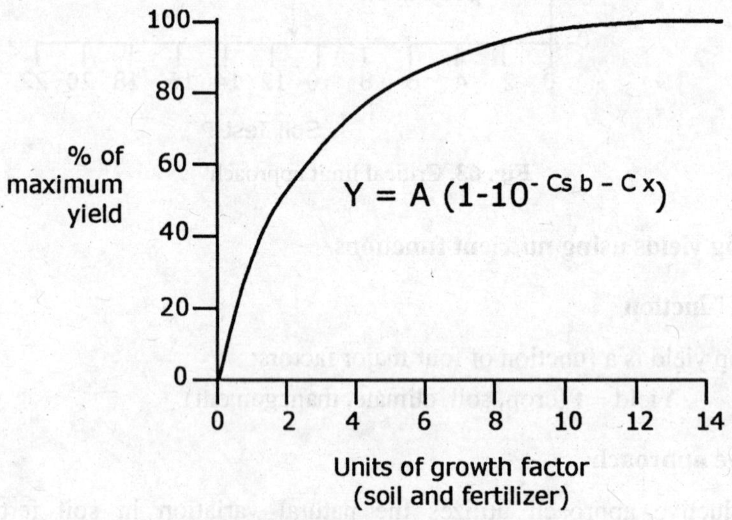

Fig. 64. Sufficiency concept

The modified Mitscherlich–Bray equation is:

$$\text{Log}\,(A-Y) = \text{Log}\,A - C_s b - Cx$$

Where, A = Calculated maximum yield

Y = Percentage yield

C_s = Proportionality factor for soil nutrient

b = Soil test value

x = Dose of fertilizer added

The maximum yield (A) could be calculated by extrapolation method.

Regression model for maximum profit

The amount of fertilizer that produces the greatest profit per hectare is called the optimum dressing (Cooke 1972). Ramamoorthy (1974) established a significant relationship between soil tests, added fertilizers and crop yields by fitting a multiple regression of the quadratic form:

$$Y = A + b_1SN + b_2SP + b_3SK + b_4SN^2 + b_5SP^2 + b_6SK^2$$

$$+ B_7FN + b_8FP + b_9FK + b_{10}FN^2 + b_{11}FP^2 + b_{12}FK^2$$
$$+ b_{13}SNFN + b_{14}SPFP + b_{15}SKFK$$

Where, Y = Crop yield (kg/ ha)

A = Intercept

bi = Regression coefficient

SN, SP, SK = Available contents of soil N, soil P and soil K

FN, FP, FK = Fertilizer N, Fertilizer P, Fertilizer K

Fertilizer calibrations for varying soil test value for obtaining maximum profit per hectare could be derived where the response to added nutrient follows the law of diminishing returns (Ramamoorthy 1974).

Fertilizer adjustment equation is derived in the form:

$$FN = a - b\ SN - cR$$
$$FP_2O_5 = a - b\ SP - cR$$
$$FK_20 = a - b\ SK - cR$$

Where,

$$R = \frac{\text{Cost of fertilizer nutrient (Rs/kg)}}{\text{Value of produce (Rs/kg)}}$$

Targeted yield approach

Fertilizer recommendation must aim at providing balanced nutrition to crops. Balanced nutrition should ensure that the nutrients present in available forms are in adequate quantities and required proportion for the plant in order to produce maximum yield. The requirement of nutrient to produce the expected yield can be worked out based on nutrient uptake.

Table 80. Nutrient requirement of crops

Crop	Nutrient Requirement (kg) to produce 100 kg of economic produce		
	N	P_2O_5	K_2O
Rice	2.01	1.12	3.00
Wheat	2.45	0.86	3.28
Maize	2.63	1.39	3.58
Sorghum	2.24	1.33	3.40
Finger millet	2.98	1.13	3.90
Chick pea	4.63	0.84	4.96
Soya bean	6.68	1.77	4.44
Ground nut	5.81	1.96	3.01
Potato	0.39	0.14	0.49
Cotton	4.45	2.83	7.47

Liebig's law (Law of Minimum) states that the growth of plants is limited by the plant nutrient element in the smallest quantity, when all others being present in adequate amounts. This forms the basis for fertilizer application for targeted yields, first advocated by Troug (1960) by significant linear relationship between the yield of grain and uptake of nutrients. Yield target can be projected within the linear region of the response function.

This approach, popularly known as Soil Test Crop Response (STCR) function, implies that for obtaining a specific yield (grain or any other economic produce) a definite quantity of the nutrient must be taken up. This value can be determined by the magnitude of the expected yield (T in q/ha) and the nutrient requirement to produce unit quantity of that target (NR in kg/ha).

Once it is known for a target yield, the fertilizer dose (FD kg/ha) can be estimated.

It is done by taking into account the efficiency of soil contribution (CS in per cent) from the soil–available nutrients (STV in kg/ha), and the efficiency of fertilizer contribution (CF in per cent) from the fertilizer nutrients (FD in kg/ha) towards the total uptake.

$$FD = \frac{NR}{CF} \times 100 \; T - \frac{CS}{CF} \times STV$$

Where, F and S stand for fertilizer and soil nutrient in Kg/ha and T is yield target in q/ha.

25.2. NUTRIENT DEFICIENCY SYMPTOMS OF PLANTS

An abnormal appearance of the growing plant may be caused by a deficiency of one or more nutrient elements. If a plant is lacking in a particular element, more or less characteristic symptoms may appear. This visual method of evaluating soil fertility is unique in that it requires no expensive or elaborate equipment, and can be used as a supplement to other diagnostic techniques.

- Nutrient-deficiency symptoms may be classified as follows:
- Complete crop failure at seedling stage
- Severe stunting of plants
- Specific leaf symptoms appearing at varying times during the season
- Internal abnormalities such as clogged conductive tissues
- Delayed or abnormal maturity
- Obvious yield differences with or without leaf symptoms
- Poor quality of crops viz., differences in keeping /storage quality, taste etc. (including that of the unseen chemical composition differences in oil, sugar etc.)
- Poor root development

Deficiencies are actually relative; and a deficiency of one element implies adequate or excessive quantities of another. e.g., Lime induced iron chlorosis; K induced Mg deficiency etc. In addition, a sufficient supply under one condition may become deficient in another as other elements become more abundant. In other words, once the first limiting factor is eliminated the second limiting factor appears. A short supply of two or more elements will also complicate the diagnosis. Similarly, the symptoms appear only after the supply of an element is so low that the plant can no longer function properly and in such cases it would have been profitable to have applied fertilizer long before the symptoms appeared.

Hidden hunger

Hidden hunger refers to a situation in which a crop needs more of a given element, yet has shown no deficiency and the nutrient content is not sufficient to give the top profitable yield.

Fertilization with the 'sure' rate rather than the bare economic optimum for an average year helps to obtain the top profitable yield. With most nutrients on most crops, significant responses can be obtained even though no recognisable symptoms have observed. It should be emphasized that the wise use of deficiency symptoms in conjunction with other methods of diagnosis such as plant or soil analysis can do much to promote proper fertilization for profitable yields.

25.3. PLANT ANALYSIS

Two general types of plant analysis have been used. One is the tissue test, which is customarily made on fresh tissue in the field, and the other is the total analysis performed in the laboratory with precise analytical techniques.

Plant analysis is based on the premise that the amount of a given element in a plant is an indication of the supply of that particular nutrient and as such is directly related to the quantity in the soil.

Critical levels have been suggested for a number of nutrient elements in a number of plants. Critical level is the content of an element below which the crop yield or performance is decreased below optimum. However, it is difficult to choose a specific level because the content of other nutrients in the plant may affect the critical point for a particular element.

1. Tissue tests

Rapid tests for the determination of nutrient elements in the plant sap of fresh tissue have found an important place in the diagnosis of the needs of growing plants. In these tests, the sap from ruptured cells is tested for unassimilated N, P, K, Ca, Mg etc. and rated as very low, low, medium or high. The purpose of the test is to assess the general levels of the nutrients in the plants.

One of the tissue test methods suggested was to chop the plant parts and extracted with reagents. The intensity of colour developed is compared with standards.

In another more rapid method, plant sap is transferred to filter paper by squeezing the plant with pliers and tested for the nutrients by using various reagents. This is a very simple test and easy to perform. Semi–quantitative values for the plant nutrients can be obtained in about a minute. The best part of the plant to use for testing is generally that showing the greatest range of levels as the nutrient goes from deficient to adequate levels.

Table 81. Plant tissue testing methods

Nutrient	Plant part to be tested	Methods
Nitrogen	Main stem	Diphenylamine test (1% DPA in conc. H_2SO_4) Blue colour
	Leaf mid ribs	Brays modified test (Chemical reagent involving $BaSO_4$, Zn, $MnSO_4$, CH_3COOH, Sulphanitric acid and alpha Napthylamine) (Std: 20 ppm NO_3-N)
Phosphorus	Leaf mid ribs near Petioles, leaf tissues	PO_4 – Test reagent: (Amm. Molybdate, conc. HCl stannous oxalate) No colour – Light yellow – greenish Blue – light blue – medium blue – dark blue
Potassium	Blade tissue, petioles Mid ribs near Upper third of stem	Cobaltinitrite test: (Sod. Cobaltinitrate, $NaNO_2$, acetic acid, ethyl alcohol) – degree of turbidity indicates the K content.

The most critical stage of growth for tissue testing is at the time of bloom or from bloom to early fruiting stage and the time of testing can be either in the early morning hours or late in the afternoon.

2. Total analysis

Total analysis is performed on the whole plant or plant parts after the plant material are dried, grounded and ash. Much smaller differences can be detected by this method than by tissue tests and more elements can be determined. Recently–matured plant parts can be selected for the analysis. Quantitative plant analysis is employed extensively in research to obtain another measure of the effect of treatment.

3. Diagnosis and Recommendation Integrated System (DRIS)

DRIS is a system that identifies all the nutritional factors limiting crop production. Index values measure how far particular nutrients in the leaf or plant are from optimum levels. Index values are used in the calibration to classify yield factors in the order of limiting importance. To develop a DRIS for a given crop, the following criteria are to be well considered:

- All factors having effect on yield
- Relationship among factors
- Calibration norms
- Continually–refined recommendations

DRIS was developed based on nutrient ratios. When compared to concentration that normally varies with season, nutrient ratio does not vary much. When a nutrient ratio has an optimal value, optimum yield occurs unless some other limiting factor reduces the yield.

	N→			N↑			N↓	
	P→			P↑			P↓	
Both numerator and denominator optimal			Both numerator and denominator excessive			Both numerator and denominator insufficient		

When the ratio is too low a response in the numerator will be obtained if it is limiting. If the nutrient in the denominator is excessive, a yield response may or may not occur depending on the level of other yield factors.

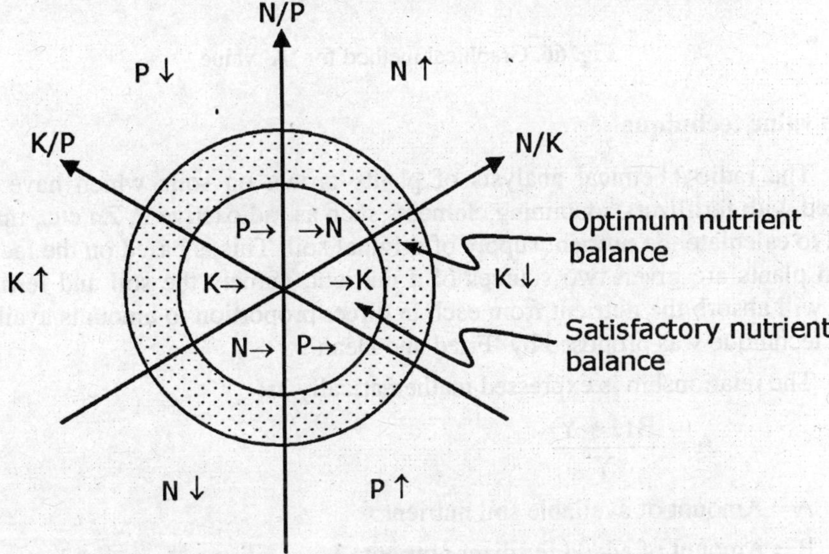

Fig. 65. General diagram for a DRIS; constructed on nutrient ratios for N-P-K

When the ratio is too high the reverse is true. Usually, N/S, K/Mg, K/Ca, Ca + Mg /K, N/P ratios are commonly used. Initially, relationship among N-P-K is calibrated.

Thus, DRIS has several advantages to integrate much nutrient concentration at various stages, in different seasons suitable for many cultivars of a crop. DRIS has been found suitable for several grain crops and perennial fruit trees.

Crop logging

An excellent example of the use of plant analysis in crop production operations is the crop logging carried out for sugarcane in Hawaii. In this method, periodical monitoring of the nutrient status of the plant as well as other plant characteristics are done, and based on the results the adjustments in the management practices are introduced as and when needed.

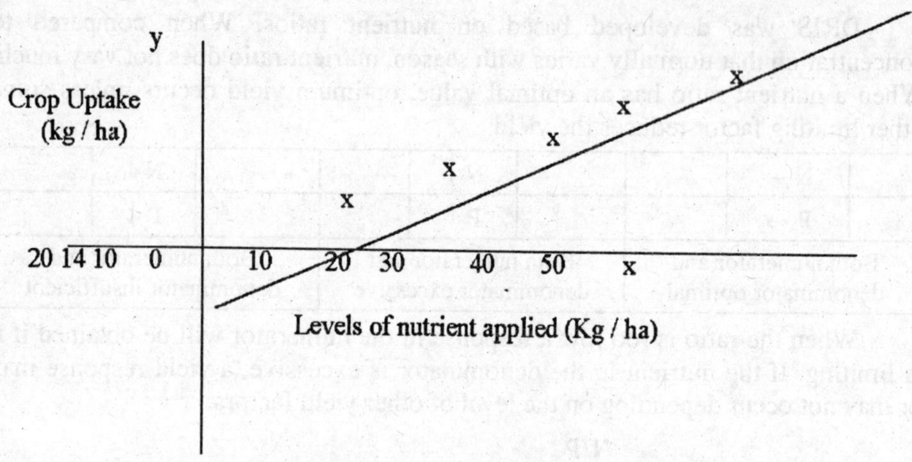

Fig. 66. Graphical method for 'A' value

'A'- value technique

The radio–chemical analysis of plants grown on soils which have been treated with fertilizers containing elements such as radioactive P, Zn etc., may be used to calculate the nutrient supply of original soil. This is based on the fact that when plants are given two sources of a nutrient, namely the soil and fertilizer; they will absorb the nutrient from each in direct proportion to amounts available. This technique was proposed by 'Fried and Dean'.

The relationship is expressed mathematically as:

$$A = \frac{B(1-Y)}{Y}$$

A = Amount of available soil nutrient

B = Amount of added fertilizer nutrient

Y = Fraction of the nutrient in plant, which is derived from the fertilizer.

An additional development related to the 'A' value method is the graphical method in which the data on the uptake of the nutrient by plants at several rates of applied nutrients are plotted in a graph and the extrapolation of the graphical line back to the X-axis will give a value corresponding to the A-value. This is an alternate method in the absence of the use of radioactive materials and the equipments necessary for the analysis.

25.4. BIOLOGICAL TESTS

Use of the growing plants or microorganisms understandably has much appeal in the study of fertilizer requirements and serves as a tool for measuring the fertility status of the soils. This includes a) Field tests b) Laboratory / Green house tests, and c) Microbiological methods.

1. Field tests

The field plot method is one of the oldest and best known of the biological tests. Treatments involving various levels of fertilizer nutrients are selected and randomly allotted in field strips of different localities and the response of the crop is observed. From the results the variations in soil and management practices are accounted for. These experiments help for formulating a general recommendation for a particular crop. Stewart (1951) and Mukkerjee (1951) were the larger contributors in this type of field experimentation. Blanket recommendations were evolved through this type of simple fertilizer experiments and trials in farmers' fields. The Blanket recommendation became popular during the 1950's.

2. Laboratory and green house tests

Simpler and more rapid biological techniques, which still involve higher plants but, utilize small quantities of soil have been developed.

a. Mitscherlich pot culture

Oats are grown to maturity in pots containing 6 lb of soil each in 10 pots involving the N, P, K treatment combinations. The percentage yields of the various treatments are worked out in comparison with the maximum yield i.e., the yield of NPK plot. Using the yield data, the plant nutrient reserve in the unfertilized soil can be worked out. From the data, the expected yield for a given level of fertilizer nutrients can also be obtained.

b. Neubauer seedling Method

This is based on the uptake of nutrients by a large number of plants grown on a small amount of soil. The roots of densely–grown plants thoroughly penetrate the soil exhausting the available nutrient supply within a short time. The nutrients taken by plants or removed are quantitatively assessed and the minimum values for satisfactory yields of various crops have been calculated and tabulated which can be used as a ready reckoner.

c. Sunflower pot culture technique

This is more specifically used to assess the boron status of the soils. Sunflower seedlings are grown in a small amount of soil and the number of days interval taken for the appearance of B deficiency in the seedlings are accounted for assessing the B status.

< 28 days	: Marked deficiency of B in soil
28 – 36 days	: Moderate
> 36 days	: Little or no deficiency

3. Microbiological tests

Winogradsky was one of the first to observe that in the absence of mineral elements, certain microorganisms exhibit a behaviour similar to that of higher plants. It was shown that the growth of Azotobacter served as an indicator of the limiting mineral nutrients in soils especially Ca, P and K. In comparison with methods that utilize higher plants, microbiological methods are rapid, simple and require little space.

a. Sackett and Stewart technique - Cultures are prepared using the different soil extracts, added with P, K and P+ K nutrients, and Azotobacter is grown. Based on the amount of colonies developed after 72 hours, the soils are rated from very deficient to not deficient.

b. Aspergillus Niger test - Soils are incubated for a period of four days in flasks containing appropriate nutrient solutions. The weight of the Aspergillus niger mycelial pad (or) the amount of nutrient absorbed by the pad is used as a measure of the nutrient deficiency. This technique was used for the assessment of the nutrient supply in soil in cases of P, K, Ca, Mg, Mn and Cu.

c. Mehlich's Cunning hamella – Plaque method - The soil is mixed with the nutrient solution; a paste is made and spread uniformly on the inner wall of a clay dish, inoculated at the centre with *Cunning hamella* and allowed to incubate for $4^1/_2$ days. The diameter of the mycelial growth on the dish is used to estimate the amount of nutrient in the soil. This is particularly recommended for the assessment of P supply in soil.

26 FERTILIZERS AND BIOFERTILIZERS

Fertilizers are the materials either natural or manufactured, containing nutrients essential for normal growth and development of plants. It may also be defined that any material or substance intended for use as a nutrient carrier in soil or crops for boosting crop yields.

In India, the use of artificial fertilizers was first initiated in 1896 when imported Chilean nitrate was used as a fertilizer. By about 1905 calcium nitrate, calcium cyanamide, ammonium sulphate, super phosphate and potassium sulphate were also imported and used.

After the World War I, the Imperial Chemical Industries carried out valuable field experiments on different crops particularly on rice with ammonium sulphate during the period 1920-30, which established the general superiority over other nitrogenous fertilizers. The manufacture of ammonium sulphate in India was first started at **Belegolla in Mysore** in 1938 on a small scale.

Later in 1947, its manufacture was started at **Alwaye**. In 1951, the Government of India set up fertilizer factory at **Sindri** for the production of ammonium sulphate in the public sector. Gradually several factories were established and use of fertilizers become increasingly popular.

Presently India is the **fourth** producer of N and P fertilizers in the world.

Table 82. Share of total fertilizer consumption (%) in India

	Share of total fertilizer consumption (%) in India
Rice	40.5
Wheat	24.2
Sugarcane	8.7
Cotton	6.7
Groundnut	2.2
Other crops	17.7

Classification of fertilizers

The fertilizer materials may be classified in several ways based on various factors

Based on fertilizer mixture behavior

Straight or simple-nitrogenous, phosphatic or Potassic fertilizers;

Complex fertilizers-having more than one nutrient in single material

Fertilizer mixtures -They are home mixed or factory mixed materials.

Based on their chemical reaction and nature

Acidic, neutral, and basic fertilizers

Based on nutrients present

Simple fertilizers -which contain only one nutrient (Single carrier)

Compound fertilizers-which contain more than one nutrient (Multinutrient carrier)

Physical properties of fertilizers

Physical properties of fertilizers have bearing on bagging storage, transportation, handling and distribution or application in the field. The properties include – hygroscopic nature, free moisture content, particle size, melting point, solubility, specific gravity, segregation, granule hardness, angle of response, drill ability etc.

Chemical properties of fertilizers

Chemical properties include the nature and amount of nutrients present, associated elements, their chemical reaction, and salt index. The chemical properties of fertilizers make a basis for their selection by the cultivators so that they may achieve highest productivity with least damage to soil fertility and least unit cost of fertilizers.

NITROGENOUS FERTILIZERS

Synthetic or chemical fertilizers are important sources of N. Over last 20 years, annual world N consumption has increased from 22 to 29 million tonnes. In India, N is consumed to a tune of 2 million tonnes. N Fertilizers can be grouped into four classes depending on the form of N present in them, such as readily available amide, ammonium, nitrate, combined/ mixed forms and slow release materials.

Classification of N fertilizers

I. Fertilizers containing N in ammonium (NH_4^+) form

Table 83. Ammonium form of fertilizers

Fertilizer	Composition	% N	% S	Reaction
Ammonium Sulphate	$(NH_4)_2SO_4$	21	24	Acidic
Ammonium Chloride	NH_4Cl	25	-	Acidic
Anhydrous ammonia	NH_3	82	-	Acidic

Characteristics

- Readily soluble in water, as such adsorbed on soil colloids, and thus protected from being washed away by run off or leaching.
- Solid ammonium fertilizers are not hygroscopic, but they are acidic in their residual effect on the soil.
- Anhydrous ammonia is a boiling liquid hence stored in high-pressure vessel.
- Nitrate form is more readily utilized by upland crops than ammonium form.
- Rice prefers ammonium form of N in early stage.
- The ammonium N has to be nitrified to nitrate in the soil before it can be absorbed by the plants, except for rice.

II. Fertilizers containing N in nitrate (NO_3^-) form

Table 84. Nitrate form of fertilizers

Fertilizer	Composition	% N	% K_2O	Reaction
Sodium nitrate (Chilean nitrate)	$NaNO_3$	16	-	Basic
Calcium nitrate	$Ca(NO_3)_2$	15	-	Basic
Potassium nitrate	KNO_3	13	44	Basic

Characteristics

- Nitrate fertilizers are obtained both as natural salt deposits (Chilean nitrate) and as manufactured products from synthetic ammonia.
- Nitrate fertilizers rapidly dissociate into soil solution as NO_3^- ions, which is readily absorbed by the plants.
- Nitrate fertilizers are hygroscopic.

- Nitrate ions being highly reactive and mobile are susceptible to losses due to leaching and (by denitrification) under water logged conditions and profuse irrigation.

III. Fertilizers containing N in ammonium and nitrate forms

Table 85. Ammonium and nitrate fertilizers

Fertilizer	Composition	% N		Reaction
Ammonium nitrate	NH_4NO_3	33	-	Acidic
Ammonium Sulphate Nitrate (ASN)	$(NH_4)_2SO_4 + NH_4NO_3$	26	6	Acidic
Calcium ammonium nitrate (CAN)	$Ca NH_4(NO_3)_3$	25	-	Neutral

Characteristics

- Contain N in both nitrate and ammonium form.
- The fertilizers of this group are hygroscopic.
- Nitrate N is readily available to plants for rapid growth and ammonium N resists leaching loss and provides a steady source of N.
- All are acidic in reaction; calcium ammonium nitrate is neutral.

IV. Fertilizers containing N in amide (NH_2) form

Table 86. Amide form of fertilizers

Fertilizer	Composition	% N	Reaction
Urea	$CO(NH_2)_2$	46	Acidic
Calcium cynamide	$CaCN_2$	21	Basic

Characteristics

- Amide fertilizers are synthetic organic fertilizers, however rapidly hydrolyzed to inorganic compounds in soils.
- Amide fertilizers are readily soluble in water and hygroscopic
- After application, amide fertilizers are quickly converted to ammonium, and then to nitrate due to action of soil microorganisms. Complete conversion of amide into inorganic forms takes in about 3-7 days.
- Leaching loss is very less because once amide is converted to ammonium form, it is absorbed by soil colloids and slowly released and nitrified to nitrates.

V. Other N fertilizers

Table 87. Other N carriers

Fertilizer	% N
Ammonia anhydrous (NH_3)	82
Ammonia solution (NH_4OH)	20-25
Calcium nitrate (Nitrate of lime) $Ca(NO_3)_2$	16-25
Calurea (Calcium nitrate + Urea)	34
Urea - Ammonia solution	35-45

VI. Slow release chemical N fertilizers

Table 88. Slow release N fertilizers

Fertilizer	% N
Urea Formaldehyde	38
Isobutylidine Diurea (IBDU)	32
Oxamide $H_2NCO–CONH_2$	32
Crotonylidene Diurea (CDU)	32
Dicyandiamide (DCD)	66

SLOW RELEASE UREA PRODUCTS

Urea formaldehyde (ureaform or UF)

The fertilizer condensates of urea formaldehyde are reaction products of urea and formaldehyde, containing at least 35% N, largely in almost insoluble slowly available form. Actually, ureaform covers many products, which are mixtures of methylene ureas. They range from short chain water-soluble materials to long chain water-insoluble materials.

Crotonylidene Diurea (2 oxo –4 methyl -6 ureido hexa hydro pyrimide).

This fertilizer contains 28% N. One-fourth of the N is from a nitrate source. The product decomposes slowly in the soil and the N release extends over a growing season.

Oxamide $(CONH_2)_2$

It contains about 31.8 per cent N. The availability of N decreases as the particle size increases. It is the diamide of oxalic acid and is a white crystalline solid. It is not hygroscopic and non-toxic, but very costly.

Isobutylidene Diurea (IBDU)

It is widely used in Japan. It is single component chemical obtained by the condensation reaction of urea and Isobutyraldehyde and contains 31-32% N.

Guanyl Urea sulphate (GUS)

It is the sulphate salt of Guanyl Urea produced from calcium cynamide by acid treatment. Though it is soluble, it is absorbed in soil colloids; as a result leaching is small.

OTHER N FERTILIZERS

Calcium nitrate is a crystalline white hygroscopic solid containing 15.5% N. It is produced by reacting nitric acid with crushed limestone.

$$CaCO_3+2\ HNO_3 \rightarrow Ca\ (NO_3)_2+CO_2\ +H_2O$$

Sodium nitrate used for fertilizer purposes is a natural product mined and processed in Chile. However, synthetically it is produced by treating sodium chloride salt with 70% HNO_3

Calcium cyanamide

It is one of the oldest methods employed for the fixation of atmospheric N for agricultural purposes is the nitrification of calcium carbide to form calcium cyanamide.

The first step in this process is the production of calcium carbide by fusion of lime and coke, carried out in an electric arc furnace to provide the high level of temperature needed:

$$CaO\quad +3C \quad \rightarrow \quad CaC_2 \quad +CO$$

$$CaC_2 \quad +N_2 \quad \rightarrow \quad CaCN_2+C$$

The second step of the process is initiated at approximately 1000°C and once started, it is self-sustaining. The use of calcium cyanamide as a nitrogen fertilizer has decreased markedly in the last few years as lower cost N products have come on the market.

SOLUTIONS CONTAINING NITROGEN

Water solution of any solid N fertilizer may be used for direct application or as the N source for blended or granulated fertilizers.

Aqua ammonia

Water solutions of ammonia are commonly referred to as "aqua" and the most widely used concentration is about 25% ammonia, which corresponds to 20% N.

Ammonium nitrate solution

Ammonium nitrate-water solutions are used almost exclusively for blending mixed fertilizer solutions and as the N source for granulated fertilizer.

PHOSPHATE FERTILIZERS

Nutrient content of P in Phosphatic fertilizers is usually expressed in terms of **phosphorus pentoxide (P_2O_5)**. **Bones** were the only known source of phosphate until the **rock phosphate (RP)** deposits were not known to human beings. RP contains more P than bones. Liebig (1857) found out that the complex form of phosphate present in RP could be reacted with sulphuric acid and made as a soluble P fertilizer.

Types of phosphate fertilizers

Water and citric acid insoluble phosphate fertilizers

Table 89. Fertilizer containing water and citric acid **insoluble** Tricalcium Phosphates [$Ca_3(PO_4)_2$] **[TCP]**

Fertilizer	Composition	P_2O_5 (%)	Reaction
Rock phosphate			
Purulia	$Ca_3(PO_4)_2CaF_2$	23	
Mussorie		23-24	-
Bone meal	$Ca(PO_4)_3CaF_2$	20-25	Alkaline

Citric acid soluble fertilizers

Table 90. Fertilizers containing citric acid soluble Dicalcium Phosphate [$Ca_2(H_2PO_4)_2$] **[DCP]**

Fertilizer	Composition	P_2O_5 (%)	Reaction
Basic Slag (India)	$(CaO)_3P_2O_5SiO_2$	3-8	Alkaline
Dicalcium Phosphate	$CaHPO_4$	35-40	Acidic

Water-soluble phosphatic fertilizers:

Table 91. Fertilizers containing water soluble monocalcium Phosphate [$Ca(H_2PO_4)_2$]

Fertilizer	Composition	P_2O_5 (%)	Reaction
Single Super phosphate	$Ca(H_2PO_4)_2$	16-20	Neutral
Double Super phosphate	$2Ca(H_2PO_4)_2$	30-35	Neutral
Triple Super phosphate	$3Ca(H_2PO_4)_2$	45-50	Neutral or Acidic

Water and citric acid insoluble phosphatic fertilizers

Rock phosphate

Rock phosphate (RP) is mainly used for the manufacture of phosphatic fertilizer such as Superphosphate, Dicalcium phosphate (DCP), Tricalcium phosphate (TCP), and calcium metaphosphate. Further, it is used also for reinforcement of manures.

The general formula apatite minerals comprising the pure RP is $Ca_{10}(PO_4)_6(X)_2$, where X is F^-, OH^-, or Cl^-. The common RP mined is fluorapatite or chlorapatites. High reactivity of some mined RP is due to the occurrence of primary mineral **francolite** (Carbonate-fluorapatite). The major deposits of RP are found in US (40%), Morocco (50%), Russia, South Africa, and China. RP contains 12-18% total P as P_2O_5 depending on the source of raw rock phosphate.

Total P in Rock-phosphate generally largely constituted by (55-80%) tricalcium phosphate, which is neither water-soluble nor citrate-soluble, hence requires acid treatment.

Citric acid soluble phosphatic fertilizers

Some grades of RP containing more **citrate soluble P** (18-25%) can be used as fertilizer straightway to supply P to the plants. The material varies in colour from grey to brown.

For most efficient use the material must be finely ground and to be applied to soil. Fine particles have increased surface area for more reaction. Best results obtained when particles are less than 300 mesh size. P supplying power increases greatly in the presence of decaying OM, which releases organic acids, particularly the carbonic acid formed by CO_2 dissolution. For the most of industrial manufacture of soluble P fertilizer the raw RP is imported.

Advantages of RP use

- It is a chief source of P. Besides P, Mg, and S it contains micronutrients like as Cu, Mo and B.
- It substitutes effectively superphosphate in acid soil with pH <6.
- Legumes respond well to RP due to their high calcium requirement. It should be thoroughly mixed into the top soil by ploughing with country plough.
- RP is commonly also used in rubber, tea and cacao plantations

Few indigenous sources of RP are available in Bihar (Mussorie-phos) and Trichirappalli, which are only suitable for direct application. Mussorie-phos is an economical and effective natural carbonate type rock phosphatic mineral for direct application in acidic, neutral, and alkaline soils. It is highly reactive variety of sedimentary origin and can be a very good source of phosphorus for direct application. Use of indigenous RP would save foreign exchange needed for the import of raw materials for manufacture of soluble P fertilizers.

Water-soluble phosphatic fertilizers

RP and Sulphur are the basic raw materials for production of phosphatic fertilizers. Sulphuric acid and phosphoric acids are intermediate products.

In the sulphuric acid plant, S is burnt to SO_2, which is further oxidized to SO_3 and dissolved in water to form sulphuric acid. In the manufacture of phosphoric acid, RP is attacked by sulphuric acid. Calcium sulphate formed along with phosphoric acid is removed by filtration.

While manufacturing phosphatic fertilizers a number of products are formed which vary in their solubility in the following order:

- Ammonium phosphates
- Monocalcium phosphates (MCP)
- Calcium metaphosphates
- Dicalcium phosphates (DCP)
- Hydroxy apatites

Other P carriers

Bone meal

Bone meal is prepared by crushing bones. It is available in two forms, raw bone meal (20% P_2O_5), and steamed bone meal (22% P_2O_5). The availability of P in bone-meal increase as the materials is crushed to a finer powder. Citric acid soluble P is 8 % in raw and 16% in steamed bone meal.

Bone meal is suitable for acid soils and long-duration crops, like fruit-trees. 150 kg of P in bone meal (0.5 mm) is equal to 100 kg P in super phosphate in acidic soils with low available P.

Basic slag

Basic slag is obtained as a by-product from steel industry and ranks next to super phosphate as the leading phosphatic fertilizer. In the manufacture of high-quality steel, the impurities like Si, S, and P combine with calcium, and rise to the top in the furnace, which are poured off. When solidified, they are finely ground and marketed as basic slag. Some grades of slag may contain 17% P_2O_5. Its chemical formula is $(CaO)_5\ P_2O_5\ SiO_2$. Basic slag also contains traces of Zn, Cu, Mn, and B.

In basic slag as 80% of the total P is citrate-soluble, the crop response is likely to be much greater than RP. Basic slag contains varying quantities of magnesium and manganese and it may be used for correcting deficiencies.

100 kg of basic slag is equivalent to 100 kg of limestone containing 80% of $CaCO_3$ in neutralizing capacity.

POTASSIUM FERTILIZERS

Extensive deposits of soluble K salts are found underground and in brines in many parts of world. The Saskatchewan deposit is the world's largest high-grade deposit between Manitoba, North Dakota, and Montana in North America. Next, it is found in New Brunswick in Canada. Germany and Russia also have large deposits.

The raw materials used for manufacturing K fertilizers are Carnallite (KCl. $MgCl_2$, 6 H_2O), Kainite (KCl. $MgSO_4$. 3 H_2O), Langbeinite (K_2SO_4. 2 $MgSO_4$), and Sylvite (KCl). KCl is extracted to smaller extent from salt bittern.

Classification

Table 92. Potassium fertilizers

Fertilizers	Composition	K_2O %	Reaction
Chloride form			
Muriate of Potash	KCl	50-60	Neutral
Non-chloride form			
Sulphate of Potash	K_2SO_4	48-52	Neutral
Potassium Magnesium Sulphate	K_2SO_4. 2 $MgSO_4$	22	Neutral
Potassium Schoenite	K_2SO_4 2 $MgSO_4$. 6 H_2O	22-24	Neutral

Practically all K fertilizers are water-soluble salts, but not hygroscopic in nature. The content of K fertilizers is usually expressed as potassium oxide (K_2O). The nutrients in K fertilizers are readily available to plants. K fertilizers have little or no effect on soil pH, as they are neutral in reaction. As a mobile cation, K^+ is readily susceptible for leaching in low CEC soils.

Most of the Indian soil contains a substantial amount of soil K. Potassic fertilizers, should therefore, be applied only to such soils that are definitely known to be K deficient or to soils which respond to K application, such as sandy soils.

Potassium chloride

Potassium chloride is most common K fertilizer under the commercial term **muriate of potash**. Muriate represents the muriatic acid (HCl). It is prepared mainly from Sylvite, Carnallite minerals, Kainite etc. by crystallization and flotation method. It contains 60 to 63% K_2O. The whole of K is readily available to plants. It is highly soluble in water and ionizes to K^+ and Cl^- ions in soil.

Muriate of potash is recovered from minerals by hot leaching or flotation. In leaching process, crushed ore is mixed with sufficient quantity of brine saturated with NaCl and heated almost to boiling to dissolve KCl. This KCl rich brine is then cooled by vacuum evaporation. This produces KCl crystals, which are centrifuged, washed, dried, and packed.

Potassium sulphate

Potassium sulphate is manufactured by treating potassium chloride with concentrated sulphuric acid.

$$2\,KCl + H_2SO_4 \rightarrow K_2SO_4 + 2\,HCl$$

It is also manufactured by dissolving Langbeinite in water and adding concentrated solution of KCl. The solubility of potassium sulphate is more than magnesium chloride. The potassium sulphate is separated by fractional crystallization.

$$K_2SO_4 2\,MgSO_4 + 4\,KCl \rightarrow 3\,K_2SO_4 + 2\,MgCl_2$$

(Langbeinite)

The fractioned product of potassium sulphate is then separated from the mother-liquor and dried. This product contains 48-50 per cent K_2O. The commercial product contains not more than 2.5 per cent chlorine.

Potassium nitrate

Potassium nitrate is also known as **saltpeter**. It is a white crystalline solid. It contains 14% N and 46-58% K_2O. Potassium nitrate is produced domestically for many years by the double decomposition reaction between $NaNO_3$ and KCl.

$$NaNO_3 + KCl \rightarrow KNO_3 + NaCl$$

In this process, solid KCl is reacted in a hot solution of $NaNO_3$. Sodium chloride crystallized is separated from hot KNO_3 liquor. The solid KNO_3 is produced by subsequent crystallization and filtration

Potassium phosphates

Potassium metaphosphate (KPO_3) contains water-soluble 40% K_2O, 60% P_2O_5. Other forms like $K_4P_2O_7$, KH_2PO_4, K_2HPO_4 are also available. They are preferred for their high analysis, low salt index, clarity in solution, suitability in fertigation units and tubes, low halide concentration, etc.

Potassium schoenite

It is the double sulphate of potassium and magnesium (K_2SO_4. $MgSO_4$. 6 H_2O) produced economically from the bitterns. The proportion of potassium chloride and magnesium sulphate in the mixed salt are in the ratio required for the formation of schoenite, but the presence of sodium chloride hinders the formation of direct schoenite. Hence, it is eliminated by the process of floatation using octadecylamine acetate and highly sulphonated castor oil. The floated product containing potassium schoenite and magnesium sulphate on treatment with water under constant agitation yields schoenite.

SECONDARY NUTRIENT FERTILIZERS

Calcium Fertilizers

Table 93. Calcium carterining fertilizers

Sl.No.	Fertilizers	Calcium Content (%)
1	Calcium Ammonium Nitrate	8.1
2	Calcium Nitrate	19.5
3	Calcium Cyanamide	39.1
4	Single super phosphate	20.0

5	Dicalcium Phosphate	23.0
6	Tricalcium Phosphate	13.6
7	Calcium EDTA	3-5
8	Rock Phosphate	38.0
9	Gypsum	23.0
10	Ground limestone	32.0
11	Calcium Chloride	18.0-36.0
12	Quicklime	71.0
13	Hydrated lime	54.0
14	Dolomite limestone	14.0-31.5
15	Basic slag	15.0-33.0
16	Bone meal	23.0
17	FYM	0.2-0.7

Primary sources: Calcite, dolomite, hydrated lime, blast furnace slag which are applied to neutralize soil acidity.

Gypsum: If Calcium is required without the need for correcting soil acidity gypsum is the favorite source.

Magnesium Fertilizers

Table 94. Magnesium containing fertilizers

S.No	Fertilizers	MgO content (percent)
1	Dolomitic limestone(Ca and Mg carbonate)	5-20
2	Ground burnt Magnesium lime (Ca and Mg oxide)	9-33
3	Serpentine	
4	Magnesia, MgO	55
5	Kieserite($MgSO_4$,H_2O)	27
6	Epsom salt($MGSO_4$,$7H_2O$)	16
7	Magnesium chloride ($MgCl_2 \cdot 6H_2O$)	20
8	Magnesium chloride solution, $MgCl.10H_2O$	8-9
9	Magnesite($MgCO_3$)	45
10	Magnesium nitrate $Mg(NO_3)_2$	16
11	Magnesium silicate	3-4
12	Synthetic chealets	2-4

Suspension formulae

Magnesium silicate, Potassium Magnesium Sulfate, Magnesium Chelates, Magnesium nitrate, Magnesium sulfate, Magnesium.chloride and Magnesium

Oxide - provide upto 3% Magnesium. Potassium Magnesium sulfate is most widely used as additive in suspension.

SULPHUR FERTILIZERS

Table 95. Sulphur containing fertilizers

Fertilizer	Sulphur content (%)
Ammonium sulphate [$(NH_4)2\ SO_4$]	24.2
Single superphosphate [$CaH4(PO_4)2, 2H_2O$	13.9
Gypsum [$CaSO_4,2H_2O$]	18.6
Magnesium sulphate [$MgSO_4$, $7H_2O$]	13.0
Pyrites [FeS_2]	53.5
All micronutrients [Sulphates]	13-19
Poultry manure	0.77
Farmyard manure	0.04
Basic slag	3.40
Elemental sulphur	100.0

There is numerous sulphur containing fertilizer materials. Elemental sulphur can be placed on the surface of the soil. Sulphur-Bentonite contains 90% elemental sulphur and 10% Bentonite.

Solid materials

Ammonium Phosphate – sulphur – 15.5%
Ammonium Polyphosphate, Ammonium Phosphate – sulphur – 2.2%
} For direct application and bulk blending several months before beginning of growing season

Ammonium sulfate (24.2%) – For direct application to some extends, bulk blends should be effective at almost any time.

Calcium sulfate (23.5%), $ZnSO_4$ – 17.8% } – For direct application

MICRONUTRIENT FERTILIZERS

Zinc fertilizers

Table 96. Zinc sources

Zinc Fertilizers	Zn content
Zinc Sulfate, $ZnSO_4H_2O$	35%
$ZnSO_4.7H_2O$	23%
Zinc Oxide ZnO	78%

Zinc Carbonate $ZnCO_3$	52%
Zinc Sulfide ZnS	67%
Zinc Phosphate $Zn_3(PO_4)_2$	51%
Zinc Chelates $Na_2ZnEDTA$	14%

Zn Frits are melted glass particles containing measured amounts of the minor nutrients, Cu, Fe, B, Zn, Mn and/or Mo. This impregnated powdered material it usually mixed with N, P and K fertilizers to form more porous and to make application more efficient.

For most field and vegetable crops 10 lb/A (12 kg/ha) is recommended in clay and loam soils and 3 to 5 lb/A (3 to 6 kg/ha) in sandy soils. In most cropping situation 10 lb/A(12 kg/ha) is recommended.

The methods of application

- soil application
- foliar application
- tree injection

Foliar application is effective and they are used primarily for tree crops. Chelates and natural organic complexes are particularly suitable for foliar sprays. These can be used in high analysis liquid fertilizers. Chelates such as Zn EDTA are mobile and can be surface applied under irrigated conditions.

Copper fertilizers

Table 97. Copper sources

Copper Fertilizers	Cu content	H_2O solubility
Copper metal Cu	100	Ins
Cuprite Cu_2O	89	Ins
Covellite CuS	66	Ins
ChalcociteCu_2S	80	Ins
Chalcopyrite $CuFeS_2$	35	I
Cu Sulphate $CuSO_45H_2O$	25	S
Cu Sulphate $CuSO_4H_2O$	35	S
Copper Nitrate $Cu(NO_3)_23H_2O$	32	S

Copper-Chelates

Disodium Cu EDTA - 13%

Monosodium Cu EDTA - 9%

Copper deficiency is corrected through foliar application. To avoid the scorching effect $CuSO_4$ solution can be mixed with lime. Sprayings are required biweekly intervals. The number of sprays will be 3 to 4 depending upon the intensity. If there is a Copper toxicity it can be over come by applying lime, liberal application of P fertilizers.

Iron Fertilizers

Table 98. Iron sources

Source	Formula	Percentage
Ferrous sulfate	$FeSO_4.7H_2O$	19
Ferric sulfate	$Fe_2(SO_4)_3\ 4H_2O$	23
Ferrous oxide	FeO	77
Ferric oxide	Fe_2O_3	69
Fe. Amm. Phosphate	$Fe(NH_4)\ PO_4H_2O$	29
Fe. Amm. sulphate	$(NH_4)_2SO_4.FeSO_4.6H_2O$	14
IronAmmPolyphosphate	$Fe(NH_4)HP_2O_7$	22
Iron chelates	Na Fe EDTA	5-14
	Na Fe HEDTA	5-9
	Na Fe EDDHA	6
	Na Fe DTPA	10

Iron chlorosis – Most difficult deficiency to correct in the field. Soil application is not satisfactory because of rapid oxidation to much less soluble ferric iron.(Ferrous $\rightarrow$ ferric)

Iron chlorosis is mainly corrected by foliar sprays. One dressing of 2-3% ferrous sulphate solution is usually sufficient to alleviate mild chlorosis. Several sprays 7 to 14 days apart may be needed to remedy severe Iron deficiencies.

Pears and Plums: Injection of iron salts directly into trunks and limbs controls, iron deficiency / chlorosis. Ferrous sulphate is commonly used

Others: Synthetic chelates are water soluble. Used for soil application and foliar sprays. The stability of Iron – Chelates depends on the pH and other soil conditions.

HEDTA - Effective in acid soils

EDTA - effective in slightly acid soils

CDTA, DTPA, EDDHA- Used above pH 7.0

Local acidification of small portions of the root zone is an effective method for correcting Iron deficiencies in calcareous and high pH soils.

The use of Elemental S, Ammonium thiosulfate, H_2SO_4, Ammonium bisulfite, SO_2 and Ammonium polysulfide will lower soil pH and simultaneously act as reducing agents to convert ferric iron more readily available ferrous form.

Application of gypsum may improve Iron availability by reducing the interference of Mo and plant uptake of Iron. This can be used in basic soils.

Mn fertilizers

Table 99. Manganese sources

Source	Manganese content (per cent)
Manganese sulphate	26-28
Manganese oxide (Mn O)	41 - 68
Manganese Chloride	17
Manganese Carbonate	31
Mn EDTA	12
Manganese frits	10 - 25

Boron fertilizers

Sodium tetraborate (Borax) is the most commonly used B source.

Table 100. Boron sources

	Sources	B content
Borax	$Na_2B_4O_7.10H_2O$	11.1
Boric acid,	H_3PO_3	17%
Colemanite,	$Ca_2B_6O_{11}.5H_2O$	10-16
Sodium Penta borate,	$Na_2B_{10}O_{16}.10H_2O$	18%
Boron frits		2-6%

Methods of application

Basal application is common. Common methods are broadcasting, banded application and foliar spray / dust spray. Boric acid can be used for foliar spray. Solubor is highly concentrated and completely soluble source of B. It can be applied as a spray / dust directly to foliage of fruit trees and vegetables. B-frits are extremely less soluble in H_2O and very slow release of B suited for long duration crops.1-3 kg B/ha is recommended for high B required crops and 0.6-1.2 kg B/ha for low B requirement. Foliar application is preferred for fruit trees than soil (immobile) and frequency should be more.

Molybdenum carriers

Table 101. Molybdenum sources

Source	Molybdenum content (%)
Sodium molybdate	39
Ammonium molybdate	54
Molybdenum trioxide	66
Molybdenite	60
Molybdenum frits	3

BIO FERTILIZERS

A biofertilizer (also *bio-fertilizer*) is a substance which contains living microorganisms which, when applied to seed, plant surfaces, or soil, colonizes the rhizosphere or the interior of the plant and promotes growth by increasing the supply or availability of primary nutrients to the host plant. Bio-fertilizers add nutrients through the natural processes of nitrogen fixation, solubilizing phosphorus, and stimulating plant growth through the synthesis of growth-promoting substances. Bio-fertilizers can be expected to reduce the use of chemical fertilizers and pesticides. The microorganisms in bio-fertilizers restore the soil's natural nutrient cycle and build soil organic matter. Through the use of bio-fertilizers, healthy plants can be grown, while enhancing the sustainability and the health of the soil.

Use of biofertilizers is one of the important components of integrated nutrient management, as they are cost effective and renewable source of plant nutrients to supplement the chemical fertilizers for sustainable agriculture. Several microorganisms and their association with crop plants are being exploited in the production of biofertilizers. They can be grouped in different ways based on their nature and function.

Table 102. Classification of Biofertilizers

S. No.	Groups	Examples
N2 fixing Biofertilizers		
1.	Free-living	*Azotobacter, Beijerinkia, Clostridium, Klebsiella, Anabaena, Nostoc*
2.	Symbiotic	*Rhizobium, Frankia, Anabaena azollae*
3.	Associative Symbiotic	*Azospirillum*
P Solubilizing Biofertilizers		
1.	Bacteria	*Bacillus megaterium* var. *phosphaticum, Bacillus subtilis, Bacillus circulans, Pseudomonas striata*
2.	Fungi	*Penicillium* sp, *Aspergillus awamori*
P Mobilizing Biofertilizers		
1.	Arbuscular mycorrhiza	*Glomus* sp.,*Gigaspora* sp., *Acaulospora* sp., *Scutellospora* sp. & *Sclerocystis* sp.
2.	Ectomycorrhiza	*Laccaria* sp., *Pisolithus* sp., *Boletus* sp., *Amanita* sp.
3.	Ericoid mycorrhizae	*Pezizella ericae*
4.	Orchid mycorrhiza	*Rhizoctonia solani*
Biofertilizers for Micro nutrients		
1.	Silicate & Zinc solubilizers	*Bacillus* sp.
Plant Growth Promoting Rhizobacteria		
1.	Pseudomonas	*Pseudomonas fluorescens*

Different types of biofertilizers

Rhizobium

Rhizobium is a soil habitat bacterium, which can able to colonize the legume roots and fixes the atmospheric nitrogen symbiotically. The morphology and physiology of Rhizobium will vary from free-living condition to the bacteroid of nodules. They are the most efficient biofertilizer as per the quantity of nitrogen fixed concerned. They have seven genera and highly specific to form nodule in legumes, referred as cross inoculation group. *Rhizobium* inoculant was first made in USA and commercialized by private enterprise in 1930s and the strange situation at that time has been chronicled by Fred (1932).

Table 103. Quantity of biological N fixed by Liqiud *Rhizobium* in different crops

Host Group	*Rhizobium* Species	Crops	N fix kg/ha
Pea group	*Rhizobium leguminosarum*	Green pea, Lentil	62- 132
Soybean group	*R.japonicum*	Soybean	57- 105
Lupini Group	*R. lupine orinthopus*	Lupinus	70- 90
Alfafa Group	*R. melliloti, Medicago Trigonella*	Melilotus	100- 150
Beans group	*R. phaseoli*	Phaseoli	80- 110
Clover group	*R. trifoli*	Trifolium	130
Cowpea group	*R. species*	Moong, Redgram, Cowpea, Groundnut	57- 105
Cicer group	*R. species*	Bengal gram	75- 117

Azotobacter

Of the several species of Azotobacter, A. chroococcum happens to be the dominant inhabitant in arable soils capable of fixing N_2 (2-15 mg N_2 fixed /g of carbon source) in culture media. The bacterium produces abundant slime which helps in soil aggregation. The numbers of A. chroococcum in Indian soils rarely exceeds 105/g soil due to lack of organic matter and the presence of antagonistic microorganisms in soil.

Azospirillum

Azospirillum lipoferum and *A. brasilense* (*Spirillum lipoferum* in earlier literature) are primary inhabitants of soil, the rhizosphere and intercellular spaces of root cortex of graminaceous plants. They perform the associative symbiotic relation with the graminaceous plants. The bacteria of Genus *Azospirillum* are N_2 fixing organisms isolated from the root and above ground parts of a variety of crop plants. They are Gram negative, *Vibrio* or *Spirillum* having abundant accumulation of polybetahydroxybutyrate (70 %) in cytoplasm.

Five species of *Azospirillum* have been described to date *A. brasilense*, *A. lipoferum*, *A. amazonense*, *A. halopraeferens* and *A. irakense*. The organism proliferates under both anaerobic and aerobic conditions but it is preferentially micro-aerophilic in the presence or absence of combined nitrogen in the medium. Apart from nitrogen fixation, growth promoting substance production (IAA), disease resistance and drought tolerance are some of the additional benefits due to *Azospirillum* inoculation.

Cyanobacteria

Both free-living as well as symbiotic cyanobacteria (blue green algae) have been harnessed in rice cultivation in India. A composite culture of BGA having heterocystous *Nostoc*, *Anabaena*, *Aulosira etc.* is given as primary inoculum in trays, polythene lined pots and later mass multiplied in the field for application as soil based flakes to the rice growing field at the rate of 10 kg/ha. The final product is not free from extraneous contaminants and not very often monitored for checking the presence of desired algal flora.

Once so much publicized as a biofertilizer for the rice crop, it has not presently attracted the attention of rice growers all over India except pockets in the Southern States, notably Tamil Nadu. The benefits due to algalization could be to the extent of 20-30 kg N/ha under ideal conditions but the labour oriented methodology for the preparation of BGA biofertilizer is in itself a limitation. Quality control measures are not usually followed except perhaps for random checking for the presence of desired species qualitatively.

Azolla

Azolla is a free-floating water fern that floats in water and fixes atmospheric nitrogen in association with nitrogen fixing blue green alga *Anabaena azollae*. *Azolla* fronds consist of sporophyte with a floating rhizome and small overlapping bi-lobed leaves and roots. Rice growing areas in South East Asia and other third World countries have recently been evincing increased interest in the use of the symbiotic N_2 fixing water fern *Azolla* either as an alternate nitrogen sources or as a supplement to commercial nitrogen fertilizers. *Azolla* is used as biofertilizer for wetland rice and it is known to contribute 40-60 kg N/ha per rice crop.

Phosphate solubilizing microorganisms (PSM)

Several soil bacteria and fungi, notably species of *Pseudomonas, Bacillus, Penicillium, Aspergillus etc.* secrete organic acids and lower the pH in their vicinity to bring about dissolution of bound phosphates in soil. Increased yields of wheat and potato were demonstrated due to inoculation of peat based cultures of *Bacillus polymyxa* and *Pseudomonas striata*. Currently, phosphate solubilizers are manufactured by agricultural universities and some private enterprises and sold to farmers through governmental agencies.

AM fungi

The transfer of nutrients mainly phosphorus and also zinc and sulphur from the soil *milleu* to the cells of the root cortex is mediated by intracellular obligate fungal endosymbionts of the genera *Glomus, Gigaspora, Acaulospora, Sclerocysts* and *Endogone* which possess vesicles for storage of nutrients and arbuscles for funneling these nutrients into the root system. By far, the commonest genus appears to be *Glomus*, which has several species distributed in soil. Availability for pure cultures of AM (Arbuscular Mycorrhiza) fungi is an impediment in large scale production despite the fact that beneficial effects of AM fungal inoculation to plants have been repeatedly shown under experimental conditions in the laboratory especially in conjunction with other nitrogen fixers.

Silicate solubilizing bacteria (SSB)

Microorganisms are capable of degrading silicates and aluminum silicates. During the metabolism of microbes several organic acids are produced and these have a dual role in silicate weathering. They supply H+ ions to the medium and promote hydrolysis and the organic acids like citric, oxalic acid, Keto acids and hydroxy carbolic acids which from complexes with cations, promote their removal and retention in the medium in a dissolved state.

The studies conducted with a *Bacillus* sp. isolated from the soil of granite crusher yard showed that the bacterium is capable of dissolving several silicate minerals under *in vitro* condition. The examination of anthrpogenic materials like cement, agro inputs like super phosphate and rock phosphate exhibited silicate solubilizing bacteria to a varying degree. The bacterial isolates made from different locations had varying degree of silicate solubilizing potential. Soil inoculation studies with selected isolate with red soil, clay soil, sand and hilly soil showed that the organisms multiplied in all types of soil and released more of silica and the available silica increased in soil and water. Rice responded well to application of organic sliceous residue like rice straw, rice husk and black ash @ 5 t/ha. Combining SSB with these residues further resulted in increased plant growth and grain yield. This enhancement is due to increased dissolution of silica and nutrients from the soil.

Plant Growth Promoting Rhizobacteria (PGPR)

The group of bacteria that colonize roots or rhizosphere soil and beneficial to crops are referred to as plant growth promoting rhizobacteria (PGPR).

The PGPR inoculants currently commercialized that seem to promote growth through at least one mechanism; suppression of plant disease (termed Bioprotectants), improved nutrient acquisition (termed Biofertilizers), or phytohormone production (termed Biostimulants). Species of *Pseudomonas* and *Bacillus* can produce as yet not well characterized phytohormones or growth regulators that cause crops to have greater amounts of fine roots which have the effect of increasing the absorptive surface of plant roots for uptake of water and nutrients. These PGPR are referred to as Biostimulants and the phytohormones they produce include indole-acetic acid, cytokinins, gibberellins and inhibitors of ethylene production.

Recent advances in molecular techniques also are encouraging in that tools are becoming available to determine the mechanism by which crop performance is improved using PGPR and track survival and activity of PGPR organisms in soil and roots. The science of PGPR is at the stage where genetically modified PGPR can be produced. PGPR with antibiotic, phytohormone and siderophore production can be made.

Despite of promising results, biofertilizers has not got widespread application in agriculture mainly because of the variable response of plant species or genotypes to inoculation depending on the bacterial strain used. Differential rhizosphere effect of crops in harbouring a target strain or even the modulation of the bacterial nitrogen fixing and phosphate solubilizing capacity by specific root exudates may account for the observed differences. On the other hand, good competitive ability and high saprophytic competence are the major factors determining the success of a bacterial strain as an inoculant.

Studies to know the synergistic activities and persistence of specific microbial populations in complex environments, such as the rhizosphere, should be addressed in order to obtain efficient inoculants. In this regards, research efforts are made to obtain appropriate formulations of microbial inoculants incorporating nitrogen fixing, phosphate- and silicate- solubilizing bacteria and plant growth promoting rhizobacteria which will help in promoting the use of such beneficial bacteria in sustainable agriculture.

27 SOIL ORGANIC MATTER

Organic matter in the soil comes from the remains of plants and animals. As new organic matter is formed in the soil, a part of the old becomes mineralized (mineral form such as sulphur, phosphorus etc). The original source of the organic matter in the soil is plant tissue. Under natural conditions, the tops and roots of trees, grasses and other plants annually supply large quantities of organic residues. Thus, higher plant tissue is the primary source of organic matter. Animals are usually considered as secondary sources of organic matter. Various organic manures such as farm yard manure, compost, green manure etc. that are added to the soil time to time, further add to the store of soil organic matter.

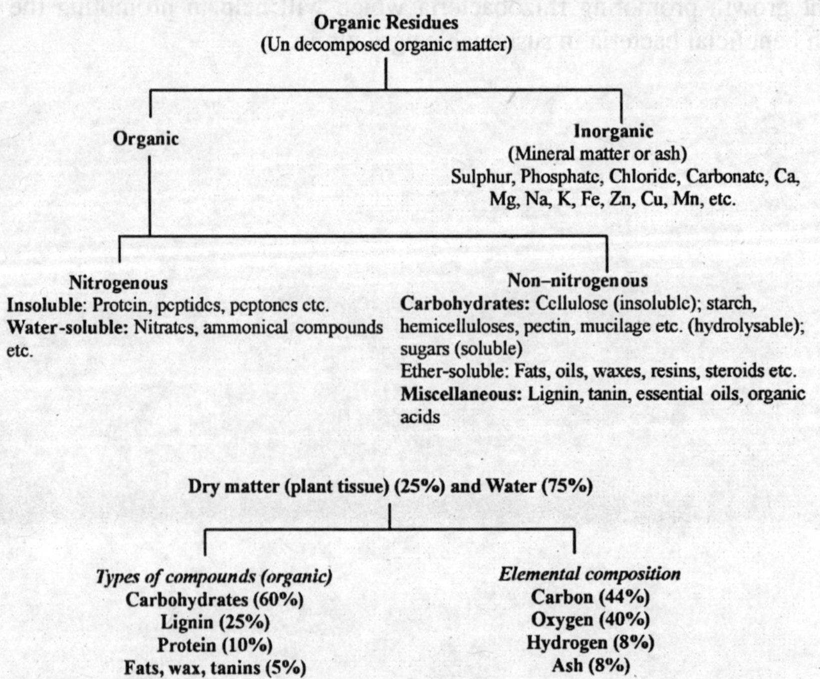

Fig. 67. List of compounds constituting the plant residues

27.1. COMPOSITION OF PLANT RESIDUES

Composition of organic residues have undecomposed soil organic matter (mainly plant residues together with animal remains i.e., animal excreta etc.). The moisture content of plant residues varies from 60–90 per cent (average 78 per cent) and 25 per cent dry matter (solid). Plant tissues (organic residues) may be divided into - organic, and inorganic (elemental) composition.

A list of compounds constituting the plant residues or undecomposed soil organic matter is shown in the diagram:

27.2. DECOMPOSITION OF ORGANIC MATTER

The organic materials (plant and animal residues) incorporated in the soil do not remain as such very long. They are at once attacked by a great variety of microorganisms, worms and insects present in the soil especially if the soil is moist.

The microorganism, for obtaining their food, break up the various constituents of the organic residues are composed, and convert them into new substances. Some of these new substances are very simple in composition and some are highly complex. The whole of the organic residues are not decomposed all at once or as a whole. Some of the constituents are decomposed very rapidly, some less readily, and some very slowly.

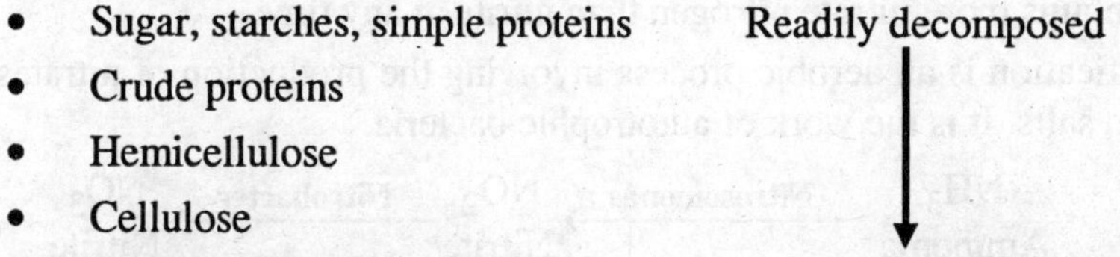

27.2.1. Decomposition of Soluble Substances

Sugar and water–soluble nitrogenous compounds are the first to be decomposed as they offer a very readily available source of carbon, nitrogen and energy for the microorganisms. Thus, when glucose is decomposed under aerobic conditions, the reaction is as under:

$$\text{Sugar} + \text{Oxygen} \longrightarrow \text{Carbon dioxide} + \text{Water}$$

When the nutrient or oxygen supplies are restricted, different partially–oxidized compounds are formed:

$$\text{Sugar} + \text{Oxygen} \longrightarrow \text{Aliphatic acids (acetic, formic etc.)}$$

Or

Hydroxy acids (citric, lactic etc.)

Or

Alcohols (ethyl alcohol etc.)

1. Ammonification

Soluble nitrogenous compounds, viz., amino acids, amides, ammonium compounds, nitrates etc., are also attacked by the microorganisms. The transformation of organic nitrogenous compounds into ammonia is called ammonification. During the course of action under aerobic conditions by heterotrophic organisms, oxygen is taken up and carbon dioxide is released. Ammonification process involves a gradual simplification of complex compounds.

Protein ⟶ Polypeptides ⟶ Amino acids ⟶ Ammonia or ammonium salts

Or

Organic Nitrogen ⟶ NH_3

The ammonification occurs as a result of action of enzymes produced by microorganisms. Their action is chiefly hydrolytic and oxidative (in presence of air).

2. Nitrification

The process of conversion of ammonia to nitrite (NO_2) and then to nitrate (NO_3) is known as nitrification. The production of nitrate is more rapid than that of nitrite, while the formation of ammonia is the slowest process. Therefore, soil usually contains more nitrate nitrogen than nitrite at any time.

Nitrification is an aerobic process involving the production of nitrates from ammonium salts. It is the work of autotrophic bacteria.

$$\underset{\text{Ammonia}}{NH_3} \xrightarrow{\text{Nitrosomonas}} \underset{\text{Nitrite}}{NO_2} \xrightarrow{\text{Nitrobacter}} \underset{\text{Nitrite}}{NO_3}$$

The net reactions are as follows:

$$NH_4 + O_2 \longrightarrow NO_2^- + 2H^+ + H_2O + \text{energy (66 k cal)}$$

$$NO_2^- + O_2 \longrightarrow NO_3^- + \text{energy (18 k cal)}$$

The process which involves conversion of soil nitrate into gaseous nitrogen or nitrous oxide is called denitrification.

$$\text{Nitrate} \xrightarrow{\text{Pseudomonas/Bacillus}} \text{Nitrogen gas} \uparrow$$

Waterlogging (e.g., Rice field) and high pH increases nitrogen loss by denitrification.

27.2.2. Decomposition of Insoluble Substances

1. Breakdown of proteins

Proteins are complex organic substances containing nitrogen, sulphur, and sometimes phosphorus, in addition to carbon, hydrogen and oxygen. During the course of decomposition of plant materials, the proteins are first hydrolyzed to a number of intermediate products known as polypeptides such as proteoses, peptones, peptides etc. The changes may be represented as follows:

Proteins $\xrightarrow{\text{hydrolysis}}$ Proteoses, Peptones, Peptides → Amino acids / Amides → Ammonia

The process of conversion of proteins to amino acids is known as amminization.

2. Breakdown of cellulose

Cellulose is the most abundant carbohydrate present in plant residues. The microorganisms break up cellulose into cellobiose and glucose. Glucose is further attacked by organisms and converted into organic acids.

The decomposition of cellulose in acid soils proceeds more slowly than in neutral and alkaline soils. It is quite rapid in well–aerated soils and comparatively slow in poorly–aerated soils.

$$\text{Cellulose} \xrightarrow[\text{(Cellulase)}]{\text{hydrolysis}} \text{Cellobiose} \xrightarrow[\text{(Cellobiase)}]{\text{hydrolysis}} \text{Glucose}$$

$$\text{Glucose} \xrightarrow{\text{Oxidation}} \text{Organic acids} \xrightarrow{\text{Oxidation}} CO_2 + H_2O$$

3. Breakdown of hemicellulose

When subjected to microbial decomposition, hemicelluloses are first hydrolyzed to their component sugars and uronic acids. The sugars are further attacked by microorganisms and are converted to organic acids, alcohols, carbon dioxide and water.

The uronic acids are broken down to pentose and carbon dioxide. Thus, the newly–synthesized hemicelluloses form a part of the humus. Hemicelluloses decompose faster than cellulose.

4. Breakdown of starch

Chemically, starch is a glucose polymer. It is first hydrolysed to maltose by the action of enzymes (amylases). Maltose is then converted to glucose by another enzyme (maltase). Glucose being soluble in water is utilized for growth and other metabolic activities.

27.2.3 Decomposition of Ether–soluble Substances

Fats are first broken down by microorganisms through the enzyme lipase into glycerol and fatty acids. Glycerol is next oxidized to organic acids, which along with the other fatty acids are finally oxidized to carbon dioxide and water.

27.2.4. Decomposition of Lignin

Lignin is deposited on the cell wall to impart strength to the skeleton framework of plant. Lignin decomposes slowly, much slower than cellulose. Complete oxidation of lignin gives rise to carbon dioxide and water.

27.2.5 Simple Decomposition Products

As the enzyme changes the soil organic matter proceed, simple products begin to form themselves. Some of these, especially carbon dioxide and water, formed immediately. Others such as nitrate nitrogen accumulate only after the peak of the vigorous decomposition is over. The simple products resulting from the activity of the soil microorganisms are as follows:

Carbon	:	CO_2, CO_3^-, HCO_3^-, elemental carbon
Nitrogen	:	NH_4^+, NO_2^-, NO_3^-, gaseous nitrogen
Sulphur	:	S, H_2S, SO_3^{--}, SO_4^{--}, CS_2
Phosphorus	:	$H_2PO_4^-$, HPO_4^-
Others	:	H_2O, O_2, H_2, H^+, OH^-, K^+, Ca^{++}, Mg^{++} etc.

Mineralization of organic sulphur

Many organic compounds especially those of a nitrogenous nature, carry sulphur. Heterotrophic bacteria simplify the complex organic compounds, and then autotrophic bacteria (sulphur bacteria) oxidize them into sulphate form.

$$\underset{\text{Sulphur}}{S + 2O_2} \xrightarrow[\text{Oxidation}]{\text{Enzyme}} \underset{\text{Sulphate}}{SO_4^{=}}$$

Mineralization of organic phosphorus

A large proportion of the soil phosphorus is carried in organic combinations. The organic phosphorus compounds are mineralized by the microorganisms i.e., they are changed to inorganic combinations. It depends upon the soil pH. As the pH goes up from 5.5 to 7.5, the available phosphorus changes from $H_2PO_4^-$ to HPO_4^-. Both of these forms are available to higher plants.

27.3. FACTORS AFFECTING ORGANIC MATTER DECOMPOSITION

The most important conditions that alter the accumulation of soil humus are:

- **Temperature** - Warmer temperature favours biomass production and makes the decomposition faster.
- **Soil moisture** - Both, plant growth and organic matter decomposition, require sufficient moisture. Extremes of both arid (dry) and anaerobic

(waterlogged) conditions reduce plant growth and microbial decomposition.

- **Nutrient levels -** Lack of nutrients particularly N, reduce plant growth and also it slows down decomposition reaction because microbes require sufficient N for their growth and activity. Materials with narrow C:N ratio (legumes) degrade faster than those with wider C:N ratio (non–legumes)
- **Soil pH -** Decomposition of organic matter proceeds faster at pH range of 6–8, since most of the microbes grow well at this range.
- **Soil texture -** Degradation and depletion of organic matter is faster in lighter–textured soils because of higher aeration, higher microbial activity and higher leaching. Clay soils tend to accumulate organic matter due to the formation of clay–humus complex and limited leaching.
- **Other factors -** Presence of toxic substances, excess soluble salts, cultivation practices, cropping pattern etc. influence the OM degradation and its accumulation in soils.

27.4. MAINTENANCE OF SOIL ORGANIC MATTER

27.4.1. Carbon:Nitrogen (C:N) Ratio

The ratio of the weight of organic carbon (C) to the weight of total nitrogen (N) in a soil (or organic material), is known as C:N ratio. When fresh plant residues are added to the soil, they are rich in carbon and poor in nitrogen. The content of carbohydrates is high. This results in wide C:N ratio, which may be 40:1. Upon decomposition, the organic matter of soil changes to humus and has an approximate C:N ratio of 10:1. The C:N ratio in the arable (cultivated) soils commonly ranges from 8:1–15:1. The C:N ratio in plant material is variable, ranging from 20:1–30:1. Low C:N ratios (10:1) in soil organic matter generally indicate an average stage of decomposition and resistance to further microbiological decomposition. A wide C:N ratio (35:1) indicates little or no decomposition; it is susceptible to further and rapid decomposition and slow nitrification.

27.4.2. Significance of C:N Ratio

1. Keen competition for available nitrogen when organic residues (with high C:N ratio) are added to soils. When organic residues with a wide C:N ratio (50:1) are incorporated in the soil, decomposition occurs quickly and carbon dioxide is produced in large quantities. Under these conditions, nitrate nitrogen disappears from the soil because of the insistent microbial demand for this element. And for the time being, little (or no) nitrogen is available to plants; consequently plant growth is affected. As decay (decomposition) occurs, the C:N ratio of the plant material decreases since carbon is being lost and nitrogen is conserved. Nitrogen, in the form of nitrate, again becomes available for plant growth.

2. Consistency of C:N ratio - As the decomposition process continues, both carbon and nitrogen are subjected to loss; the carbon as carbon dioxide and the nitrogen as nitrates, which are leached or absorbed by plants. At a point, C:N ratio becomes more or less constant and generally stabilizes at 10:1 or 12:1.

It has always been considered profitable to mix the highly carbonaceous material such as straw with a small quantity of ammonium sulphate. With this available source of nitrogen from ammonium sulphate, the decomposition of fresh organic material is hastened and thus, the release of available nitrogen is affected in a shorter period of time.

27.4.3. Role of Soil Organic Matter (SOM) in Soil Fertility

1. Improvement in physical properties of soil

- **Bulk density** - Addition of OM to soil lowers the bulk density of the soil and makes the soil lighter.
- **Aggregation** - OM improves the aggregate formation in soil by acting as binding agent. The stability of the aggregates is also improved; thereby reducing the soil erosion.
- **Pore space** - The pore spaces of the soil get increased because of the increased aggregation.
- **Soil aeration** - Because of better aggregation and pore spaces, the soil aeration gets improved favouring respiration and growth of plant roots.
- **Soil temperature** - Higher OM content of the soil helps to keep the soil warmer during winter and cooler during summer, because of its lower heat conductivity.
- **Soil Colour** - OM gives darker colour to the soil.
- **Water–holding capacity** - Soils with high OM imbibe and retain more water, and help conservation of soil moisture and also to increase the water use efficiency.

2. Improvement in Chemical properties

- **Buffering Capacity of soils** - Addition of OM reduces the degree of acidity or alkalinity of soils and increases the buffering capacity of soils.
- **CEC of soils** - Soil OM increases the CEC of soils thereby helping the adsorption / retention and availability of nutrients in the soil. Soil humus has the highest CEC value of about 300 meq$^+$/100g. The loss of

applied nutrients is minimized. It acts as a store house of nutrients in the soil.

- **Solubility of soil nutrients** - The organic acids released during organic matter decomposition helps for the solubilization of soil mineral nutrients making them more available to crops.
- **Source of nutrients for plants** - OM or humus largely contains C, N, P, S, which forms a major source of nutrients for plants. In addition, it releases other secondary micronutrients into the soil during decomposition.
- **Chelation** - OM helps in the formation of mineral or chelates with the plant nutrients, more particularly metallic cations such as Cu, Zn, Mn, Fe making them more available to crops, and it also prevents them from precipitation or oxidation reactions.
- **Transformation of soil nutrients** - OM helps for the activity of soil microbes, which are responsible for the transformation and availability of various nutrients such as N, P, S and other micronutrients.
- **Sink for heavy metal pollutants** - The heavy metal pollutants received from the various Industrial wastes are made complex by the OM; thereby preventing their toxicities to crop plants.
- **Metabolism of soil applied pesticides** - SOM combines with the organic molecules of applied soil pesticides and affects their bioactivity persistence and biodegradation.
- **Soil Formation** - SOM plays an important role in the soil formation by the process called 'Biogeochemical processes' which include the processes such as podzolization, laterization, melanization etc.

3. Biological properties

- **Growth and activity of microbes** - SOM serves as the energy or substrate for the growth and activity of soil microbes, which are involved in the biological N fixation, mineralization and immobilization of nutrients, degradation of crop residues and pesticides etc.
- **Growth and activity of earthworms** - SOM favours the activity of earthworms in soil, which helps in improving the soil aggregates, aeration, water retention and movement etc.
- **Soil Enzyme activity** - The enzyme concentration and enzyme activity are higher in soils having higher OM due to increased microbial activity. This helps in the transformation of applied nutrients in soil viz., urea hydrolysis.

27.5. HUMUS - SYNTHESIS AND NATURE

Humus is a heterogeneous mixture of complex organic compounds. The bulk of these compounds have resulted from decomposition and synthesis. The organic residues, which undergo breakdown and transformation are complex in nature, containing carbohydrates, proteins and other nitrogenous compounds, liginins, fats etc.

Humus is a complex mixture of amorphous and colloidal organic substances, which is more or less a biologically–stable product.

Based on the above facts, humus may be defined as follows:

Humus is a complex and rather resistant mixture of brown or dark brown amorphous and colloidal substances that result from microbial decomposition and synthesis, and has chemical and physical properties of great significance to soils and plants. — *Nyle* C. *Brady*

The humic and non–humic compounds collectively form humus.

1. **Humic substances** - This comprises about 60–80 per cent of the soil organic matter. Two more complex substances - polyphenols and polyquinones are formed by decomposition, synthesis and polymerization processes.

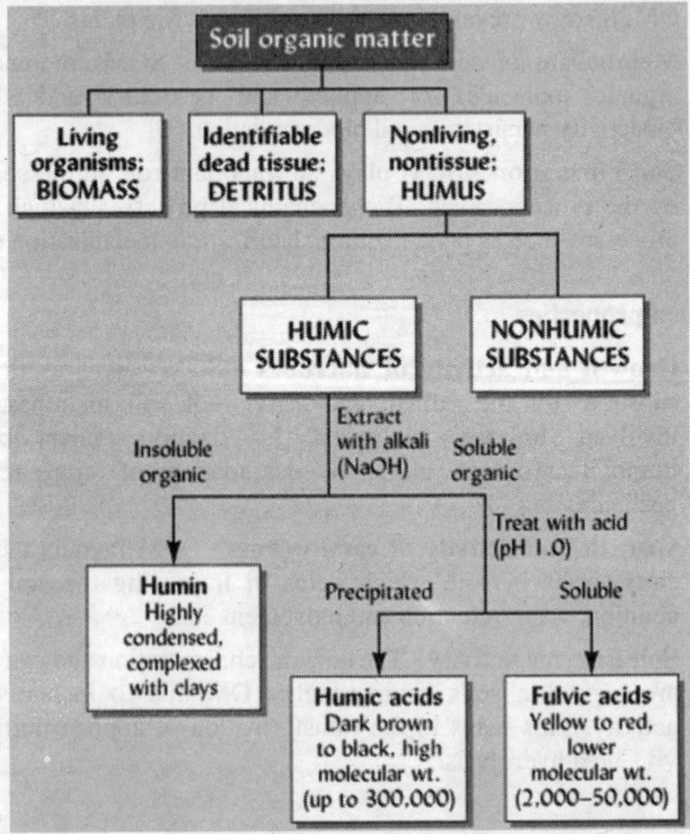

Fig. 68. Fractionation of Soil organic matter

Humic substances are dark in colour, amorphous and have high molecular weight. Humic substances can be classified into three chemical groups:

a. Humin - Highest in molecular weight, dark in colour, insoluble in both acid and alkali, and more resistant to microorganism attack

b. Humic acid - Medium molecular weight, medium–dark in colour, soluble in alkali, insoluble in acid, and moderate resistant to microbial degradation

c. **Fulvic acid** - Lowest molecular weight, light in colour, soluble in both acid and alkali, and most susceptible to microorganism attack.

2. Non–humic group - This makes up about 20–30 per cent of the organic matter in soils. The non–humic substances consist of polymers, polysaccharides and polyuronides. Polysaccharides are especially effective in enhancing soil aggregate stability.

Non–humic group also have some simpler compounds such as organic acids and some protein–like materials. Non–humic substances are less complex and less resistant to microbial attack than humic substances.

Humus is more resistant to decay than most of the compounds found in plant residues. Humus may remain in soils even for hundreds of years. This resistant property of humus is important in maintaining organic matter level in soils and this protects nitrogen and other essential nutrients that are found in humus complex.

27.5.1. Colloidal Characteristics of Humus

Important colloidal properties are as follows:

Colloidal humus particles are composed of carbon, hydrogen and oxygen (probably in the form of polyphenols, polyquinones, polyuronides and polysaccharides).

- Colloidal humus particles are black in colour.
- Humus helps in aggregate formation and provides stability.
- Humus and silicate clay are similar in cation exchange reaction. At high pH values, the cation exchange capacity on a mass basis far exceeds that of most silicate clays.
- The water–holding capacity of humus is more than silicate clays.
- The surface area of humus colloids is higher than silicate clays.
- The colloidal surfaces of humus are negatively charged.

27.6. DIFFERENCE BETWEEN ORGANIC MATTER AND HUMUS

Humus is a complex and rather resistant mixture of brown (or dark brown) amorphous and colloidal substances modified from the original tissues or synthesized by the various soil organisms. Soil organic matter may be distinguished from humus as soil organic matter consists of any substances of organic origin (living or dead), while humus is fairly stable and amorphous, brown to black material which is formed as a result of decomposition of plant and animal residues, with no trace of the structure of the material from which it is derived.

Humus, in reality, is not merely a decomposition product since decomposition implies transformation to simpler compounds, whereas humus is more complex than original organic matter. Besides decomposition of organic matter, synthesis of organic substances is involved in humus formation. On an average, humus contains 58 per cent carbon, which is somewhat greater than the carbon content of plant tissues (45–50 per cent).

Humus possesses a definite C:N ratio, which varies from 10:1–12:1. The actual variation in C:N ratio of soil organic matter is very wide (40:1–12:1). Carbon content of humus is less variable than nitrogen. Another characteristic, which distinguishes humus from non–humified organic matter, is its behaviour with hydrogen peroxide. Three per cent hydrogen peroxide has been found to oxidize humus leaving the undecomposed matter unattacked.

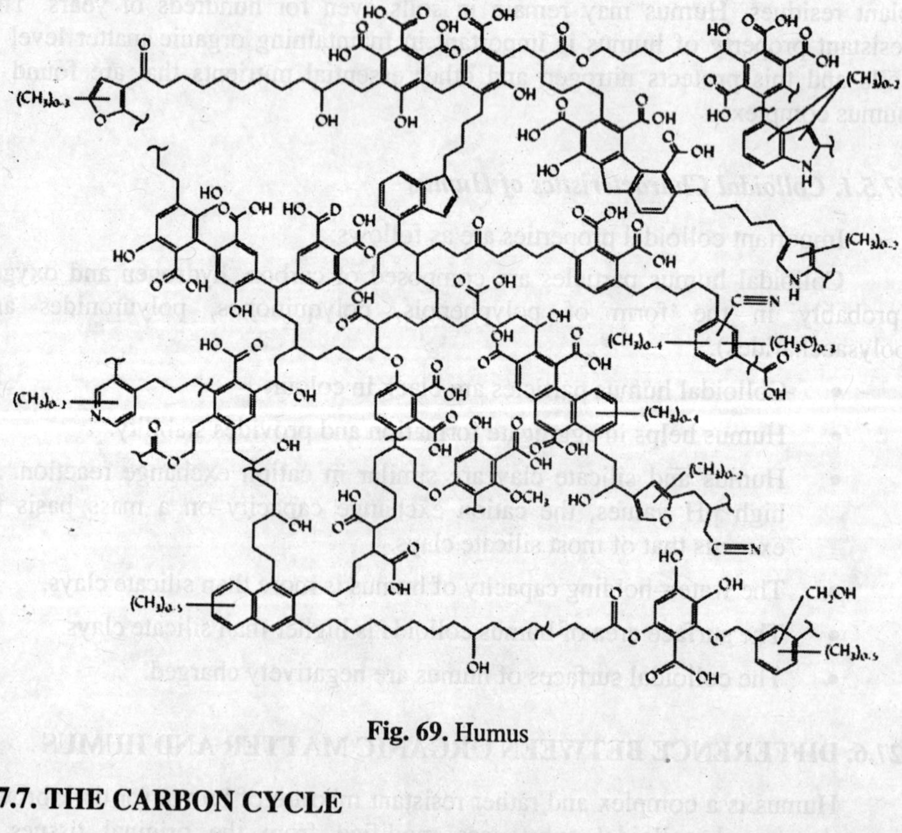

Fig. 69. Humus

27.7. THE CARBON CYCLE

The concentration of carbon in living matter (18 per cent) is almost 100 times greater than its concentration in the Earth (0.19 per cent). So, living things extract carbon from their non–living environment. For life to continue, this carbon must be recycled.

Carbon exists in the non–living environment as

- Carbon dioxide (CO_2) in the atmosphere and dissolved in water (forming HCO_3^-)
- Carbonate rocks (limestone and coral = $CaCO_3$)
- Deposits of coal, petroleum, and natural gas derived from once living things
- Dead organic matter, e.g., humus in the soil

Carbon enters the biotic world through the action of autotrophs

- Primarily, photoautotrophs such as plants and algae use the energy of light to convert carbon dioxide to organic matter.
- To a small extent, chemoautotrophs such as bacteria and archaea do the same but use the energy derived from oxidation of molecules in their substrate.

Carbon returns to the atmosphere and water by

- Respiration (as CO_2)
- Burning
- Decay (as CO_2 if oxygen is present, and as methane (CH_4) if oxygen is not present).

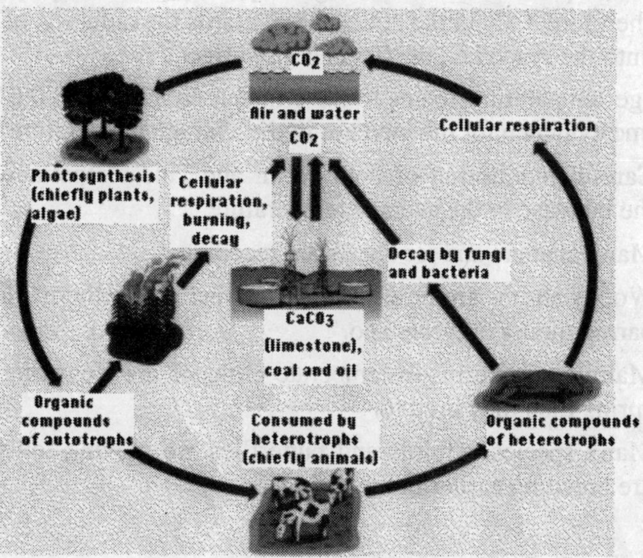

Fig. 70. Carbon cycle

The uptake and return of CO_2 are not in balance

The carbon dioxide content of the atmosphere is gradually and steadily increasing. The increase in CO_2 probably began with the start of the industrial revolution. This increase is surely anthropogenic i.e., caused by human activities such as:

- Burning of fossil fuels (coal, oil, natural gas), which returns to the atmosphere; carbon that has been locked within the Earth for millions of years.
- Clearing and burning of forests, especially in the tropics. In recent decades, large areas of the Amazon rainforest have been cleared for agriculture and cattle grazing.

Where is the missing carbon?

Curiously, the increase in atmospheric CO_2 is only about one–half of what would have been expected from the amount of fossil fuel consumption and forest burning. Where has the rest of CO_2 gone? Research has shown that increased CO_2 levels lead to increased net production of biomass by photoautotrophs. There is some evidence that the missing CO_2 has been incorporated by:

- Increased growth of forests, especially in North America
- Increased amounts of phytoplankton in the oceans

27.7.1. The Greenhouse Effect and Global Warming

Despite these sinks (forests and phytoplankton) for our greatly increased CO_2 production, the concentration of atmospheric CO_2 continues to rise. Carbon dioxide is a transparent medium for light rays but, it is an opaque medium for heat rays. Therefore, CO_2 in the atmosphere retards the radiation of heat from the Earth back into the space i.e., the 'greenhouse effect'.

Average temperature seems to have increased slightly (~0.6°C) in the last century. Some evidence for the increase in the average temperature:

- Careful monitoring of both ocean and land temperatures is showing the increase in the average temperature.
- Many glaciers and ice sheets are receding.
- Woody shrubs are now growing in areas of Northern Alaska that were barren tundra 50 years ago.
- Many angiosperms in temperate climates are flowering earlier in the spring than they used to.
- Many species of birds and butterflies are moving towards north and are breeding earlier in the spring.

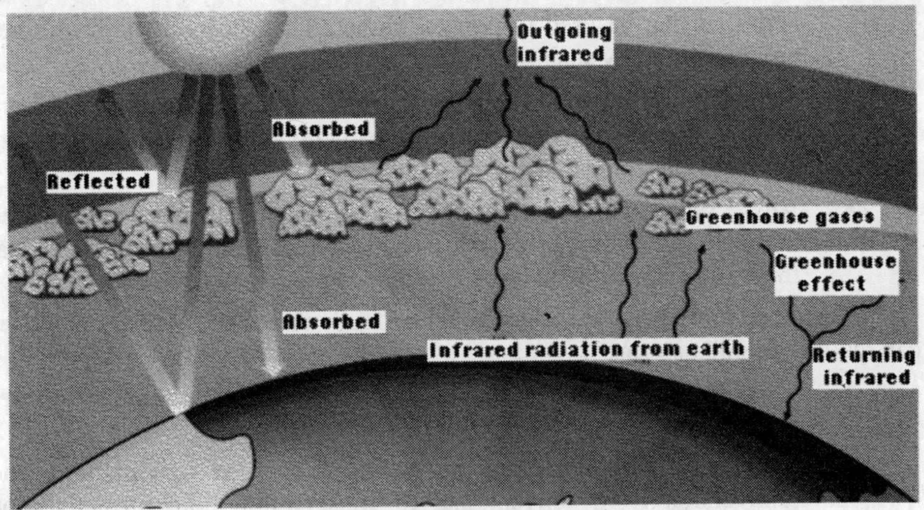

Fig. 71. Terrestrial carbon

Carbon storage in the biosphere is influenced by a number of processes on different time-scales while net primary productivity follows a diurnal and seasonal cycle. Carbon can be stored up to several hundreds of years in trees and up to thousands of years in soils. Thus, changes in these long term carbon pools (e.g. through deforestation or afforestation or through temperature–related changes in soil respiration) can directly affect global warming.

Sequestration

Terrestrial carbon sequestration is the process through which carbon dioxide (CO_2) from the atmosphere is absorbed by trees, plants and crops through photosynthesis, and stored as carbon in biomass (tree trunks, branches, foliage and roots) and soils. The term 'sinks' is also used to refer to forests, croplands and grazing lands, and their ability to sequester carbon. Agricultural and forestry activities can also release CO_2 to the atmosphere. Therefore, a carbon sink occurs when carbon sequestration is greater than carbon releases over some time period.

Agricultural and forestry sequestration activities

- Forests and soils have a large influence on atmospheric levels of carbon dioxide (CO_2), which is the most important global warming gas emitted by human activities.
- Tropical deforestation is responsible for about 20 per cent of the world's annual CO_2 emissions on a global scale. However, these emissions are more than the offset by the uptake of atmospheric CO_2 by forests and agriculture.
- Therefore, both agricultural and forestry activities can contribute to the accumulation of greenhouse gases in our atmosphere, as well as these

activities can be used to prevent climate change, by avoiding further emissions and by sequestering additional carbon.

- Sequestration activities can be carried out immediately. These are relatively cost–effective emission reduction opportunities, and may generate environmental co–benefits.
- At the same time, it is important to recognize that carbon sequestered in trees and soils can be released back to the atmosphere, and that there is a finite amount of carbon that can ultimately be sequestered.

There are three general means by which agricultural and forestry practices can reduce greenhouse gases:

1. Avoiding emissions by maintaining existing carbon storage in trees and soils.
2. Increasing carbon storage by activities such as tree planting and conversion from conventional to conservation tillage practices on agricultural lands.
3. Substituting bio–based fuels and products for fossil fuels such as coal and oil, and energy–intensive products that generate greater quantities of CO_2 when used.

Carbon sequestered by agricultural and forestry practices

- Carbon sequestration rates vary as per tree species, regional climate, topography and management practice.
- Soil carbon sequestration rates vary as per soil type and cropping practice and are less well–documented. But, information and research in this area is growing rapidly.
- Carbon accumulation in forests and soils eventually reaches a saturation point, beyond which additional sequestration is not possible.
- This happens when trees reach maturity, or when the organic matter in soils builds back up to original levels before losses are occurred.
- Even after saturation, the trees or agricultural practices would need to be sustained to maintain the accumulated carbon and prevent subsequent losses of carbon back to the atmosphere.

27.7.2. Carbon Trading

Carbon trading is a practice which is designed to reduce overall emissions of carbon dioxide, along with other greenhouse gases, by providing a regulatory and economic incentive. The carbon trade came about in response to the Kyoto Protocol. Signed in Kyoto, Japan, by 180 countries in December 1997, the Kyoto Protocol calls for 38 industrialized countries to reduce their greenhouse gas emissions between the years 2008 to 2012 to levels that are 5.2% lower than those of 1990.

Carbon is an element stored in fossil fuels such as coal and oil. When these fuels are burned, carbon dioxide is released and acts as what we term a "greenhouse gas". In fact, the term "carbon trading" is a bit misleading, as a number of greenhouse emissions can be regulated under what are known as cap and trade systems. For this reason, some people prefer the term "emission trading," to emphasize the fact that far more than just carbon is being traded.

The idea behind carbon trading is quite similar to the trading of securities or commodities in a marketplace. Carbon would be given an economic value, allowing people, companies or nations to trade it. If a nation bought carbon, it would be buying the rights to burn it, and a nation selling carbon would be giving up its rights to burn it. The value of the carbon would be based on the ability of the country owning the carbon to store it or to prevent it from being released into the atmosphere. (The better you are at storing it, the more you can charge for it.)

A market would be created to facilitate the buying and selling of the rights to emit greenhouse gases. The industrialized nations for which reducing emissions is a daunting task could buy the emission rights from another nation whose industries do not produce as much of these gases. The market for carbon is possible because the goal of the Kyoto Protocol is to reduce emissions as a collective measure.

Right now, there is a market opportunity for India—but only till 2012. Closer to that clean-up date prices of carbon credits will rise and in the years leading up to it there will be a scramble to buy credits cheap. The World Bank has built itself a role in this market as a referee, broker and macro-manager of international fund flows. The scheme has been entitled Clean Development Mechanism (CDM) in 2000 or more commonly CARBON TRADING.

Emissions trading (or cap and trade)

It is an administrative approach used to control pollution by providing economic incentives for achieving reductions in the emissions of pollutants.

In such a plan, a central authority (usually a government agency) sets a limit or cap on the amount of a pollutant that can be emitted. Companies or other groups that emit the pollutant are given credits or allowances, which represent the right to emit a specific amount. The total amount of credits cannot exceed the cap, limiting total emissions to that level. Companies that pollute beyond their allowances must buy credits from those who pollute less than their allowances or face heavy penalties. This transfer is referred to as a trade.

- In effect, the buyer is being fined for polluting, while the seller is being rewarded for having reduced emissions. Thus, companies that can easily reduce emissions will do so and those for which it is harder will buy credits, which reduce greenhouse gasses at the lowest possible cost to society.
- There are currently several trading systems in place with the largest being the European Union's Scheme.

- The carbon market makes up the bulk of these and is growing in popularity. Many businesses have welcomed emissions trading as the best way to mitigate climate change.
- Enforcement of the caps is a problem, but unlike traditional regulation, emissions trading markets can be easier to enforce because the government overseeing the market does not need to regulate specific practices of each pollution source. However, monitoring (or estimating) and verifying of actual emissions is still required, which can be costly.
- Critics doubt whether these trading schemes can work as there may be too many credits given by the government such as in the first phase of the European Union's scheme. Once a large surplus is discovered, the price for credits will be bottomed out and they will effectively collapse, with no noticeable reduction of emissions.

28 SOIL ORGANISMS

Soil flora and soil fauna are the terms used to describe the organisms that live within the soil. The flora is composed primarily of microorganisms such as bacteria, actinomycetes, algae and fungi (though some fungi are visible to the unaided eye). The fauna is composed of microfauna (such as protozoans), mesofauna (such as nematodes, and arthropods such as mites, centipedes, springtails), and macrofauna (such as insects and worms, mice, moles).

In agriculture and horticulture, the soil flora and fauna play important roles such as decomposing organic materials, moving nutrients through soil layers, fixing atmospheric nitrogen, and improving soil structure.

Some parts of the soil flora and fauna are plant pathogens, while others are antagonists. Without the activities of soil organisms, organic materials would accumulate and litter the soil surface, and there would be no food for living organisms. The soil biota includes:

- Megafauna - Size range 20 mm upwards e.g., moles, rabbits, and rodents
- Macrofauna - Size range 2–20 mm e.g., woodlice, arañas, earthworms, beetles, centipedes, slugs, snails, ants, and harvestmen
- Mesofauna - Size range 100 μm–2 mm e.g., tardigrades, mites and springtails
- Microfauna and Microflora - Size range 1–100 μm e.g., yeasts, bacteria, fungi, protozoa, roundworms and rotifers

Of these, bacteria and fungi play key role in maintaining a healthy soil. They act as decomposers that break down organic materials to produce detritus and other breakdown products. Soil detritivores such as earthworms, ingest detritus and decompose it. Saprotrophs, well represented by fungi and bacteria, extract soluble nutrients from detritus.

Bacteria

- Bacteria are single–celled organisms, and are the most numerous denizens of the soil, with populations ranging from 100 million–3 billion in a gram.
- They are capable of very rapid reproduction by binary fission (dividing into two) in favourable conditions.
- One bacterium is capable of producing 16 million bacteria in just 24 hours. Most soil bacteria live in close proximity to plant roots and are often referred to as rhizobacteria. Bacteria live in soil water including the film of moisture surrounding soil particles, and some are able to swim by means of flagella.
- The majority of the beneficial soil–dwelling bacteria need oxygen (and are thus termed aerobic bacteria), while those that do not require air are referred to as anaerobic, and tend to cause putrefaction of dead organic matter.
- Aerobic bacteria are most active in a moist soil (but not saturated, as this will deprive aerobic bacteria of the air that they require), and in neutral soil pH, and where there is plenty of food (carbohydrates and micronutrients from organic matter) available.

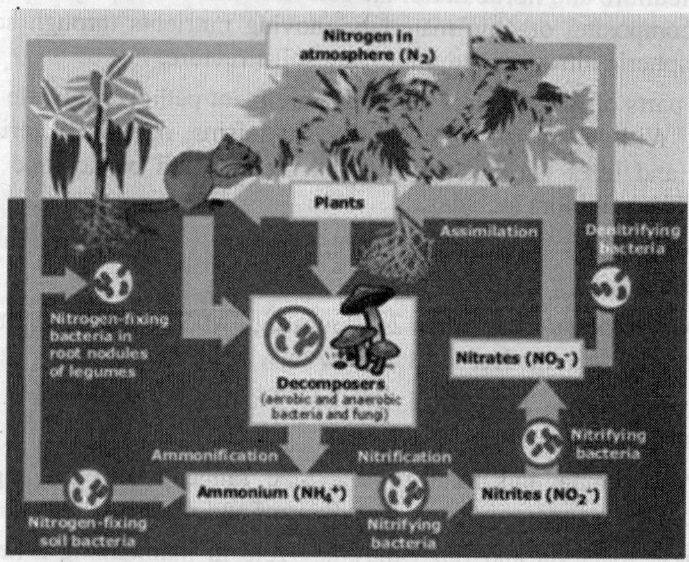

Fig. 72. The nitrogen cycle

- The important roles that bacteria play are:

Bacteria take on many roles in the environment. They act as decomposers at the end of food chains and food webs. During decompositions, they also liberate advantageous gases and nutrients which are used by other living beings. Some bacteria also participate in the nitrogen cycle, making fixation of nitrogen, nitrification, and denitrification almost always in mutualism ecological interaction with plants. Excessive proliferation or mass destruction of bacteria can impact entire ecosystems. For example, when a river is polluted by organic material the population of aerobic bacteria increases since the organic material is food for them, the great number of bacteria then exhausts the oxygen dissolved in water and other aerobic beings (like fishes) undergo mass death.

28.1. THE NITROGEN CYCLE

Nitrification

Nitrification is a vital part of the nitrogen cycle wherein certain bacteria are able to transform nitrogen in the form of ammonium, (which is produced by the decomposition of proteins), into nitrates. These nitrates are available to growing plants, and once again converted to proteins.

Nitrogen fixation

In another part of the cycle, the process of nitrogen fixation constantly puts additional nitrogen into biological circulation. This is carried out by free–living, nitrogen–fixing bacteria in the soil or water such as azotobacter, or by those, which live in close symbiosis with leguminous plants such as rhizobia. These bacteria form colonies in nodules they create on the roots of peas, beans, and related species. These are able to convert nitrogen from the atmosphere into nitrogen–containing organic substances.

Denitrification

While nitrogen fixation converts nitrogen from the atmosphere into organic compounds, a series of processes called denitrification returns an approximately equal amount of nitrogen to the atmosphere. Denitrifying bacteria tend to be anaerobes, or facultatively anaerobes (can alter between the oxygen dependent and oxygen independent types of metabolisms), including *Achromobacter* and *Pseudomonas*. The putrefaction process, caused by oxygen–free conditions, converts nitrates and nitrites from soil into nitrogen gas or into gaseous compounds such as nitrous oxide or nitric oxide.

Actinobacteria

Actinobacteria are critical in the decomposition of organic matter and in humus formation, and their presence is responsible for the sweet ‘earthy’ aroma, which is associated with a good healthy soil. They require plenty of air and a pH between 6.0 and 7.5, but are more tolerant of dry conditions than most other bacteria and fungi.

Fungi

- A gram of garden soil can contain around one million fungi such as yeasts and moulds. Fungi have no chlorophyll, and are not able to photosynthesize; besides, they cannot use atmospheric carbon dioxide as a source of carbon. Therefore, they are chemoheterotrophic, which means like animals, they require a chemical source of energy as they are not able to use light as an energy source as well as organic substrates to get carbon for growth and development.
- Many fungi are parasitic, often causing disease to their living host plant, although some have beneficial relationships with living plants. In terms of soil and humus creation, the most important fungi tend to be saprotrophic i.e., they live on dead or decaying organic matter; they break it down and convert it to the forms, which are available to the higher plants. A succession of fungi species will colonise the dead matter, beginning with those that use sugars and starches, which are succeeded by those that are able to break down cellulose and lignins.
- Fungi spread underground by sending long thin threads known as mycelium throughout the soil; these threads can be observed throughout many soils and compost heaps. From the mycelia, the fungi is able to throw up its fruiting bodies, the visible part above the soil (e.g., mushrooms, toadstools and puffballs), which may contain millions of spores. When the fruiting body bursts, these spores are dispersed through the air to settle in fresh environments, and are able to lie dormant for up to years until the right conditions for their activation arise or the right food is made available.

Mycorrhizae

- Those fungi that are able to live symbiotically with living plants, creating a relationship that is beneficial to both, are known as Mycorrhizae (*myco* meaning fungal and *rhiza* meaning root).
- Plant root hairs are invaded by the mycelia of the mycorrhiza, which lives partly in the soil and partly in the root, and may either cover the length of the root hair as a sheath or be concentrated around its tip. The mycorrhiza obtains the carbohydrates that it requires from the root, in return providing the plant with nutrients including nitrogen and moisture. Later the plant roots will also absorb the mycelium into its own tissues.
- Beneficial mycorrhizal associations are found in many of our edible and flowering crops. These include at least 80 per cent of the *brassica* and *solanum* families (including tomatoes and potatoes), as well as the majority of tree species, especially in forest and woodlands.

28.2. ROLE OF SOIL MICROORGANISMS IN THE DECOMPOSITION OF ORGANIC MATTER

Soil microorganism plays an important role in completing the carbon cycle. They convert organic matter into simple substances, liberate CO_2 and water, increase the concentration of nitrogen content and bring down the C:N ratio in soil. This improves the soil fertility. The released CO_2 goes back to complete the cycle and this ensures that there is no major lock–up in organic tissues.

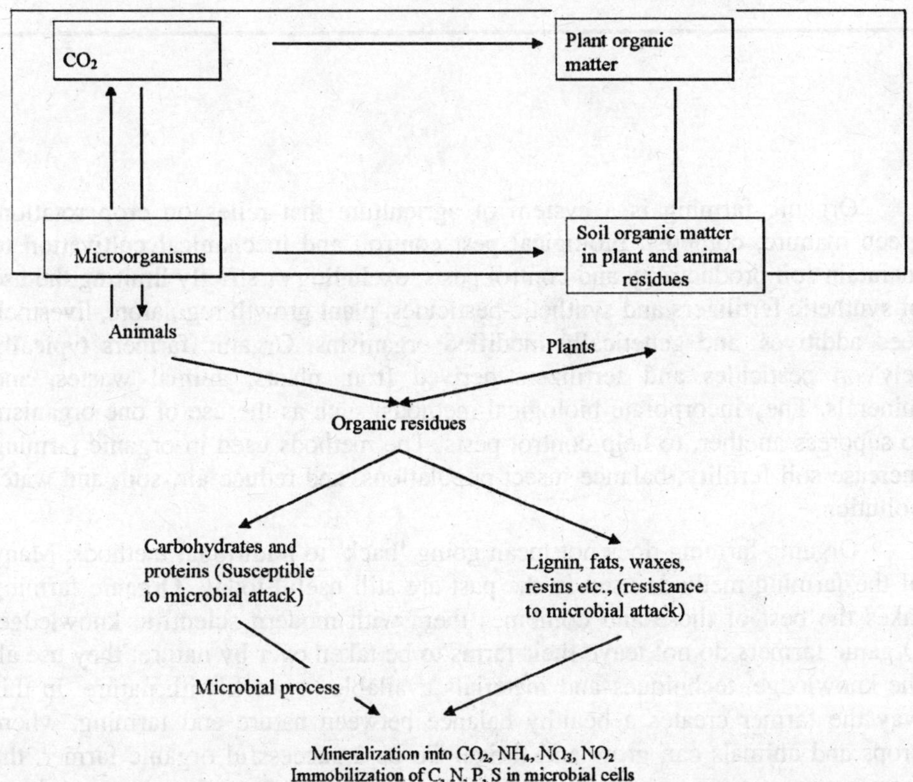

Fig. 73. The C: N ratio is ~ 10:1. C: N ratio – is a tool to predict the rate of decomposition. If the C: N ratio is wider, decay will be slower.

29 ORGANIC FARMING

Organic farming is a system of agriculture that relies on crop rotation, green manure, compost, biological pest control, and mechanical cultivation to maintain soil productivity and control pests, excluding or strictly limiting the use of synthetic fertilizers and synthetic pesticides, plant growth regulators, livestock feed additives, and genetically modified organisms. Organic farmers typically rely on pesticides and fertilizers derived from plants, animal wastes, and minerals. They incorporate biological methods, such as the use of one organism to suppress another, to help control pests. The methods used in organic farming increase soil fertility, balance insect populations, and reduce air, soil, and water pollution.

Organic farming does not mean going 'back' to traditional methods. Many of the farming methods used in the past are still useful today. Organic farming takes the best of these and combines them with modern scientific knowledge. Organic farmers do not leave their farms to be taken over by nature; they use all the knowledge, techniques and materials available to work with nature. In this way the farmer creates a healthy balance between nature and farming, where crops and animals can grow and thrive. To be a successful organic farmer, the farmer must not see every insect as a pest, every plant out of place as a weed and the solution to every problem in an artificial chemical spray. The aim is not to eradicate all pests and weeds, but to keep them down to an acceptable level and make the most of the benefits that they may provide.

Need of Organic farming

Several reasons have been emphasized for the need of organic farming like limited land holdings, poor socio-economic conditions of farmers, rise in input cost etc. The broadest view shows two major reasons. They are mainly, population and environment. They have emphasized the ultimate need for eco friendly technologies. The current methods of farming are damaging to the natural resources and the environment. This type of agronomy also requires the farmers to invest heavily into inputs to improve yields and productivity. In addition to this many research studies have proved the presence of pesticide residues in agricultural products.

- There are reports for problems in reproductive processes as a consequence of consumption of the chemical methyl bromide through vegetables.
- About one million people suffer from pesticides poisoning.
- 20000 die every year due to the toxic effect of the chemicals used in agriculture.
- The crop loss due to pest is still 15000 crores.

In order to avoid the deleterious effects of synthetic chemical fertilizers and pesticides, organic agriculture is needed as an alternative to provide ecologically safe methods of farming. Organic farming provides long-term benefits to people and the environment.

Organic farming aims to:

- increase long-term soil fertility.
- control pests and diseases without harming the environment.
- ensure that water stays clean and safe.
- use resources which the farmer already has, so the farmer needs less money to buy farm inputs.
- produce nutritious food, feed for animals and high quality crops to sell at a good price.

Modern, intensive agriculture causes many problems, including the following:

- Artificial fertilisers and herbicides are easily washed from the soil and pollute rivers, lakes and water courses.
- The prolonged use of artificial fertilisers results in soils with a low organic matter content which is easily eroded by wind and rain.
- Dependency on fertilisers. Greater amounts are needed every year to produce the same yields of crops.
- Artificial pesticides can stay in the soil for a long time and enter the food chain where they build up in the bodies of animals and humans, causing health problems.
- Artificial chemicals destroy soil micro-organisms resulting in poor soil structure and aeration and decreasing nutrient availability.
- Pests and diseases become more difficult to control as they become resistant to artificial pesticides. The numbers of natural enemies decrease because of pesticide use and habitat loss.

Objectives of Organic farming

The main principles of organic farming were laid down by IFOAM in 1992.

- To produce food of high nutritional quality in sufficient quantity.
- To interact in a constructive and life enhancing way with all natural systems and cycles.
- To encourage and enhance biological cycles within the farming system, involving micro-organisms, soil flora and fauna, plants and animals.
- To maintain and increase long term fertility of soils.
- To use, as far as possible, renewable resources in locally organised agricultural systems.
- To work, as far as possible, within a closed system with regard to organic matter and nutrient elements. This aims to reduce external inputs.
- To work, as far as possible, with materials and substances which can be reused or recycled, either on the farm or elsewhere?

Components of Organic Farming

There are assumptions throughout the organic literature of differences between organic and conventional systems with respect to their effects on soil physical properties, soil insect fauna and nutrient flow within the soil, crop health and nutritional value of the harvested crop. Different components of organic farming are as follows:

1. Crop and Soil Management

Organic farming system encourages the use of rotations and measures to maintain soil fertility. Soil carefully managed with a high production of humus offer essential advantages with respect to water retention, ion exchange, soil erosion and animal life in the soil. Green manuring and inter-cropping of legumes is another important aspect for biological farming systems not only in regard to weed control but also in reducing the leaching of nutrients and in reducing soil erosion. A green cover throughout most of the year is one of the main goals of such farming methods. Depending on the green manure mixture or the legumes used for under sowing, there may be an increased soil organic matter and soil N_2 as well as in other nutrients.

2. On-farm Waste Recycling

Increase price of chemical fertilizers have enables organic wastes to regain an important role in the fertilizer practices on the farm. Good manure management means improved fertilizers value of manure and slurry and less nutrient losses. Composting of all organic wastes in general and of Farm Yard Manure (FYM) or feedlot manure in particular is important in organic farming.

3. Non-chemical Weed Management

Weed management is one of the main concerns in organic agriculture. Generally, all aspects of arable crop production play an important role in a system approach to problems. The elements to consider in preventing weed problems are crop rotation, green manuring, manure management and tillage. Mulching on a large scale by using manure spreaders may also be useful in weed control.

4. Domestic and Industrial Waste Recycling

Sewage and sludge use for crop production can form an important component of organic farming if treatment and application methods are improved further.

5. Energy Use

In the energy requirement for production measured per rupees of produce for organic farms is only one third of what it is for their conventional counterparts. Because N-fertilizer and pesticides are not used by biological farmers, the comparison of total energy input/ha. with total energy output favours biological farming systems.

6. Food Quality

Food quality is one of the main issues, which concerns both scientists and consumers. Nitrates in water and farm produce, desirable components, pesticides residues, keeping quality and physiological imbalances are some of the important aspects of food quality.

7. Ecological Agriculture

The growing concern about environmental degradation, dwindling natural resources and urgency to meet the food needs of the increasing population are compelling farm scientist and policy makers to seriously examine alternative to chemical agriculture. If one gets a premium price for organically grown rice, the ecońomic returns from the ecological farming system will highly encouraging.

8. Integrated Intensive Farming System (IIFS)

IIFS involves intensive use of farm resources. To be ecologically sustainable, such intensification should be based on techniques which are knowledge intensive and which replace to the extent possible, market purchased chemical inputs with farm grown biological inputs.

Nutrient Management in Organic Farming

Introduction of high yielding varieties, extension of irrigated areas, use of high analysis NPK fertilizers and increase in cropping intensity during the era of Green Revolution propelled India towards self sufficiency in food production. In the process, relative contribution of Organic manures as a source of plant nutrients vis – a – vis chemical fertilizers declined substantially. With increase in cost of inputs, inorganic fertilizers became increasingly more expensive. Another issue of great concern is the sustainability of soil productivity as lands began to be intensively tilled to produce higher yields under multiple and intensive cropping systems. It is distressing to note that organic carbon present in soil is fast declining. The present level of organic carbon in arable lands in India varies between 0.3 to 0.5%. Frequently tilling the lands, use of chemicals for plant protection measures also minimize soil microbial activity, fauna and flora development.

Organic farming system solely depends on the use of crop residues, animal manures, green manures, use of crop residues, off- farm organic wastes, crop rotation incorporating legumes and biological pest control to maintain soil productivity. The organic manures viz., cattle dung, pig manure, poultry manure, rural and urban composts, crop residues, green manure, oil seed cakes, coir pith and sewage sludge are commonly used. Bulky organic manures such as FYM, compost and crop residues supply low quantities of major plant nutrients, concentrated organic manures such as oil cakes, slaughters house wastes, fish meal, poultry manure etc., are comparatively richer in NPK contents.

Organic manures undergo microbial decomposition. In this process, the nutrients held in organic combinations are slowly released in available forms besides improving the availability of nutrient elements present in the soil. In addition, the organic carbon level of the soil also slowly increases when the manures are used continuously.

Fertilization practices

- Organic farming uses a variety of methods to improve soil fertility, including crop rotation with high and low biomass crops, cover cropping, application of vermicompost, biocompost and mulching.
- Biodegradable materials of microbial, plant and animal origin should be returned to the soil to increase or at least maintain the its fertility and the biological activity.
- Organic farming heavily relies on the natural breakdown of organic matter, using techniques like green manure and composting, to replace nutrient taken from the soil by previous crops.
- Recycling of nutrients from livestock manures or crop operations (trimmings), or usage of naturally mined fertilizer nutrients can be adopted.

- Organic farmers also use processed natural fertilizers such as bone meal, blood meal and various mineral powders such as rock phosphate and green sand (a naturally occurring form of potash)
- Emphasis should be given to generate and use own on-farm organic fertilizers.

Benefits of Soil Organic Matter

- It serves as a slow release of nitrogen, phosphorus and sulphur for plant nutrition and microbial growth
- It posses considerable water holding capacity and thereby helps to maintain the water regime of the soil.
- It acts as a buffer against pH of the soil.
- Its dark color contributes to absorption of energy from the sun and heating of the soil
- It acts as "cement" for holding clay and silt particles together, thus contributing to the crumb structure of the soil and resistance against soil erosion
- It binds micro-nutrients metal ion in the soil that otherwise might be leached out of the surface soil
- Organic constituents in the humic substances may act as plant growth stimulants

Organic Farming with Sustainability

Organic farming is imperative rather than a choice; it is much nearer to nature and natural processes; it can achieve greater sustainability than chemical farming. The primary motto is to improve soil quality in terms of physical and biological characters. The sources of organic farming are able to naturally fit themselves into an agro eco – system than chemicals and therefore, are able to mould the soil system to long term sustainability. The biological bases of sustainable crop yield lies in sustainable fertility and sustainable eco – system.

Status of organic farming in India

India is bestowed with lot of potential to produce all varieties of organic products due to its various agro climatic regions. In several parts of the country, the inherited tradition of organic farming is an added advantage. This holds promise for the organic producers to tap the market which is growing steadily in the domestic market related to the export market. Currently, India ranks 33rd in terms of total land under organic cultivation and 88th position for agriculture land under organic crops to total farming area. The cultivated land under certification is around 2.8 million ha (2007-08, 1.9% of the GCA). This includes

1 million ha under cultivation and the rest is under forest area (wild collection) (APEDA, 2010).

The organic land area has increased substantially between 2005 and 2006 and is now more than 500,000 hectares. India produced around 3,96,997 MT (2007-08) of certified organic products which includes all varieties of food products namely Basmati rice, Pulses, Honey, Tea, Spices, Coffee, Oil Seeds, Fruits, Processed food, Cereals, Herbal medicines and their value added products (Table 104). The production is not limited to the edible sector but also produces organic cotton fiber, garments, cosmetics, functional food products, body care products etc. The commodity wise details of quantity exported and its value is presented in Table 105.

Table 104. Organic food productions in India, 2007-08

Total production	9,76,646 M.T.
Total quantity exported	37,533 M.T
Value of total export	USD 100.4 million
Total area under certification (including wild harvest)	2.8 million hectares
Total area under certified organic cultivation	0.45 million hectares
Share of exports to total production	4% approx.
Increase in export value over previous year	30% approx.

Source: APEDA, 2010

Table 105. Commodity wise organic food exports, 2007- 08

Commodity	Export Contribution (of volume)	Export Contribution (of value)	Export Contribution (Rs. Cr)
Cotton	43%	25%	123.88
Basmati Rice	15%	13%	59.20
Honey	11%	10%	46.41
Tea	8%	20%	92.13
Dry fruits	7%	18%	84.31
Processed food	5%	4%	17.99
Sesame	4%	2%	9.13
Spices	3%	4%	20.09
Medicinal & Herbal plants/products	2%	2%	10.59
Others	2%	2%	5.05

Source: APEDA, 2010

India exported 86 items last year (2007-08) with the total volume of 37533 MT. The export realization was around 100.4 million US $ registering a 30%

growth over the previous year. Organic products are mainly exported to EU, US, Australia, Canada, Japan, Switzerland, South Africa and Middle East. Cotton leads among the products exported (16,503 MT). According to the Indian Centre for Organic Agriculture (ICCOA), a major reason for the growth in organic farming is increased awareness among consumers in the country, even though until recently food was mainly being exported. But over the last couple of years, the domestic market has started growing.

The Indian government is also implementing a National Project on Organic Farming (NPOF) for production, promotion, certification and market development of organic farming in the country. Financial assistance is being provided for the capacity building through service providers, setting up of organic input production units, promotion of organic farming through training programs, field demonstrations, setting up of model organic farms and market development.

India's Organic Input potential

Bhattacharya and Chakraborty (2005) estimated the current status of organic farming in India and other countries. They noticed various problems in the conventional farming in India and opined that the integration of organic and inorganic farming would be an ideal model. Based on their results the industrial nitrogen fixation (INF) is 40 mt/year which accounts for only 15.3% of total nitrogen fixation. On the other hand, the quantity of biological nitrogen fixation (BNF) is 175 mt/year contributes for 67.3% of the total amount. Plant also uses nutrients from organic sources through mineralization and billions of microorganisms are available in soil for this job. Excess and indiscriminate use of inorganic fertilizer has deteriorated soil badly with deficiency of macro and micronutrients.

India is endowed with various types of naturally available organic form of nutrients in different parts of the country and which will help for organic cultivation of crops substantially. According to Bhattacharya (2006) the major sources of organic inputs in India are presented below:

Sources of organic inputs in India

Live stock = 2.47 million ton
Crop residues = 2.00 million ton
Biogas slurry = 0.12 million ton
Bio-fertilizer = 0.20 million ton
Green manure = 0.10 million ton
City refuse = 0.68 million ton
Others* = 1.00 million ton
Total = 6.57 million ton
Bio-pesticide = 1000 ton

* Rural compost, vermi-compost, agricultural wastes

There is enough scope for production of sufficient organic inputs in India. Almost 7 million ton of production potential currently is having through different sources. Among different sources, livestock accounts for lion share (nearly 40 per cent). It is followed by crop residues (30 per cent) and other sources (15 per cent). Other sources include the rural compost, vermi-compost and agricultural wastes. India is also having a huge potential of nearly 1000 tons of bio-pesticide preparations.

Government initiatives in promotion and regulation of Organic agriculture

National Programme of Organic Production (NPOP) launched by the Ministry of Commerce during 2001 and National Project on Organic farming (NPOF) launched during 2004 by the Department of Agriculture and Cooperation, Ministry of Agriculture were the two milestones towards institutionalization of organic farming in the country. The details of the two programs are:

National Programme of Organic Production (NPOP)

To provide a focused and well directed development of organic agriculture and quality products, Ministry of Commerce and Industry, Government of India, launched a National Program on Organic Production (NPOP) in the year 2000, which was formally notified in October 2001 under the Foreign Trade & Development Act (FTDR Act). It provides information on standards for organic production, systems criteria, and procedures for accreditation of inspection and certification bodies, the national organic logo and the regulations governing its use. The standards and procedures have been formulated in harmony with international standards such as those of Codex and IFOAM.

The NPOP proposes to provide an institutional mechanism for the implementation of National Standards for Organic Production, through a National Accreditation Policy and Program. The aim of the national program for organic production, inter alia, includes the following:

a. To provide the means of evaluation of certification programmes for organic agriculture and products as per the approved criteria.
b. To accredit certification programmes
c. To facilitate certification of organic products in conformity to the National Standards for Organic Products.
d. To encourage the development of organic farming and organic processing

Scope

The National Programme for Organic Production shall, among others, include:

(a) Policies for development and certification of organic products
(b) National standards for organic products and processes
(c) Accreditation of programmes to be operated by inspection and certification agencies
(d) Certification of organic products

National Project on Organic Farming (NPOF)

In 2000, Ministry of Agriculture also set up a task force on organic farming under the chairmanship of Kunwarjee Bhai Yadav. Based on their recommendations, the Department of Agriculture and Cooperation, Ministry of Agriculture, Govt. of India has launched a Central Sector Scheme "National Project on Organic Farming during Xth Five year plan on pilot basis with an out lay of Rs. 57.05 crore w.e.f. 1st October, 2004. The scheme is continuing in XIth Five year plan with an allocation of Rs. 150 crore. Main objectives of this scheme are as follows:

1. Capacity building through service provider
2. Financial support to different production units engaged in production of bio-fertilizers, compost, vermi-compost etc.
3. Human Resource Development through organizing training on certification and inspection, production and quality control of organic inputs, training of extension officers / field functionaries, farmers training on organic farming etc.
4. Field demonstration on organic inputs and enriched biogas slurry
5. Setting up of model organic farms
6. Market development for organic produce
7. Development of domestic standards
8. Support to new initiatives on technology related to organic farming
9. Awareness programmes etc.
10. Quality Control of various Bio-fertilizers and Organic fertilizers as per Fertilizer Control Order (FCO)

Organic Certification

It is a certification process for producers of organic food and other organic agricultural products. In general, any business directly involved in food production can be certified, including seed suppliers, farmers, food processors, retailers and restaurants. Requirements vary from country to country, and generally involve a set of production standards for growing, storage, processing, packaging and shipping that include:

- avoidance of synthetic chemical inputs (e.g. fertilizer, pesticides, antibiotics, food additives, etc) and genetically modified organisms;

- use of farmland that has been free from chemicals for a number of years (often, three or more);
- keeping detailed written production and sales records (audit trail);
- maintaining strict physical separation of organic products from non-certified products;
- undergoing periodic on-site inspections.

In some countries, certification is overseen by the government, and commercial use of the term *organic* is legally restricted. Certified organic producers are also subject to the same agricultural, food safety and other government regulations that apply to non-certified producers.

Advantages of organic farming

- Conservation of Natural resources (soil, water, petroleum products, energy etc)
- Enhances soil productivity
- Prevent damage to ecosystem(soil, water, atmospheric pollution)
- Reduce- prevent entry of toxicants in to the Food chain.
- Promote exports of Organic foods, fiber
- Generates on-farm rural employment in developing countries.
- Encourages- promotes adoption of self-sustaining production systems.

Drawbacks of Organic Farming

There are not many drawbacks in this farming practice. But they need to be considered to have a good yield. Following are some of the drawbacks of this method of agronomy:

- **Not nutrient specific:** Most organic manures are not nutrient specific.
- **Labour:** Ample labour is required; otherwise the land goes a waste.
- **Timely vigilance:** Regular vigil is required. As in the case of chemical farming, the pesticides once sprayed last for certain duration. Since Organic agronomy requires the complete discard of pesticides, a regular inspection of land is necessary.

Despite the drawbacks of this practice of agronomy, it can be regarded as a sustainable farming method.

30 SOIL POLLUTION

Soil pollution is defined as the build–up in soils of persistent toxic compounds, chemicals, salts, radioactive materials, or disease causing agents, which have adverse effects on plant growth and animal health. Soil is the thin layer of organic and inorganic materials that covers the Earth's rocky surface. The organic portion, which is derived from the decayed remains of plants and animals, is concentrated in the dark uppermost topsoil. The inorganic portion made up of rock fragments was formed over thousands of years by physical and chemical weathering of bedrock. Productive soils are necessary for agriculture to supply the world with sufficient food.

30.1. NATURAL LAND POLLUTION

Land pollution occurs massively during earthquakes, landslides, hurricanes and floods, creating a mess which is hard and expensive to clean mess, and may sometimes take years to restore the affected area. These kinds of natural disasters are not only a problem in that they cause pollution but, also because they leave many victims homeless.

The wars that hit the Earth are probably the immediate cause of soil pollution. Not talking in the sense of how many people died but in that it is through this period that many countries found the necessity to improve their living standards.

After the World War II, many countries suffered from food shortage and this facilitated the introduction of fertilizers and other agricultural chemicals. Although Nitrogen, Phosphorus, and Potassium fertilizers have not led to soil pollution, the application of trace elements has. Pesticides such as DDT [dichloro diphenyl trichloroethane], a colourless chemical pesticide, which is a potent nerve poison in insects, was first used to combat diseases such as yellow fever and malaria. It was later used to control and / or eradicate disease–carrying and crop–eating insects. DDT was later on discovered to cause endangerment of species in the same food chain as the controlled insects, particularly birds. DDT prevents the shelling of bird eggs and in humans causes health threats.

In yet another famous war of Vietnam in 1970, another chemical substance, Dioxin was introduced, which had a more adverse effect than that of DDT. Dioxin is a chemical impurity resulting from the production of the auxin 2,4,5T. Dioxin is a toxic chemical and was used as a defoliant by the American army. Dioxin was a major constituent of Agent Orange, which was applied on trees, which would then fall off revealing enemy camps. After the war, it was found that the chemical cause congenital deformities and mental effects to the children born to the American soldiers and in the area over which it was applied. In minute amount dioxin has the ability to cause cancer, chloracne, miscarriage and foetal abnormalities.

Glass industries have also been responsible for soil pollution.

The glass industries use arsenic to eliminate a green colour caused by impurities of iron compounds. Arsenic is a violent poison, yet it is widely used and therefore, is a frequent contaminant. James Marsh, supplies a simple method for detecting traces of arsenic so minute that they would escape discovery in ordinary analysis. Arsenic is sometimes added to lead to harden it and is also used in the manufacture of such military poison gases as lewisite and adamsite. Until the introduction of penicillin, arsenic was of great importance in the treatment of syphilis. In other medicinal uses, it has been displaced by sulpha drugs or antibiotics. Lead arsenate, calcium arsenate, and Paris green are used extensively as insecticides. Pollution of land by heavy metals is a result of the mining of ores to extract metals such as tin, silver, nickel, lead, iron, chromium and copper. Most of these metals occur naturally as ions in the soils.

Though some metals such as copper, iron, and zinc, are necessary for plant growth, it is the high concentration if these ions that renders the land unsuitable for plant growth. Soil pollution is widely linked to chemical substances but irrigation is also somehow linked to it as well because irrigation can transport some chemicals from field to field.

30.2. STATUS OF SOIL CONTAMINATION IN INDIA

In India, soil pollution is mainly due to pollution concentration in urban areas and unplanned development of industries. It has been estimated that the urban population of India produces about 15 million tonnes of solid waste per annum causing pollution.

- 125 major contaminated sites across the country
- 175 million hectare (out of 329 million ha) are considered degraded
- 40 per cent of chemical fertilizers leached into soil
- 14 States are affected by Fluoride contamination
- 65 per cent of India's villages are exposed to residual pesticides risk
- Heavy metals beyond permissible limits affecting ground water of 40 districts from 13 States

30.3. CAUSES OF SOIL POLLUTION

Soil pollution is a result of many activities and experiments done by mankind which end up contaminating the soil. Here are some of the leading soil pollution causes:

- Most common causes of soil pollution - Industrial wastes such as harmful gases and chemicals, agricultural pesticides, fertilizers and insecticides
- Ignorance towards soil management and related systems
- Unfavourable and harmful irrigation practices
- Improper septic system, and management and maintenance of the same
- Leakages from sanitary sewage
- Acid rains, when fumes released from industries get mixed with rains
- Fuel leakages from automobiles that get washed away due to rain and seep into the nearby soil
- Unhealthy waste management techniques, which are characterized by release of sewage into the large dumping grounds and nearby streams or rivers

30.4. TYPES OF SOIL POLLUTION

- Agricultural soil pollution
  - Pollution of surface soil
  - Pollution of underground soil
- Soil pollution by industrial effluents and solid wastes
  - Pollution of surface soil
  - Disturbances in soil profile
- Pollution due to urban activities
  - Pollution of surface soil
  - Pollution of underground soil

Pollution in soil is associated with

- Indiscriminate use of fertilizers
- Indiscriminate use of pesticides, insecticides and herbicides
- Dumping of large quantities of solid waste
- Deforestation and soil erosion

Indiscriminate use of fertilizers

Soil nutrients are important for plant growth and development. Plants obtain carbon, hydrogen and oxygen from air and water. But other necessary nutrients such as nitrogen, phosphorus, potassium, calcium, magnesium, sulphur and more must be obtained from the soil. Generally, farmers use fertilizers to correct soil deficiencies. Fertilizers contaminate the soil with impurities, which come from the raw materials used for their manufacture. Mixed fertilizers often contain ammonium nitrate (NH_4NO_3), phosphorus as P_2O_5, and potassium as K_2O. For instance, As, Pb and Cd present in traces in rock phosphate mineral get transferred to super phosphate fertilizer. Since the metals are not degradable, their accumulation in the soil above their toxic levels due to excessive use of phosphate fertilizers becomes an indestructible poison for crops. The overuse of NPK fertilizers reduce quantity of vegetables and crops grown on soil over the years. It also reduces the protein content of wheat, maize, grams, etc., grown on that soil. The carbohydrate quality of such crops also gets degraded. Excess potassium content in soil decreases Vitamin C and carotene content in vegetables and fruits. The vegetables and fruits grown on over–fertilized soil are more prone to attacks by insects and disease.

Indiscriminate use of pesticides, insecticides and herbicides

Plants on which we depend for food are under attack from insects, fungi, bacteria, viruses, rodents and other animals, and they must compete with weeds for nutrients. To kill unwanted populations living in or on these crops, farmers use pesticides. The first widespread use of insecticide began at the end of World War II and included DDT (dichloro diphenyl trichloroethane) and gammaxene. Insects soon became resistant to DDT and as the chemical did not decompose readily, it persisted in the environment. Since it was soluble in fat rather than water, it biomagnified up to the food chain and disrupted calcium metabolism in birds, causing eggshells to be thin and fragile. As a result, large birds of prey such as the brown pelican, ospreys, falcons and eagles became endangered. DDT has now been banned in most western countries. Ironically many of them including USA, still produce DDT for export to other developing nations whose needs outweigh the problems caused by it. The most important pesticides are DDT, BHC, chlorinated hydrocarbons, organophosphates, aldrin, malathion, dieldrin, furodan, etc. The remnants of such pesticides used on pests may get adsorbed by the soil particles, which then contaminate root crops grown in that soil. The consumption of such crops causes the pesticides remnants to enter human biological systems, affecting them adversely. An infamous herbicide used as a defoliant in the Vietnam War called Agent Orange (dioxin), was eventually banned. Soldiers' cancer cases, skin conditions and infertility have been linked to exposure to Agent Orange. Pesticides not only bring toxic effect on human and animals but also decrease the fertility of the soil. Some of the pesticides are quite stable and their bio- degradation may take weeks and even months. Pesticide problems such as resistance, resurgence, and health effects have caused scientists to seek alternatives. Pheromones and hormones have been suggested to attract or repel insects and use of natural enemies or sterilization by radiation have also been suggested.

Dumping of solid wastes

In general, solid waste includes garbage, domestic refuse and discarded solid materials such as those from commercial, industrial and agricultural operations. They contain increasing amounts of paper, cardboards, plastics, glass, old construction material, packaging material and toxic or otherwise hazardous substances. Since a significant amount of urban solid waste tends to be paper and food waste, the majority is recyclable or biodegradable in landfills. Similarly, most agricultural waste is recycled and mining waste is left on site. The portion of solid waste that is hazardous such as oils, battery metals, heavy metals from smelting industries and organic solvents are the ones we have to pay particular attention to. These can, in the long run, get deposited to the soils of the surrounding area and pollute them by altering their chemical and biological properties. They also contaminate drinking water aquifer sources. More than 90 per cent of hazardous waste is produced by chemical, petroleum and metal–related industries and small businesses such as dry cleaners and gas stations contribute as well. Solid Waste disposal was brought to the forefront of public attention by the notorious Love Canal case in USA in 1978. Toxic chemicals leached from oozing storage drums into the soil underneath homes, causing an unusually large number of birth defects, cancer and respiratory, nervous and kidney diseases.

Deforestation

Soil erosion occurs when the weathered soil particles are dislodged and carried away by wind or water. Deforestation, agricultural development, temperature extremes, precipitation including acid rain, and human activities contribute to this erosion. Humans speed up this process by construction, mining, cutting of timber, over cropping and overgrazing. It results in floods and cause soil erosion. Forests and grasslands are an excellent binding material that keeps the soil intact and healthy. They support many habitats and ecosystems, which provide innumerable feeding pathways or food chains to all species. Their loss would threaten food chains and the survival of many species. During the past few years quite a lot of vast green land has been converted into deserts. The precious rain forest habitats of South America, tropical Asia and Africa are coming under pressure of population growth and development (especially timber, construction and agriculture). Many scientists believe that a wealth of medicinal substances including a cure for cancer and aids, lie in these forests. Deforestation is slowly destroying the most productive flora and fauna areas in the world, which also form vast tracts of a very valuable sink for CO_2.

Pollution due to urbanization

Pollution of surface soils

Urban activities generate large quantities of city wastes including several biodegradable materials (e.g., vegetables, animal wastes, papers, wooden pieces, carcasses, plant twigs, leaves, cloth wastes as well as sweepings) and many non–biodegradable materials (e.g., plastic bags, plastic bottles, plastic wastes, glass bottles, glass pieces, stone / cement pieces). On a rough estimate Indian cities are producing solid city wastes to the tune of 50,000–80,000 metric tons every day. If left uncollected and decomposed, they are a cause of several problems such as:

- Clogging of drains - Causing serious drainage problems including the burst / leakage of drainage lines leading to health problems.
- Barrier to movement of water - Solid wastes have seriously damaged the normal movement of water thus, creating problem of inundation, damage to foundation of buildings as well as public health hazards.
- Foul smell - It is generated by dumping the wastes at a place.
- Increased microbial activities - Microbial decomposition of organic wastes generate large quantities of methane besides many chemicals to pollute the soil and water flowing on its surface.
- Hospital wastes - When such solid wastes are hospital wastes they create many health problems as they may have dangerous pathogen within them besides dangerous medicines and injections.

Pollution of underground soil

Underground soil in cities is likely to be polluted by:

- Industrial wastes and chemicals released through industrial wastes
- Decomposed and partially–decomposed materials of sanitary wastes
- Many dangerous chemicals such as cadmium, chromium, lead, arsenic, selenium products are likely to be deposited in underground soil.
- Similarly, underground soil polluted by sanitary wastes generates many harmful chemicals. These can damage the normal activities and ecological balance in the underground soil.

30.5. EFFECTS OF SOIL POLLUTION

The effects of pollution on soil are quite alarming and can cause huge disturbances in the ecological balance and health of living creatures on the Earth. Some of the most serious soil pollution effects are:

Agricultural

- Decrease in soil fertility and reduced nitrogen fixation and therefore, decrease in the crop yield. How can one expect contaminated soil to produce healthy crops?
- Loss of soil and natural nutrients present in it. Plants also would not thrive in such soil, which would further result in soil erosion.
- Increase in salinity of the soil, which therefore makes it unfit for vegetation, thus making it useless and barren.
- Generally crops cannot grow and flourish in polluted soil. Yet, if some crops manage to grow, they would be poisonous enough to cause serious health problems in people consuming them.
- Soil pollutants would bring in alteration in the soil structure, which would lead to death of many essential organisms in it. This would also affect the larger predators and compel them to move to other places, once they lose their food supply.
- Increased erodibility and larger loss of soil and nutrients.

Industrial

- Polluted water discharged from factories
- Runoff from pollutants (paint, chemicals, rotting organic material) leaching out of landfill
- Acid rain (fumes from factories mixing with rain)
- Sewage discharged into rivers instead of being treated properly
- Purposeful injection into groundwater as a disposal method
- Interconnections between aquifers during drilling (poor technique)
- Dangerous chemicals entering underground water
- Release of pollutant gases and radioactive rays causing health problems
- Creation of toxic dust is another potential effect of soil pollution

Urban

- Foul smell due to industrial chemicals and gases might result in headaches, fatigue, nausea, etc., in many people.
- Clogging of drains and inundation of urban areas leads to public health problems
- Septic tank seepage and pollution of drinking water sources
- Waste management problems

- Oil and petroleum leaks from vehicles washed off the road by the rain into the surrounding habitat

Environmental

- Disturbance in the balance of flora and fauna residing in the soil
- In addition, the pollutants change the make–up of the soil and the types of microorganisms living in it.
- Lagoon seepage, which, in turn, can cause even more harm because a lack of plants on the soil will cause more erosion, spreading the contaminants onto land that might not have been tainted before. Soil that has been contaminated should no longer be used to grow food, because the chemicals can leach into the food and harm people who eat it.
- If certain organisms die off in the area, the larger predator animals also have to move away or die because they lost their food supply. Thus, it is possible for soil pollution to change whole ecosystems.

30.6. CONTROL OF SOIL POLLUTION

To prevent soil erosion, we can limit construction in sensitive area. In general, we would need less fertilizer and fewer pesticides. We need to adopt the three R's: Reduce, Reuse, and Recycle. This would give us less solid waste.

Reducing chemical fertilizer and pesticide use

Applying biofertilizers and manures can reduce chemical fertilizer and pesticide use. Biological methods of pest control can also reduce the use of pesticides and thereby, minimize soil pollution.

Reusing of materials

Materials such as glass containers, plastic bags, paper, cloth etc. can be reused at domestic levels rather than being disposed, reducing solid waste pollution.

Recycling and recovery of materials

This is a reasonable solution for reducing soil pollution. Materials such as paper, some kinds of plastics and glass can be recycled. This decreases the volume of refuse and helps in the conservation of natural resources e.g., recovery of one ton of paper can save 17 trees.

Reforesting

Control of land loss and soil erosion can be attempted through restoring forest and grass cover to check wastelands, soil erosion and floods. Crop rotation or mixed cropping can improve the fertility of the land.

Solid waste treatment

Proper methods should be adopted for management of solid waste disposal. Industrial wastes can be treated physically, chemically and biologically until they become less hazardous. Acidic and alkaline wastes should be first neutralized; the insoluble material if biodegradable should be allowed to degrade under controlled conditions before being disposed. As a last resort, new areas for storage of hazardous waste should be investigated such as deep well injection and more secure landfills. Burying the waste in locations situated away from residential areas is the simplest and most widely used technique of solid waste management. Environmental and aesthetic considerations must be taken into consideration before selecting the dumping sites. Incineration of other wastes is expensive and leaves a huge residue and adds to air pollution. Pyrolysis is a process of combustion in the absence of oxygen or the material burnt under controlled atmosphere of oxygen. It is an alternative to incineration. The gas and liquid, thus, obtained can be used as fuels. Pyrolysis of carbonaceous wastes such as firewood, coconut, palm waste, corn combs, cashew shell, rice husk paddy straw and saw dust, yields charcoal along with products such as tar, methyl alcohol, acetic acid, acetone and a fuel gas. Anaerobic or aerobic decomposition of biodegradable municipal and domestic waste is also being done and gives organic manure. Cow dung, which releases methane into the atmosphere, should be processed further in 'gobar gas plants' to produce 'gobar gas' and good manure.

31 SOIL EROSION AND CONSERVATION

Soil erosion is a natural process and a major problem in Agriculture. Human activity causes it to occur much faster than under natural conditions. Balanced ecosystems comprising soil, water and plant environments are essential for the survival and welfare of mankind. However, ecosystems are being disturbed due to over exploitation in many parts of the world, including some parts of India. The resulting imbalance in the ecosystem is revealed through various undesirable effects, such as degradation of soil surfaces, frequent occurrence of intense floods etc. Tonnes of soil are lost from fields every year. This not only reduces crop production, the soil acts as a pollutant to rivers, lakes, and other water systems. Soil erosion can be controlled with the proper techniques.

Soil erosion is the deterioration of soil by the physical movement of soil particles from a given site. Wind, water, ice, animals, and the use of tools by man are usually the main causes of soil erosion. It is a natural process which usually does not cause any major problems. It becomes a problem when human activity causes it to occur much faster than under normal conditions

Soil is naturally removed by the action of water or wind: such 'background' (or 'geological') soil erosion has been occurring for some 450 million years, since the first land plants formed the first soil. Even before this, natural processes moved loose rock, or regolith, off the Earth's surface.

In general, background erosion removes soil at roughly the same rate as soil is formed. But 'accelerated' soil erosion — loss of soil at a much faster rate than it is formed — is a far more recent problem. It is always a result of mankind's unwise actions, such as overgrazing or unsuitable cultivation practices. These leave the land unprotected and vulnerable. Then, during times of erosive rainfall or windstorms, soil may be detached, transported, and (possibly travelling a long distance) deposited.

Accelerated soil erosion by water or wind may affect both agricultural areas and the natural environment, and is one of the most widespread of today's environmental problems. It has impacts which are both on-site (at the place where the soil is detached) and off-site (wherever the eroded soil ends up). Soil erosion is just one form of soil degradation.

Worldwide, farmers are losing an estimated 24 billion tonnes of topsoil each year. Although soil erosion is a physical process, it also affects productivity and growth. Reductions in yield of up to 50% have been documented on severely eroded soils. Losses of soil take place much faster than new soil can be created. It takes thousands of years to form just a few centimetres of soil. The difference between creation and loss represents an annual loss of 7.5 to 10 tonnes per acre worldwide.

Topsoil contains most of the soil's nutrients, organic matter, and pesticides. Soil erosion causes these substances to move also. What is left behind is soil with poorer structure, lower water-holding capacity, different pH values, and low nutrient levels. Therefore, fertilizers and organic matter must be added in an attempt to restore the soil to its original composition. The soil also has a lower resistance to drought.

Soil Erosion in India

Soil Erosion in India or degradation of soil is one of the prime concerns of the nation as agriculture is adversely affected to a great extent. Soil Erosion in India is amongst the leading areas of concern to the Government.

In India almost 130 million hectares of land, that is, 45% of total geographical surface area, is affected by serious soil erosion through gorge and gully, shifting cultivation, cultivated wastelands, sandy areas, deserts and water logging. India is losing 5,334 million tonnes of soil every year due to soil erosion because of indiscreet and excess use of fertilisers, insecticides and pesticides over the years. About one millimetre of top soil is being lost each year and the rate of loss is 16.4 tonnes per hectare every year. 20% of this is transported by river to sea, 10% deposited in reservoir results in 1 to 2% loss off storage capacity. It affects cultivation and farming in the country in adverse and unfavourable ways.

Impacts of Erosion

Soil erosion leads to deprivation of physical characteristics of soils and damages plant and crops. Soil erosion by rain and river that takes place in hilly areas causes severe landslides and floods, while cutting trees for agricultural implements, firewood and timber; grazing by a large number of livestock over and above the carrying capacity of grass lands, traditional agricultural practices, construction of roads, indiscriminate quarrying and other activities, have all led to the opening of hill-faces to extreme soil erosion.

Wind erosion causes development of deserts, dust, storms, whirlwinds and destruction of crops, while moving sand covers the land and makes it sterilised. Excessive soil erosion with resultant high rate of sedimentation in the reservoirs and decreased fertility has become solemn environmental problems for the country with disastrous economic consequences.

Impacts of erosion on soil quality

Organic matter is a small fraction (2% to 4%) of soil mainly present on the soil surface. Organic matter contributes to productivity through its effect on the physical, chemical, and biological properties of the soil. Erosion gradually depletes organic matter and decreases soil productivity.

When organic matter is lost, soils tend to lose their physical structure. The degradation of soil structure makes the soil hard, compact and cloddy. The soil aeration, water-holding capacity and permeability are also decreased. Decreased aeration means less oxygen available for plant roots to grow. Decreased water availability also means less water available for healthy plant growth. When soil permeability decreases, less water will soak into the soil and more will run-off.

Beneficial organisms that suppress disease and break down organic residues will not function well due to reduced oxygen and water in soil. This in turn will reduce nutrient storage and supply abilities of the soil.

Impacts of erosion on air quality

Soil particles blown by the wind into the air have a major impact on human and animal health. Particles suspended in air by wind are easily inhaled and accumulate in lung tissues causing major respiratory problems. Concentrated levels of wind blown particles can also reduce visibility and increase the risk of automobile accidents.

Impacts of erosion on water quality

Sediment deposition in lakes and rivers increases water turbidity making it difficult for light to penetrate the water. This causes problems for aquatic plants that need sunlight for photosynthesis. Sediments are also rich in nutrients such as phosphorus and nitrogen. These nutrients promote the excessive growth of algae. This process is called eutrophication. Areas of excessive algae growth, called algae blooms, deplete oxygen in the water resulting in the death of aquatic animals from lack of oxygen.

Processes and mechanics of erosion

The three steps are common to both water and wind erosion:

1. Detachment of Soil Particles:
   - This action dislodges the particles from the soil by the impact energy of the rain or wind.
2. Transport of particles:
   - This action carries soil particles in the moving wind or water.
3. Deposition of particles in a new location:

- This action deposits the sediment when the wind and water energy subsides. Clay or silt size particles can be carried a great distance before deposition, while larger sand-size particles will be carried only a short distance.

The impact of raindrops shatters surface aggregates and detaches soil particles from them. Raindrop impact is the primary cause of particle detachment. Raindrops can splash soil particles, moving them up to three feet away. Some of the detached particles float into soil pore spaces. This can clog and seal soil pores and result in reduced water entry (infiltration) into the soil. If the rainfall rate exceeds the rate at which water can infiltrate the soil, the excess water runs off and often carries the detached soil particles with it.

Detached particles (sediment) are carried with flowing water down the slope. How many particles and how far they are transported depend on the velocity and volume of the running water. As the water velocity slows down it loses the energy needed to continue carrying the detached suspended soil particles, and the soil particles are then deposited in their new location.

Similarly, wind erosion is a world-wide problem that occurs when strong winds blow across dry soil on unprotected surfaces. Wind detaches soil particles from the surface. Once detached these particles are transported by either suspension into air and/or rolling along the soil surface. Fine sands, silt or clay size particles can be transported for great distances by strong winds. While larger particles rolling along the soil surface move shorter distance and also shatter other soil particles along the way. As the wind speed decreases, deposition of soil particles begins. Wind erosion most commonly occurs in arid and semi-arid regions, because of the frequent occurrence of dry and windy conditions.

Types of Water and Wind Erosion

Table 106. Types of Water and Wind Erosion

Water	Wind
splash erosion	
sheet erosion	suspension
rill erosion	saltation
gully erosion	soil creep

Types of Water Erosion

Splash Erosion (Rain Drop Impact)

Splash erosion or rain drop impact represents the first stage in the erosion process. Splash erosion results from the bombardment of the soil surface by rain drops. Rain drops behave as little bombs when falling on exposed or bare soil, displacing soil particles and destroying soil structure. It is the primary cause of soil detachment and soil disintegration Studies have shown that splashed particles may raise as high as 0.6 metres above the ground and move up to 1.5 metres horizontally. Splash erosion results in the formation of surface crusts which reduce infiltration resulting in the start of runoff.

Sheet Erosion

Sheet erosion is responsible for extensive soil loss in both cultivated and non-cultivated environments. Sheet erosion occurs as a shallow 'sheet' of water flowing over the ground surface, resulting in the removal of a uniform layer of soil from the soil surface. Sheet erosion occurs when rainfall intensity is greater than infiltration (sometimes due to crusting).

- Rarely seen but accounts for large volumes of soil loss.
- Sheet erosion rarely flows for more than a few meters before concentrating into rills.
- Soil particles are only transported not detached.
- Typically results in the loss of the finest soil particles which contain the bulk of the available nutrients and organic matter.
- Difficult to observe in the field, sheet erosion is often only recognised when eroded soil is deposited along a fenceline.

Rill Erosion

Rill erosion results from the concentration of surface water (sheet erosion) into deeper, faster-flowing channels. As the flow becomes deeper the velocity increases detaching soil particles and scouring channels up to 30cm deep. Rill erosion represents the intermediate process between sheet and gully erosion.

- Particular concern in raised bed cropping systems
- Concentrated flow is able to detach and transport soil particles
- Channels form up to 30cm deep
- Intermediate process between sheet and gully erosion.

Gully Erosion

Gully erosion is an advanced stage of rill erosion where surface channels have eroded to the point where they can not be removed by tillage operations. Gully erosion is responsible for removing vast amounts of soil, irreversibly destroying farmland, roads and bridges and reducing water quality by increasing the sediment load in streams. Gully initiation is thought to be a response to excessive water in the local environment caused by the removal of perennial vegetation.

A gully head forms as rill erosion deepens and widens creating a characteristic nick point or headwall. Most gullies extend up slope as a result of headwall migration. However, it is the collapse and slumping of the sidewalls which usually contributes the greatest proportion of soil loss.

- Water running into the gully either scours the face or undercuts the head wall resulting in gully migration.
- Widening of gully sides occurs by undercutting or slumping.
- Gully head shape indicates if erosion is due to scouring (forward slope) or dispersion (undercut).
- Believed to be a response to changed hydrological conditions

Types of Wind Erosion

1. **SUSPENSION:** Fine particles less than 0.1 mm in size are moved parallel to the surface and upward into the atmosphere by strong winds. The most spectacular of erosive processes, these particles can be carried high into the atmosphere, returning to earth only when the wind subsides or they are carried downward with precipitation. Suspended particles can travel hundreds of miles.
2. **SALTATION:** Movement of particles by a series of short bounces along the surface of the ground, and dislodging additional particles with each impact. The bouncing particles ranging in size from 0.1 to 0.5 mm usually remain within 30 cm of the surface. Depending on conditions, this process accounts for 50 to 90% of the total movement of soil by wind.
3. **SOIL CREEP:** The rolling and sliding of larger soil particles along the ground surface. The movement of these particles is aided by the bouncing impacts of the saltating particles described above. Soil creep can move particles ranging from 0.5 to 1 mm in diameter, and accounts for 5 to 25% of total soil movement by wind.

Factors Contributing to Water and Wind Erosion of Soil

Climate, soil properties, vegetation, soil cover, and land management practices are factors that influence both water and wind erosion. Soil surface roughness, unsheltered distance, and wind velocity and turbulence are additional factors influencing wind erosion, and topography is an additional factor influencing water erosion.

Table 107. Factors influencing erosion

Water Erosion	Wind Erosion
climate	climate
soil properties: aggregation /soil	soil properties

moisture	
soil cover	soil cover
land management	land management
topography	soil surface roughness
	unsheltered distance
	wind velocity and turbulence

Climate

Duration and intensity of rainfall regulates the amount of soil detachment and subsequent loss from the land. During intense storms, rainfall can detach up to 100 tons/acre.

The erosive energy of running water depends on the volume and velocity. When the erosive energy is high (*i.e.*, high volume and velocity), water can detach and transport large particles as well as smaller particles.

Soil Properties

Soil texture (proportion of clay, silt, and sand particles in a soil) has two effects on soil erosion. The first is in its influence on infiltration or entry of water into the soil. When rainfall infiltrates rapidly, runoff is minimal. For example sandy textured soils have large pores acting like large pipes that allow much of the rainfall to soak right into the soil. Sandy soils are known to have good infiltration and drainage. Clay textured soil have small pores more like narrow pipes that do not allow water to soak into the soil fast. Clay soils are known to have poor infiltration and drainage.

Second, particles vary in their ease of detachment. Silt particles are most easily detached because they are small and do not easily form aggregates.

Aggregation

Particles of fine sand, silt and clay may join together to form aggregates. The soil property which describes the character and formation of these aggregates is called soil structure. The glue that joins the soil particles together includes organic matter, clays, iron oxides, aluminum oxides, and lime. Aggregate formation in clay-textured soil improves water infiltration into the soil and drainage because it increases the number of large pores (larger pipes). In sandy textured soil, aggregate formation reduces the excessively fast drainage of water by increasing the number of small pores (narrow pipes).

Aggregates can be described by their grade. The grade is a measure of how well the aggregate is cemented together or, conversely, how easily it is broken down by the impact of water, wind, or human activity. A soil with good structure has many aggregates present which are stable, meaning well cemented together.

Such aggregates resist the forces of water, wind, and human activity. Consequently, they can maintain their porosity and allow better water and air movement.

Soil Moisture

Moist soils are less prone to wind erosion due to the cohesive or binding effects of water on soil clay and organic matter. As wind dries the soil, the risk of soil loss erosion by wind increases, especially if aggregate grades are weak. The risk peaks at or below permanent wilting point.

Soil Cover

Bare soil is exposed to the full erosive power of raindrops and runoff water. Vegetative canopy helps stabilize soil and controls runoff. The vegetative canopy intercepts raindrops and reduces the erosive energy of the raindrops. Dense canopies that cover much of the soil surface intercept a large proportion of the rainfall. The roots of vegetation, such as grass, bind soil particles together to resist erosion. Vegetation can also help lessen or deflect wind, intercept wind-borne sediment, and keep soils moist, making them less susceptible to wind-related erosion.

Crop or plant residue dissipates the energy available to cause erosion. A mulch or crop residue absorbs the energy of falling raindrop, lessens detachment, reduces sealing of pores in the soil surface, and promotes good water infiltration. The chart below shows residue cover effects from 0 to 100% on soil erosion reduction. For example, a residue cover of 20% will reduce erosion by 50% compared to field with no residue cover.

Topography

Topography, or lay of the land, is an important variable in water erosion. More specifically, the degree of steepness (percent slope), as well as the slope length, is important. Steep slopes have high runoff water velocity. This increases its erosive energy (remember that erosive energy of runoff is a function of runoff velocity and volume). When the slope is longer (length), surface area for water collection also increases and therefore increases the run-off volume.

The 'plan' is a term used to describe the distribution of water across the slope, that is, whether water flow is evenly distributed across the slope (linear); water flow is concentrated in one area (concave); or water flow is moved away from the slope (convex).The pattern and speed of water flow will influence the area of most intense soil erosion across the slope.

Land Management

Land Development projects, such as road and housing construction, can contribute to soil erosion and sedimentation both during and after the actual construction activity. Clearing, grading, and other activities disturb the soil surface, remove existing vegetation, and alter topography, thereby increasing erosion risk. Massive land clearing done by large construction equipment pulverizes the soil and clears all vegetation. This mechanical disturbance exposes the soil directly to the impact of rainfall energy. Highly disturbed soils have also lost much of the organic matter that glues them together. Also, the rate of water infiltration is usually decreased and water that does not soak into the soil runs off carrying sediment.

Irrigation practices can increase runoff by over-saturating the soil or allowing sprinklers to spray onto impervious surfaces. Under sprinkler irrigation system, water droplets can create splash erosion, if the soil is not fully covered with mulch or vegetation.

Mechanical disturbance affects soil physical properties, specifically by disrupting soil aggregation, causing increase erosion potential. For example, tillage buries soil cover or residue that otherwise would serve to protect the soil from erosion.

Soil Surface Roughness

Soil surfaces that are smooth offer little resistance to the wind. Keeping the surface rough by tilling when the soil is moist enough to form large clods will reduce wind erosion. Although, it is important to recognize that excessive and frequent tillage can gradually reduce roughness of soil by breaking clods and aggregates that resist erosion. Tilling a dry soil may cause a dry dust to form, which aggravates the erosion problem.

Unsheltered Distance

The lack of windbreaks allows the wind to put soil particles into motion over greater distances thus increasing the abrasion and soil erosion. There is no obstacle to reduce wind velocity.

Wind Velocity and Turbulence

Wind erosion potential begins when the wind speed increases to 25 km/h (15 mph). Soil movement increases by the cube of the wind velocity, so the amount of soil transported increases rapidly with wind speed. Turbulent winds carry soil particles into the atmosphere to greater altitude, and increases soil particle detachment.

CONSERVATION MEASURES

Conservation of water and conservation of soil are closely related. Water is lost from the soil in four ways: (1) Surface run-off, (2) Percolation, (3) Transpiration and (4) Evaporation. Out of these sources by which the water is lost run-off is usually the largest and most damaging as it causes erosion.

The main factors in conserving moisture relate to increasing infiltration and storage capacity of soil and reducing run-off and evaporation.

Untamed water is the main cause of soil erosion. Almost all the methods that deal with soil conservation are principle methods to control and conserve water. Measures to control soil erosion also result in the conservation of water.

Water Erosion Control: The main objectives in controlling soil erosion caused by water are:

i) To reduce surface run-off

ii) To put mechanical obstruction in the way of flowing water, and

iii) To reduce dispersion of soil particles

Conservation Methods for Water Erosion

There are two methods for the control or management of land eroded by water : (a) Biological Method, and (b) Mechanical Method.

A) **Biological Method**: i) Agronomical method, and ii) Agrostological method

1. **Agronomical Method**: 1. ***Growing Cover Crops***: Sod-like crops, such as grass, sunnhelp, berseem, groundnut offer excellent soil protection against erosion by water. Such crops are effective primarily because they provide a cusion against the raindrops. The denser the vegetation the more effective is the cushioning effect. Close growing crops also help to control erosion by providing root channels through which water moves downwards more freely. In addition, plant roots aid in making a desirable soil structure.

Vegetation also encourages the growth of earthworms and other desirable forms of soil life. These in turn help to keep the soil open and receptive to raindrops.

2. ***Strip Cropping***: This practice consists of growing erosion-permitting crops (e.g., jowar, bajra, maize), in alternate strips with erosion checking close-growing crops (e.g., grasses, pulses). The erosion checking strips check and held the flowing water and soil.

Contour strip cropping is one of the type of strip cropping: Contour strip cropping is the growing of erosion-permitting and erosion-resisting crops alternately in strips across the slope and on the contour.

3. ***Crop Rotation***: Growing of two or more different crops in sequence in a field for maintaining the soil fertility. Continuous growing of clean-cultivated crops (e.g., tobacco) causes more erosion. A good rotation should include densely planted small grains, spreading legume which may check soil erosion.

4. ***Mulching***: Mulches of different kinds minimise evaporation, and increases absorption of moisture. Stubble mulching is based on merely stirring the soil with implements that leaves considerable part of vegetative materials i.e.,

crop residue or vegetative litter on the land as a surface protection against erosion and for conserving moisture by favouring infiltration and reducing evaporation.

5. ***Organic Manures***: Incorporation of organic manures in the soil improves the soil structure. Granular and crumby structures increase infiltration and permeability in the soil and conserve soil water.

II. ***Agrostological Method***: Dense canopy and profuse root system of grasses tightly bind the soil against erosion. Grasses increase water infiltration and improve physical conditions of soil.

Grasses can be grown on lands which are otherwise not suitable for cultivation. They are also used in protecting bunds, waterways, badly eroded areas and gully control. The grasses commonly grown are:

i) Dub - *Cynodon dactylon* - Checks run-off water

ii) Kudzu vine - Checks gully erosion

iii) Kan - *Saccharum spontaneum* - Checks water erosion on the bank of rivers and steams.

iv) Dinanath - *Pennisetum pedicellatum* - Checks bunds and water outlet.

B. **Mechanical Methods:** Mechanical measures constitute various engineering techniques and structures. These practices reduce run-off velocity, impound water for a longer time and provide more absorption opportunity. Following mechanical practices are adopted for checking water erosion.

1. **Contour Tillage**: On hilly land (slope) and tillage operations and the sowing of crops should be done at right angles to the slope of the land. On long slopes, the field may be laid out in narrow strips across the slope rather than with it. In contour tillage, each furrow which lies intercept flowing water and holds it and allows it to soak into the soil.

2. **Contour Bunding**: The slope of the land is broken up into smaller, more level compartments by constructing earthen embankments of suitable size on contours. Each bund thus, holds the rain water within each compartment.

Farmers in the plains could raise level bunds around their fields for holding the rain water and conserving the soil. The maintenance of bunds is very important. Planting useful grasses, strengthening the bunds at weak points, closing rat holes and closing small breaches in time, are the most important items in bund maintenance.

3. **Terracing:** On steeper slopes terraces or flat platforms are constructed in steps like series across the slope. It is like a level benches or table top for retaining and distributing rainfall for controlling run-off.

4. **Outlet Channel**: For safe removal of excess run-off water, it is essential to provide suitable outlet structure at proper places so that no harmful effects of water-logging, eroding, gullying or damage to other conservation structures are caused. Some of the important surplusing arrangements are grass waterways and diversion ditches.

5. **Basin Listing**: Scooping out small basins at regular intervals on slopes, checks run-off.

6. **Pan Breaking and sub-soiling**: Pan breaking and sub-soiling permit infiltration and percolation. This increases the water storage capacity of the soil, thus, decreasing run-off and erosion.

7. **Water Harvest**: At suitable places within possibilities of storing water pods and tanks are recommended water conservation practices. They are also very useful to some extent as flood control measures.

Control of Gully Erosion: Gully control is necessary in order to prevent complete destruction of cultivated lands and grasslands. In the first step, water flowing into the gully should be diverted away by means of a bund. The second step is to build several obstruction in the gully such as rock dams. Invariably, the gullies should be established under permanent grass and tree vegetation. Sometimes a gully may be adequately established by converting it into a paddy field.

Wind Erosion Control

i) To reduce the erodibility of the soil

ii) To reduce wind velocity at ground level

The following principles and practices should be adopted for the management of eroded land wind:

1. **Conservation of Moisture**: Obviously, if the soil can be kept moist there is little damager of wind erosion. This can be done by ploughing of the lands during rains, levelling, bundling and terracing of the land, and water harvest.

2. **Vegetative Cover**: A vegetative cover discourages soil blowing by providing a mechanical obstruction to erosive winds thus reducing their velocity and soil carrying capacity. The roots help in binding the soil particles together, in this way they ressist erosion by wind.

3. **Mulching**: Stubble mulch has proved to be effective in controlling wind erosioin

4. **Rough Surface**: Wind is more effective on smooth surface than on rough surface. Hence, if no mulch is possible then the surface should be left rough during the period of high winds.

5. **Trash Tillage**: Tillage operations that leave the clods and trash at the surface and are desirable in controlling wind erosion. The trash should not be buried but mixed with the surface soil so that it is not easily flowing away, but will obstruct the movement of the soil in a lose condition and without a good cover is liable to lead heavy wind erosion. Opening ridges and furrows at right angles to the prevailing winds is desirable to a certain extent.

6. **Organic Matter**: This is desirable for increasing the stability of the soil structure for improving water-holding capacity of the soil and for building up its

fertility. Using green manure crops is a desirable practice for this purpose. Crop residues should be left on the soil and not removed for use as fuel.

7. **Regulating Grazing**: Probably the most important cause of wind erosion is overgrazing. There must be grazing regulations to restrict the use of grazing lands.

8. **Strip Cropping**: Strip cropping is an effective method for controlling the wind erosion. Wind strip cropping, one of the types of strip cropping is especially suitable for the land which is susceptible to wind erosion. Wind strip cropping normal crop row should be alternated with tall crop plant, at right angles to the direction of the prevailing winds.

9. **Wind Break and Shelterbelt**: Trees and shrubs, commonly planted in rows at right angles to the prevailing winds are called shelterbelts. A recommended shelterbelt to control wind erosion may be made by planting three rows of trees and shrubs, the control row to consist of tall trees and the two outside rows to consist of smaller dense shrubs.

Table 108. Example of Tall trees

i)	Acacia arabica	Babul
ii)	Dalbergia sisso	Shisham
iii)	Eucalyptus rostrata	Eucalyptus
iv)	Azadirachta indica	Neem
v)	Albizzia lebbek	Siris
		Exmple of Bushy Shrubs
i)	Cassia auriculata	
ii)	Dondoned viscosa	
iii)	Glyricidia maculata	
iv)	Saccharum munja	
v)	Zizphusspecies	

Stabilizing of Drifting Sand Dunes

The deposition of sand on cultivated land happens usually in the neighbouring of sand dunes or the tract of blowing sand, from which sand is easily blown on the good cultivated land. The only method by which this can be prevented consists likewise in establishing some soil or sand binding vegetation on these dunes. Suitable grasses, generally with a creeping habit, and trees which can grow in sandy situations have to be planted and their growth encouraged and the drifting of the sand prevented thereby. The cashewnut tree and the scre pine are common trees for this purpose. The others sand binding trees are Acacia arabica, Agave americana, Andropogon laniger etc.

32 SOIL QUALITY

32.1. SOIL QUALITY BASICS

Soil quality is the ability of a soil to perform functions that are essential to people and the environment. Soil quality is not limited to agricultural soils, although most soil quality work has been done in agricultural systems. Soil quality definitions emphasize several features such as:

Management affects on soil properties

Soil quality assessments focussed on the dynamic, or management–affected properties of soil such as nutrient status, salinity, and water–holding capacity. These properties are assessed in the context of the inherent capability of a particular soil.

Essential services provided by soil

Soils support plant growth, recycle dead material, regulate and filter water flows, support buildings and roads, and provide habitat for many plants and animals. Depending on the land use, many of these functions occur simultaneously. Soil quality assessments go beyond measuring degradation (erosion, compaction, or contamination) to focus on these soil functions and the processes that create them.

Soil meeting multiple and sometimes conflicting goals

Soil functions provide private benefits such as crop production or structural support for buildings. Simultaneously, the same soil may provide societal benefits such as carbon sequestration, water quality protection, or preservation of soil productivity for future generations. Evaluating soil quality requires that we identify and prioritize these benefits and pay attention to the interactions and trade–offs among them.

Resistance and resilience to disturbance as characteristics of a healthy soil

The ability to continue providing essential services in the face of disturbance. An ecosystem disturbance can be natural or human induced stress. An example of a natural disturbance is a hurricane or a tornado. An example of a human-induced disturbance is tillage or pesticide application., whether natural or human induced, is essential to maintain or improve soil quality over time. A soil is not considered 'healthy' if it is managed for short term productivity at the expense of future degradation (Doran et al 1994).

32.2. SUSTAINABLE AGRICULTURE

Although there is no one definition of sustainable agriculture, most incorporate the ideals encompassed in the following definitions:

"An agriculture that can evolve indefinitely towards greater human utility, greater efficiency of resource use and a balance with the environment that is favourable both to humans and to most other species." *–Harwood, 1990*

"...sustainability lies in the dynamic nature of its fundamental components: ecological (spatial and temporal relations, diversity, stability, and resilience; economic (resource distribution and allocation); and social (equity, access, stewardship and institutions)."*–Miller and Wali, 1995*

The agro–ecosystem is made up of many interacting components with multiple goals. Soil quality is one important part of sustainable agro–ecosystem management, analogous to water and air quality. Assessing soil quality may help managers identify practices that could be adapted to become more sustainable.

Soil quality is one aspect of sustainable agro–ecosystem management.

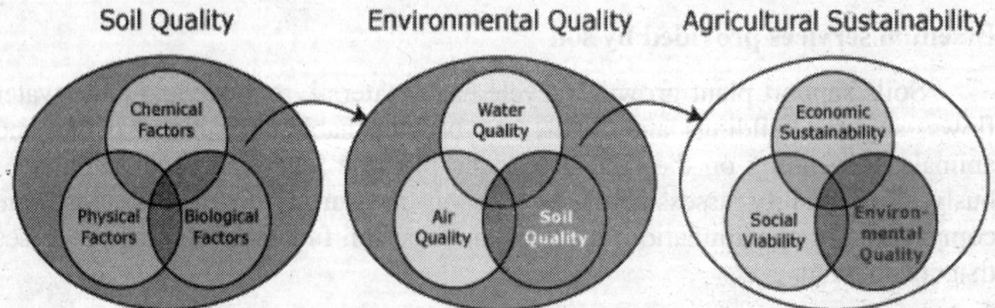

Fig. 74. Components of Soil Quality and agro-ecosystem

32.3. SOIL QUALITY INDICATORS: MEASURES OF SOIL FUNCTIONAL STATE

Scientists use soil quality indicators. A soil quality indicator is a chemical, physical or biological property of soil that is sensitive to disturbance and represents performance of ecosystem function in the soil of interest. Indicators are dynamic soil properties that evaluate how well soil functions since soil function often cannot be directly measured. Measuring soil quality is an exercise in identifying soil properties that are responsive to management, affect or correlate with environmental outcomes, and are capable of being precisely measured within certain technical and economic constraints. Soil quality indicators may be qualitative (e.g., drainage is fast) or quantitative (infiltration = 2.5 in/hr).

Ideal indicators should

- Correlate well with ecosystem processes
- Integrate physical, chemical, and biological properties and processes of soil
- Be accessible to many users
- Be sensitive to management and climate
- Be components of existing databases
- Be interpretable

(Doran and Parkin 1996)

There are three main categories of soil indicators as chemical, physical and biological. Typical soil tests only look at chemical indicators. Soil quality attempts to integrate all three types of indicators. The categories do not neatly align with the various soil functions, so integration is necessary. The table below shows the relationship between indicator type and soil function.

Table 109. Relationship between indicator type and soil function

Indicator category	Related soil function
Chemical	Nutrient Cycling, Water Relations, Buffering
Physical	Physical Stability and Support, Water Relations, Habitat
Biological	Biodiversity, Nutrient Cycling, Filtering

Organic matter, or more specifically soil carbon, transcends all three indicator categories and has the most widely recognized influence on soil quality. Organic matter is tied to all soil functions. It affects other indicators such as aggregate stability (physical), nutrient retention and availability (chemical), and nutrient cycling (biological); and is itself an indicator of soil quality.

Some indicators are descriptive and can be used in the field as part of a health card. Others must be measured using laboratory analysis. Some examples of indicators that fall into the three broad categories of chemical, physical and biological are given below:

32.4. INDICATOR CATEGORIES

Chemical indicators can give you information about the equilibrium between soil solution (soil water and nutrients) and exchange sites (clay particles, organic matter), plant health, the nutritional requirements of plant and soil animal communities, and levels of soil contaminants and their availability for uptake by animals and plants. Indicators include measures of:

- Electrical Conductivity
- Soil Nitrate
- Soil Reaction (pH)

Physical indicators provide information about soil hydrologic characteristics such as water entry and retention that influences availability to plants. Some indicators are related to nutrient availability by their influence on rooting volume and aeration status. Other measures tell us about erosional status. Indicators include measures of:

- Aggregate Stability
- Available Water Capacity
- Bulk Density
- Infiltration
- Slaking
- Soil Crusts
- Soil Structure and Macropores

Biological indicators can tell us about the organisms that form the soil food web that are responsible for decomposition of organic matter and nutrient cycling. Information about the numbers of organisms, both individuals and species, that perform similar jobs or niches, can indicate a soil's ability to function or bounce back after disturbance (resistance and resilience). Indicators include measures of:

- Earthworms
- Particulate Organic Matter
- Potentially Mineralizable Nitrogen
- Respiration
- Soil Enzymes
- Total Organic Carbon

32.5. SOIL PROBLEM–SOLVING GUIDE

The table below lists common soil problems. Each problem includes indicators to test for the problem, and practices that help or hinder.

Table 110. Soil problem–solving guide

Problem/Indicators	Possible cause of problem	To improve soil quality
Compaction	Working wet soil Excess traffic	Avoid working wet soil Reduce traffic/tillage operations

Indicators: Bulk density Penetration resistance Porosity Root growth patterns	Heavy machinery Repeated tillage at same depth Excess animal traffic Poor aggregation Low organic matter	Use controlled traffic patterns Avoid using heavy machinery Subsoil or rip when soil is not excessively wet or dry Alter tillage depth Add organic residues Diversify cropping system Use conservation tillage Add cover crops Use crop rotations Add animal manures Use non-compacting tillage (e.g., chisel vs moldboard)
Crop disease Indicators: Plant health Crop vigour Yield	Compacted layers Saturated soil Soil pathogen problems Nutrient deficiencies or imbalance Low organic matter Monoculture Low biological diversity	Soil test - correct nutrient and pH levels Check for pathogens/pests Reduce compaction following harvest Improve drainage Increase organic residue Use animal manure Add cover crops Use crop rotation Diversify cropping system
Crusting Indicators: Aggregate stability Slaking Soil Crusts	Excess sodium Low organic matter Low residues	Increase organic residues Reduce tillage depth Use animal manure Add cover crops For sodium problem - apply gypsum and flush with irrigation water
Drainage Indicators: Infiltration Hydraulic conductivity	Tillage pan High water table Poor soil structure	Subsoil to break up tillage pan Add drainage system
Soil life Indicators: Earthworms Soil respiration Microbial biomass Pitfall trapping	Low organic matter Low residues Excess pesticides or fertilizers Excess tillage Poor aeration	Use conservation tillage Use crop rotations Add cover crops
Salinity Indicators: Electrical conductivity Observe white crust	Saline seeps Saline irrigation water/well Shallow water table Poor drainage Excess evaporation	Leach excess salts Plant deep rooted crops Grow salt tolerant crops Increase vegetative cover Manage irrigation water Improve drainage

Erosion	Lack of cover and residue	Diversify crop rotations
Indicators:	Low organic matter	Reduce tillage
Observe rills, gullies	Poor aggregation	Use animal manure
Topsoil depth	Tillage pan or compacted	Use cover crops
Aggregate stability	layer	Increase surface residue or roughness
	Tillage practices that	Shorten slope length
	move soil down slope	Plant strip crops
	Excessive tillage	Use wind breaks
	Intensive crop rotation	
Infiltration	Lack of cover and residue	Add organic residue
Indicators:	Low organic matter	Add animal manure
Infiltration	Poor aggregation	Use cover crops
Aggregate stability	Tillage pan or compacted	Diversify crop rotation
Soil structure	layer	For sodium problem, apply gypsum
	Tillage practices that	and flush with irrigation water
	move soil down slope	Subsoil or rip when soil is not
	Excessive tillage	excessively wet or dry
	Intensive crop rotation	Use tillage that preserves soil structure
Organic matter/	Excess tillage	Diversify or increase crop rotations
residue	Residue burned off	Add animal manure
Indicators:	Low residue crops	Use cover crops
Organic carbon	Too much fallow	Use high residue crops
Percent residues	Insufficient additions of	Reduce tillage
	crop residue	
Soil pH	Use of ammonium	Soil test - correct pH levels
Indicators:	fertilizers	Add lime for low pH
Soil solution pH	No liming	Improve drainage

APPENDIX-I

PLATES OF IGNEOUS ROCKS

Intrusive Igneous Rocks

Gabbro

Granite

Diorite

Extrusive Igneous Rocks

Obsidian

Andesite

Basalt

PLATES OF METAMORPHIC ROCKS

Foliated

Gneiss

Schist (Muscovite Schist)

Slate

Non-Foliated

Quartzite

Amphibolite

Marble

PLATES OF SEDIMENTARY ROCKS

Conglomerate

Chert

Iron Ore

Limestone

Sandstone

Shale

APPENDIX-II

PLATES OF PRIMARY MINERALS

Ferro magnesium minerals

Ortho-ino silicates

Olivine

Pyroxene

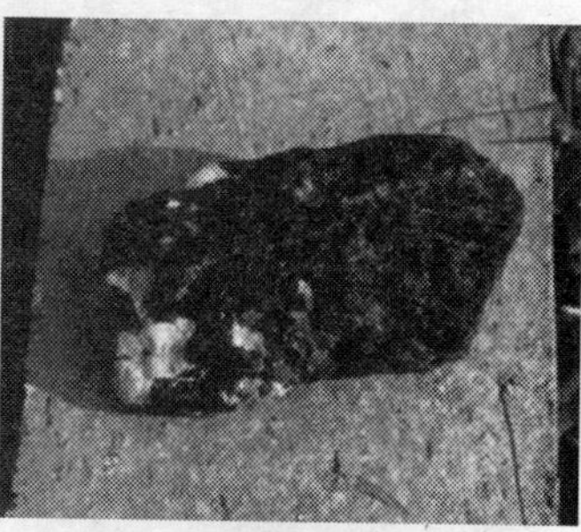

Amphibole

Phyllo Silicates (Micas)

Muscovite

Biotite

Non- Ferro magnesium minerals

Feldspars

Anorthite

Albite

Quartz Cristals

Calcite Cristals

Chlorite

Halite rock salt

APPENDIX - III

PROFILES OF SOIL ORDERS

Entisols

Inceptisol

Mollisol

Alfisol

Spodosols

Ultisols

Oxisols

Histosols

Vertisols

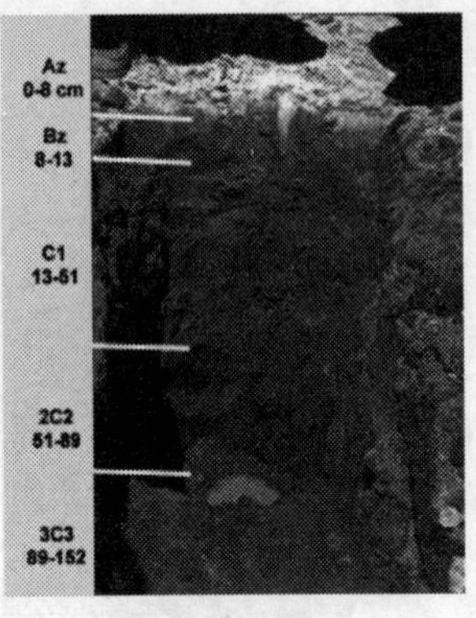

Aridosol

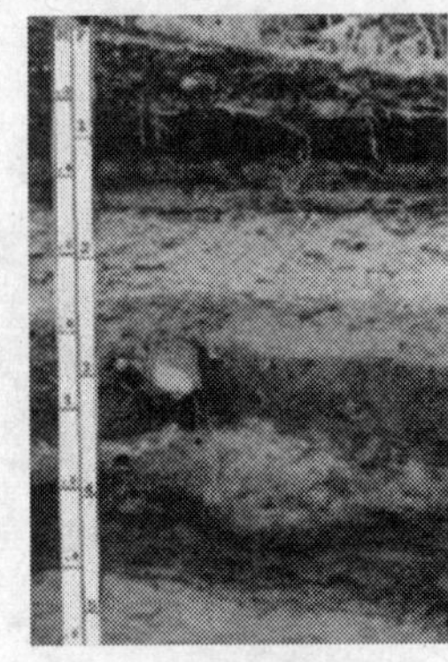

Andisols

Gelisols

GLOSSARY

ABC soil. A soil having an A, a B, and a C horizon.

Ablation till. Loose, permeable till deposited during the final down-wasting of glacial ice. Lenses of crudely sorted sand and gravel are common.

AC soil. A soil having only an A and a C horizon. Commonly, such soil formed in recent alluvium or on steep rocky slopes.

Aeration, soil. The exchange of air in soil with air from the atmosphere. The air in a well aerated soil is similar to that in the atmosphere; the air in a poorly aerated soil is considerably higher in carbon dioxide and lower in oxygen.

Aggregate, soil. Many fine particles held in a single mass or cluster. Natural soil aggregates, such as granules, blocks, or prisms, are called peds. Clods are aggregates produced by tillage or logging.

Alkali (sodic) soil. A soil having so high a degree of alkalinity (pH 8.5 or higher), or so high a percentage of exchangeable sodium (15 per cent or more of the total exchangeable bases), or both, that plant growth is restricted.

Alluvium. Material, such as sand, silt, or clay, deposited on land by streams.

Area reclaim . An area difficult to reclaim after the removal of soil for construction and other uses. Revegetation and erosion control are extremely difficult.

Association, soil. A group of soils geographically associated in a characteristic repeating pattern and defined and delineated as a single map unit.

Available water capacity (available moisture capacity). The capacity of soils to hold water available for use by most plants. It is commonly defined as the difference between the amount of soil water at field moisture capacity and the amount at wilting point. It is commonly expressed as inches of water per inch of soil. The capacity, in inches, in a 60-inch profile or to a limiting layer is expressed as:

Class	Inches/inch
Very low	0 to 3
Low	3 to 6
Moderate	6 to 9
High	9 to 12
Very high	more than 1

Basal till. Compact glacial till deposited beneath the ice (Lodgement Till is preferred).

Base saturation. The degree to which material having cation-exchange properties is saturated with exchangeable bases (sum of Ca, Mg, Na, K), expressed as a percentage of the total cation-exchange capacity.

Bedding planes. Fine stratifications, less than 5 millimeters thick, in unconsolidated alluvial, eolian, lacustrine, or marine sediments.

Bedding system. A drainage system made by plowing, grading, or otherwise shaping the surface of a flat field. It consists of a series of low ridges separated by shallow, parallel dead furrows.

Bedrock. The solid rock that underlies the soil and other unconsolidated material or that is exposed at the surface.

Bench terrace. A raised, level or nearly level strip of earth constructed on or nearly on the contour, supported by a barrier of rocks or similar material, and designed to make the soil suitable for tillage and to prevent accelerated erosion.

Bisequum. Two sequences of soil horizons, each of which consists of an illuvial horizon and the overlying eluvial horizons.

Blowout. A shallow depression from which all or most of the soil material has been removed by wind. A blowout has a flat or irregular floor formed by a resistant layer or by an accumulation of pebbles or cobbles. In some blowouts the water table is exposed.

Bog - Waterlogged, spongy ground, consisting primarily of mosses, containing acidic, decaying vegetation such as sphagnum, sedges, and heaths, that may develop into peat. Compare - fen, marsh, swamp.

Bottom land. The normal flood plain of a stream, subject to flooding.

Boulders. Rock fragments larger than 2 feet (60 centimeters) in diameter.

Broad-base terrace. A ridge-type terrace built to control erosion by diverting runoff along the contour at a nonscouring velocity. The terrace is 10 to 20 inches high and 15 to 30 feet wide and has gently sloping sides, a rounded crown, and a dish-shaped channel along the upper side. It may be nearly level or have a grade toward one or both ends.

Bulk Density - A measure of the weight of the soil per unit volume (g/cc), usually given on an oven-dry (110° C) basis

Calcareous soil. A soil containing enough calcium carbonate (commonly combined with magnesium carbonate) to effervesce visibly when treated with cold, dilute hydrochloric acid.

Caliche. A more or less cemented deposit of calcium carbonate in soils of warm-temperate, subhumid to arid areas. Caliche occurs as soft, thin layers in the soil or as hard, thick beds just beneath the solum, or it is exposed at the surface by erosion.

Capillary water. Water held as a film around soil particles and in tiny spaces between particles. Surface tension is the adhesive force that holds capillary water in the soil.

Catena. A sequence, or "chain," of soils on a landscape that formed in similar kinds of parent material but have different characteristics as a result of differences in relief and drainage.

Cation. An ion carrying a positive charge of electricity. The common soil cations are calcium, potassium, magnesium, sodium, and hydrogen.

Cation-exchange capacity. The total amount of exchangeable cations that can be held by the soil, expressed in terms of milliequivalents per 100 grams of soil at neutrality (pH 7.0) or at some other stated pH value. The term, as applied to soils, is synonymous with base-exchange capacity but is more precise in meaning.

Cement rock. Shaly limestone used in the manufacture of cement.

Channery soil. A soil that is, by volume, more than 15 percent thin, flat fragments of sandstone, shale, slate, limestone, or schist as much as 6 inches along the longest axis. A single piece is called a channer.

Chiseling. Tillage with an implement having one or more soil-penetrating points that shatter or loosen hard compacted layers to a depth below normal plow depth.

Clay. As a soil separate, the mineral soil particles less than 0.002 millimeter in diameter. As a soil textural class, soil material that is 40 percent or more clay, less than 45 percent sand, and less than 40 percent silt.

Clay film. A thin coating of oriented clay on the surface of a soil aggregate or lining pores or root channels. Synonyms: clay coating, clay skin.

Claypan. A slowly permeable soil horizon that contains much more clay than the horizons above it. A claypan is commonly hard when dry and plastic or stiff when wet.

Climax vegetation. The stabilized plant community on a particular site. The plant cover reproduces itself and does not change so long as the environment remains the same.

Coarse fragments. If round, mineral or rock particles 2 millimeters to 25 centimeters (10 inches) in diameter; if flat, mineral or rock particles (flagstone) 15 to 38 centimeters (6 to 15 inches) long.

Coarse textured soil. A soil with USDA Soil textures of loamy fine sand or coarser (loamy sand or sand).

Cobblestone (or cobble). A rounded or partly rounded fragment of rock 3 to 10 inches (7.5 to 25 centimeters) in diameter.

Colluvium. Soil material, rock fragments, or both moved by creep, slide, or local wash and deposited at the base of steep slopes.

Complex slope. Irregular or variable slope. Planning or constructing terraces, diversions, and other water-control measures on a complex slope is difficult.

Complex, soil. A map unit of two or more kinds of soil in such an intricate pattern or so small in area that it is not practical to map them separately at the selected scale of mapping. The pattern and proportion of the soils are somewhat similar in all areas.

Concretions. Grains, pellets, or nodules of various sizes, shapes, and colors consisting of concentrated compounds or cemented soil grains. The composition of most concretions is unlike that of the surrounding soil. Calcium carbonate and iron oxide are common compounds in concretions.

Congeliturbate. Soil material disturbed by frost action.

Conservation tillage. A tillage system that does not invert the soil and that leaves a protective amount of crop residue on the surface throughout the year.

Consistence, soil. The feel of the soil and the ease with which a lump can be crushed by the fingers. Terms commonly used to describe consistence are;

Loose: Noncoherent when dry or moist; does not hold together in a mass.

Friable: When moist, crushes easily under gentle pressure between thumb and forefinger and can be pressed together into a lump.

Firm: When moist, crushes under moderate pressure between thumb and forefinger, but resistance is distinctly noticeable.

Plastic: When wet, readily deformed by moderate pressure but can be pressed into a lump; will form a "wire" when rolled between thumb and forefinger.

Sticky: When wet, adheres to other material and tends to stretch somewhat and pull apart rather than to pull free from other material.

Hard: When dry, moderately resistant to pressure; can be broken with difficulty between thumb and forefinger.

Soft: When dry, breaks into powder or individual grains under very slight pressure.

Cemented: Hard; little affected by moistening.

Contour strip cropping. Growing crops in strips that follow the contour. Strips of grass or close-growing crops are alternated with strips of clean-tilled crops or summer fallow.

Control section. The part of the soil on which classification is based. The thickness varies among different kinds of soil, but for many it is that part of the soil profile between depths of 10 inches and 40 or 80 inches.

Coprogenous earth (sedimentary peat). Faecal material deposited in water by aquatic organisms.

Corrosive. High risk of corrosion to uncoated steel or deterioration of concrete.

Cover crop. A close-growing crop grown primarily to improve and protect the soil between periods of regular crop production, or a crop grown between trees and vines in orchards and vineyards.

Dense layer . A very firm, massive layer that has a bulk density of more than 1.8 grams per cubic centimeter. Such a layer affects the ease of digging and can affect filling and compacting.

Depth to rock. Bedrock is too near the surface for the specified use.

Diversion (or diversion terrace). A ridge of earth, generally a terrace, built to protect downslope areas by diverting runoff from its natural course.

Drainage class (natural). Refers to the frequency and duration of periods of saturation or partial saturation during soil formation, as opposed to altered drainage, which is commonly the result of artificial drainage or irrigation but may be caused by the sudden deepening of channels or the blocking of drainage outlets. Seven classes of natural soil drainage are recognized:

Excessively drained: Water is removed from the soil very rapidly. Excessively drained soils are commonly very coarse textured, rocky, or shallow. Some are steep. All are free of the mottling related to wetness.

Somewhat excessively drained: Water is removed from the soil rapidly. Many somewhat excessively drained soils are sandy and rapidly pervious. Some are shallow. Some are so steep that much of the water they receive is lost as runoff. All are free of the mottling related to wetness.

Well drained: Water is removed from the soil readily, but not rapidly. It is available to plants throughout most of the growing season, and wetness does not inhibit growth of roots for significant periods during most growing seasons. Well drained soils are commonly medium textured. They are mainly free of mottling.

Moderately well drained: Water is removed from the soil somewhat slowly during some periods. Moderately well drained soils are wet for only a short time during the growing season, but periodically they are wet long enough that most mesophytic crops are affected. They commonly have a slowly pervious layer within or directly below the solum, or periodically receive high rainfall, or both.

Somewhat poorly drained: Water is removed slowly enough that the soil is wet for significant periods during the growing season. Wetness markedly restricts the growth of mesophytic crops unless artificial drainage is provided. Somewhat poorly drained soils commonly have a slowly pervious layer, a high water table, additional water from seepage, nearly continuous rainfall, or a combination of these.

Poorly drained: Water is removed so slowly that the soil is saturated periodically during the growing season or remains wet for long periods. Free water is commonly at or near the surface for long enough during the growing season that most mesophytic crops cannot be grown unless the soil is artificially drained. The soil is not continuously saturated in layers directly below plow depth. Poor drainage results from a high water table, a slowly pervious layer within the profile, seepage, nearly continuous rainfall, or a combination of these.

Very poorly drained: Water is removed from the soil so slowly that free water remains at or on the surface during most of the growing season. Unless the soil is artificially drained, most mesophytic crops cannot be grown. Very poorly drained soils are commonly level or depressed and are frequently ponded. Yet, where rainfall is high and nearly continuous, they can have moderate or high slope gradients.

Drainage, surface. Runoff, or surface flow of water, from an area.

Drumlin. A low, smooth, elongated oval hill, mound, or ridge of compact glacial till. The longer axis is parallel to the path of the glacier and commonly has a blunt nose pointing in the direction from which the ice approached.

Eluviation. The movement of material in true solution or colloidal suspension from one place to another within the soil. Soil horizons that have lost material through eluviation are eluvial; those that have received material are illuvial.

Eolian soil material. Earthy parent material accumulated through wind action; commonly refers to sandy material in dunes or to loess in blankets on the surface.

Erosion. The wearing away of the land surface by water, wind, ice, or other geologic agents and by such processes as gravitational creep.

Erosion (geologic). Erosion caused by geologic processes acting over long geologic periods and resulting in the wearing away of mountains and the building up of such landscape features as flood plains and coastal plains. Synonym: natural erosion.

Erosion (accelerated). Erosion much more rapid than geologic erosion, mainly as a result of the activities of man or other animals or of a catastrophe in nature, for example, fire, that exposes the surface.

Erosion pavement. A layer of gravel or stones that remains on the surface after fine particles are removed by sheet or rill erosion.

Esker (geology). A narrow, winding ridge of stratified gravelly and sandy drift deposited by a stream flowing in a tunnel beneath a glacier.

Fallow. Cropland left idle in order to restore productivity through accumulation of moisture. Summer fallow is common in regions of limited rainfall where cereal grains are grown. The soil is tilled for at least one growing season for weed control and decomposition of plant residue.

Fen - Waterlogged, spongy ground containing alkaline decaying vegetation, characterized by reeds, that develops into peat. It sometimes occurs in sinkholes of karst regions. Compare - bog, marsh, swamp.

Fertility, soil. The quality that enables a soil to provide plant nutrients, in adequate amounts and in proper balance, for the growth of specified plants when light, moisture, temperature, tilth, and other growth factors are favorable.

Fibric soil material (peat). The least decomposed of all organic soil material. Peat contains a large amount of well preserved fiber that is readily identifiable according to botanical origin. Peat has the lowest bulk density and the highest water content at saturation of all organic soil material.

Field moisture capacity. The moisture content of a soil, expressed as a percentage of the ovendry weight, after the gravitational, or free, water has drained away; the field moisture content 2 or 3 days after a soaking rain; also called *normal field capacity, normal moisture capacity,* or *capillary capacity.*

Fine textured soil. Sandy clay, silty clay, and clay.

First bottom. The normal flood plain of a stream, subject to frequent or occasional flooding.

Flagstone. A thin fragment of sandstone, limestone, slate, shale, or (rarely) schist, 6 to 15 inches (15 to 38 centimeters) long.

Flood plain. A nearly level alluvial plain that borders a stream and is subject to flooding unless protected artificially.

Foot slope. The inclined surface at the base of a hill.

Forb. Any herbaceous plant not a grass or a sedge.

Fragipan. A loamy, brittle subsurface horizon low in porosity and content of organic matter and low or moderate in clay but high in silt or very fine sand. A fragipan appears cemented and restricts roots. When dry, it is hard or very hard and has a higher bulk density than the horizon or horizons above. When moist, it tends to rupture suddenly under pressure rather than to deform slowly.

Frost action. Freezing and thawing of soil moisture. Frost action can damage roads, buildings and other structures, and plant roots.

Genesis, soil. The mode of origin of the soil. Refers especially to the processes or soil-forming factors responsible for the formation of the solum, or true soil, from the unconsolidated parent material.

Gilgai. Commonly a succession of microbasins and microknolls in nearly level areas or of microvalleys and microridges parallel with the slope. Typically, the microrelief of *Vertisols; clayey soils having a high coefficient of expansion and contraction with changes in moisture content.*

Glacial drift (geology). Pulverized and other rock material transported by glacial ice and then deposited. Also, the sorted and unsorted material deposited by streams flowing from glaciers.

Glacial outwash (geology). Gravel, sand, and silt, commonly stratified, deposited by glacial meltwater.

Glacial till (geology). Unsorted, nonstratified glacial deposits consisting of clay, silt, sand, and boulders transported and deposited by glacial ice.

Glacial fluvial deposits (geology). Material moved by glaciers and subsequently sorted and deposited by streams flowing from the melting ice. The deposits are stratified and occur as kames, eskers, deltas, and outwash plains.

Glacial lacustrine deposits. Material ranging from fine clay to sand derived from glaciers and deposited in glacial lakes mainly by glacial meltwater. Many deposits are interbedded or laminated.

Gleyed soil. Soil that formed under poor drainage, resulting in the reduction of iron and other elements in the profile and in gray colors and mottles.

Graded stripcropping. Growing crops in strips that grade toward a protected waterway.

Grassed waterway. A natural or constructed waterway, typically broad and shallow, seeded to grass as protection against erosion. Conducts surface water away from cropland.

Gravel. Rounded or angular fragments of rock up to 3 inches (2 millimeters to 7.6 centimeters) in diameter. An individual piece is a pebble.

Gravelly soil material. Material that is 15 to 50 percent, by volume, rounded or angular rock fragments, not prominently flattened, up to 3 inches (7.6 centimeters) in diameter.

Ground water (geology). Water filling all the unblocked pores of underlying material below the water table.

Gully. A miniature valley with steep sides cut by running water and through which water ordinarily runs only after rainfall. The distinction between a gully and a rill is one of depth. A gully generally is an obstacle to farm machinery and is too deep to be obliterated by ordinary tillage; a rill is of lesser depth and can be smoothed over by ordinary tillage.

Hardpan. A hardened or cemented soil horizon, or layer. The soil material is sandy, loamy, or clayey and is cemented by iron oxide, silica, calcium carbonate, or other substance.

Hemic soil material (mucky peat). Organic soil material intermediate in degree of decomposition between the less decomposed fibric and the more decomposed sapric material.

Horizon, soil. A layer of soil, approximately parallel to the surface, having distinct characteristics produced by soil-forming processes. In the identification of soil horizons, an uppercase letter represents the major horizons. Numbers or lowercase letters that follow represent subdivisions of the major horizons. The major horizons are as follows:

O horizon. An organic layer of fresh and decaying plant residue.

A horizon. The mineral horizon at or near the surface in which an accumulation of humified organic matter is mixed with the mineral material. Also, any plowed or disturbed surface layer.

E horizon. The mineral horizon in which the main feature is loss of silicate clay, iron, aluminum, or some combination of these.

B horizon. The mineral horizon below an O, A, or E horizon. The B horizon is in part a layer of transition from the overlying horizon to the underlying C horizon. The B horizon also has distinctive characteristics, such as (1) accumulation of clay, sesquioxides, humus, or a combination of these; (2) granular, prismatic, or blocky structure; (3) redder or browner colors than those in the A horizon; or (4) a combination of these.

C horizon. The mineral horizon or layer, excluding indurated bedrock, that is little affected by soil-forming processes and does not have the properties typical of the overlying horizon. The material of a C horizon may be either like or unlike that in which the solum formed. If the material is known to differ from that in the solum, an Arabic numeral, commonly a 2, precedes the letter C.

R layer. Hard, consolidated bedrock beneath the soil. The bedrock commonly underlies a C horizon but can be directly below an A or a B horizon.

Humus. The well decomposed, more or less stable part of the organic matter in mineral soils.

Hydrologic soil groups. Refers to soils grouped according to their runoff-producing characteristics. The chief consideration is the inherent capacity of soil bare of vegetation to permit infiltration. The slope and the kind of plant cover are not considered but are separate factors in predicting runoff. Soils are assigned to four groups. In group A are soils having a high infiltration rate when thoroughly wet and having a low runoff potential. They are mainly deep, well drained, an-d sandy or gravelly. In group D, at the other extreme, are soils having a very slow infiltration rate and thus a high runoff potential. They have a claypan or clay layer at or near the surface, have a permanent high water table, or are shallow over nearly impervious bedrock or other material. A soil is assigned to two hydrologic groups if part of the acreage is artificially drained and part is undrained.

Illuviation. The movement of soil material from one horizon to another in the soil profile. Generally, material is removed from an upper horizon and deposited in a lower horizon.

Impervious soil. A soil through which water, air, or roots penetrate slowly or not at all. No soil is absolutely impervious to air and water all the time.

Infiltration. The downward entry of water into the immediate surface of soil or other material, as contrasted with percolation, which is movement of water through soil layers or material.

Infiltration capacity. The maximum rate at which water can infiltrate into a soil under a given set of conditions.

Infiltration rate. The rate at which water penetrates the surface of the soil at any given instant, usually expressed in inches per hour. The rate can be limited by the infiltration capacity of the soil or the rate at which water is applied at the surface.

Intake rate. The average rate of water entering the soil under irrigation. Most soils have a fast initial rate; the rate decreases with application time. Therefore, intake rate for design purposes is not a constant but is a variable depending on the net irrigation application. The rate of water intake in inches per hour is expressed as follows:

Less than 0.2	very low
0.2 to 0.4	Low
0.4 to 0.75	Moderately low
0.75 to 1.25	Moderate
1.25 to 1.75	Moderately high
1.75 to 2.5	High
More than 2.5	very high

Irrigation. Application of water to soils to assist in production of crops. Methods of irrigation are;

- **Border.** Water is applied at the upper end of a strip in which the lateral flow of water is controlled by small earth ridges called border dikes, or borders.
- **Basin**. Water is applied rapidly to nearly level plains surrounded by levees or dikes.
- **Controlled flooding.** Water is released at intervals from closely spaced field ditches and distributed uniformly over the field.
- **Corrugation.** Water is applied to small, closely spaced furrows or ditches in fields of close-growing crops or in orchards so that it flows in only one direction.
- **Drip (or trickle).** Water is applied slowly and under low pressure to the surface of the soil or into the soil through such applicators as emitters, porous tubing, or perforated pipe.
- **Furrow.** Water is applied in small ditches made by cultivation implements. Furrows are used for tree and row crops.
- **Sprinkler.** Water is sprayed over the soil surface through pipes or nozzles from a pressure system.
- **Subirrigation.** Water is applied in open ditches or tile lines until the water table is raised enough to wet the soil.
- **Wild flooding.** Water, released at high points, is allowed to flow onto an area without controlled distribution.

Kame (geology). An irregular, short ridge or hill of stratified glacial drift.

Karst (topography). The relief of an area underlain by limestone that dissolves in differing degrees, thus forming numerous depressions or small basins.

Lacustrine deposit (geology). Material deposited in lake water and exposed when the water level is lowered or the elevation of the land is raised.

Landslide. The rapid downhill movement of a mass of soil and loose rock, generally when wet or saturated. The speed and distance of movement, as well as the amount of soil and rock material, vary greatly.

Large stones . Rock fragments 3 inches (7.6 centimeters) or more across. Large stones adversely affect the specified use of the soil.

Leaching. The removal of soluble material from soil or other material by percolating water.

Liquid limit. The moisture content at which the soil passes from a plastic to a liquid state.

Loess. Fine grained material, dominantly of silt-sized particles, deposited by wind.

Lowmoor bog - A bog that is at or only slightly above the water table, on which it depends for accumulation and preservation of peat (chiefly the remains of sedges, reeds, shrubs, and various mosses). Compare - highmoor bog, raised bog.

Low strength. The soil is not strong enough to support loads.

Medium textured soil. Very fine sandy loam, loam, silt loam, or silt.

Metamorphic rock. Rock of any origin altered in mineralogical composition, chemical composition, or structure by heat, pressure, and movement. Nearly all such rocks are crystalline.

Mineral soil. Soil that is mainly mineral material and low in organic material. Its bulk density is more than that of organic soil.

Minimum tillage. Only the tillage essential to crop production and prevention of soil damage.

Miscellaneous area. An area that has little or no natural soil and supports little or no vegetation.

Moderately coarse textured soil. Coarse sandy loam, sandy loam, and fine sandy loam.

Moderately fine textured soil. Clay loam, sandy clay loam, and silty clay loam.

Moraine (geology). An accumulation of earth, stones, and other debris deposited by a glacier. Some types are terminal, lateral, medial, and ground.

Morphology, soil. The physical makeup of the soil, including the texture, structure, porosity, consistence, color, and other physical, mineral, and biological properties of the various horizons, and the thickness and arrangement of those horizons in the soil profile.

Moss peat [Soil Taxonomy] - An accumulation of organic material that is predominantly the remains of mosses (e.g. sphagnum moss). Compare - Herbaceous peat, sedimentary peat, woody peat, peat, muck, and mucky peat.

Mottling, soil. Irregular spots of different colors that vary in number and size. Mottling generally indicates poor aeration and impeded drainage.

Muck - Unconsolidated soil material consisting primarily of highly decomposed organic material in which the original plant parts are not recognizable (i.e. "sapric" in Soil Taxonomy). It generally contains more mineral matter and is usually darker in color, than peat. Compare - peat, mucky peat, herbaceous peat.

Mucky peat - Unconsolidated soil material consisting primarily of organic matter that is in an intermediate stage of decomposition such that a significant part of the original material can be recognized and a significant part of the material can not be recognized (i.e. "hemic" in Soil Taxonomy). Compare - peat, muck, herbaceous peat.

Munsell notation. A designation of color by degrees of three simple variables; hue, value, and chroma. For example, a notation of 10YR 6/4 has a hue of 10YR (yellow-red), value of 6, and chroma of 4.

Neutral soil. A soil having a pH value between 6.6 and 7.3.

Nutrient, plant. Any element taken in by a plant essential to its growth. Plant nutrients are mainly nitrogen, phosphorus, potassium, calcium, magnesium, sulfur, iron, manganese, copper, boron, and zinc obtained from the soil and carbon, hydrogen, and oxygen obtained from the air and water.

Organic matter. Plant and animal residue in the soil in various stages of decomposition.

Organic materials - [soil survey] Unconsolidated sediments or deposits in which carbon is an essential, substantial component. Several types of organic materials (deposits) can be identified based on the composition of the dominant fibers (grassy organic materials, herbaceous organic materials, mossy organic materials, woody organic materials). Compare - herbaceous peat, moss peat, sedimentary peat, woody peat.

Outwash, glacial. Stratified sand and gravel produced by glaciers and carried, sorted, and deposited by glacial meltwater.

Outwash plain. A landform of mainly sandy or coarse textured material of glaciofluvial origin. An outwash plain is commonly smooth; where pitted, it is generally low in relief.

Pan. A compact, dense layer in a soil that impedes the movement of water and the growth of roots. For example, *hardpan, fragipan, claypan, plowpan,* and *traffic pan.*

Parent material. The unconsolidated organic and mineral material in which soil forms.

Peat - Unconsolidated soil material consisting largely of undecomposed, or slightly decomposed, organic matter (i.e. "fibric" in Soil Taxonomy) accumulated under conditions of excessive moisture. Compare - muck, mucky peat, herbaceous peat.

Ped. An individual natural soil aggregate, such as a granule, a prism, or a block.

Pedon. The smallest volume that can be called ""a soil." A pedon is three dimensional and large enough to permit study of all horizons. Its area ranges from about 10 to 100 square feet (1 square meter to 10 square meters), depending on the variability of the soil.

Percolation. The downward movement of water through the soil.

Permafrost. Layers of soil, or even bedrock, occurring in arctic or subarctic regions, in which a temperature below freezing has existed continuously for a long time.

Permeability. The quality of the soil that enables water to move downward through the profile. Permeability is measured as the number of inches per hour that water moves downward through the saturated soil. Terms describing permeability are:

Very slow	less than 0.06 inch/hr
Slow	0.06 to 0.2 inch/hr
Moderately slow	0.2 to 0.6 inch/hr
Moderate	0.6 inch to 2.0 inches/hr
Moderately rapid	0.0 to 6.0 inches/hr
Rapid	6.0 to 20 inches/hr
Very rapid	more than 20 inches/hr

Phase, soil. A subdivision of a soil series based on features that affect its use and management. For example, slope, stoniness, and thickness.

pH value. A numerical designation of acidity and alkalinity in soil.

Piping . Formation of subsurface tunnels or pipelike cavities by water moving through the soil.

Pitting . Pits caused by melting ground ice. They form on the soil after plant cover is removed.

Plasticity index. The numerical difference between the liquid limit and the plastic limit; the range of moisture content within which the soil remains plastic.

Plastic limit. The moisture content at which a soil changes from semisolid to plastic.

Plinthite. The sesquioxide-rich, humus-poor, highly weathered mixture of clay with quartz and other diluents. It commonly appears as red mottles, usually in platy, polygonal, or reticulate patterns. Plinthite changes irreversibly to an ironstone hardpan or to irregular aggregates on repeated wetting and drying, especially if it is exposed to heat from the sun. In a moist soil, plinthite can be cut with a spade. It is a form of laterite.

Plowpan. A compacted layer formed in the soil directly below the plowed layer.

Ponding. Standing water on soils in closed depressions. Unless the soils are artificially drained, the water can be removed only by percolation or evapotranspiration.

Poor filter. Because of rapid permeability, the soil may not adequately filter effluent from a waste disposal system.

Poorly graded. Refers to a coarse grained soil or soil material consisting mainly of particles of nearly the same size. Because there is little difference in size of the particles, density can be increased only slightly by compaction.

Poor outlets . Refers to areas where surface or subsurface drainage outlets are difficult or expensive to install.

Productivity, soil. The capability of a soil for producing a specified plant or sequence of plants under specific management.

Profile, soil. A vertical section of the soil extending through all its horizons and into the parent material.

Reaction, soil. A measure of acidity or alkalinity of a soil, expressed in pH values. A soil that tests to pH 7.0 is described as precisely neutral in reaction because it is neither acid nor alkaline. The degree of acidity or alkalinity is expressed as pH:

Extremely acid	below 4.5
Very strongly acid	4.5 to 5.0
Strongly acid	5.1 to 5.5
Medium acid	5.6 to 6.0
Slightly acid	6.1 to 6.5
Neutral	6.6 to 7.3
Mildly alkaline	7.4 to 7.8
Moderately alkaline	7.9 to 8.4
Strongly alkaline	8.5 to 9.0
Very strongly alkaline	9.1 and higher

Regolith. The unconsolidated mantle of weathered rock and soil material on the earth's surface; the loose earth material above the solid rock.

Relief. The elevations or inequalities of a land surface, considered collectively.

Residuum (residual soil material). Unconsolidated, weathered or partly weathered mineral material that accumulated as consolidated rock disintegrated in place.

Rill. A stee-sided channel resulting from accelerated erosion. A rill is generally a few inches deep and not wide enough to be an obstacle to farm machinery.

Rippable. Bedrock or hardpan can be excavated using a single-tooth ripping attachment mounted on a tractor with a 200-300 draw bar horsepower rating.

Rock fragments. Rock or mineral fragments having a diameter of 2 millimeters or more; for example, pebbles, cobbles, stones, and boulders.

Rooting depth . Shallow root zone. The soil is shallow over a layer that greatly restricts roots.

Root zone. The part of the soil that can be penetrated by plant roots.

Runoff. The precipitation discharged into stream channels from an area. The water that flows off the surface of the land without sinking into the soil is called surface runoff. Water that enters the soil before reaching surface streams is called ground-water runoff or seepage flow from ground water.

Saline soil. A soil containing soluble salts in an amount that impairs growth of plants. A saline soil does not contain excess exchangeable sodium.

Salty waters. Water that is too salty for consumption by livestock.

Sand. As a soil separate, individual rock or mineral fragments from 0.05 millimeter to 2.0 millimeters in diameter. Most sand grains consist of quartz. As a soil textural class, a soil that is 85 percent or more sand and not more than 10 percent clay.

Sandstone. Sedimentary rock containing dominantly sand-size particles.

Sapric soil material (muck). The most highly decomposed of all organic soil material. Muck has the least amount of plant fiber, the highest bulk density, and the lowest water content at saturation of all organic soil material.

Saprolite . Unconsolidated residual material underlying the soil and grading to hard bedrock below.

Sedimentary rock. Rock made up of particles deposited from suspension in water. The chief kinds of sedimentary rock are conglomerate, formed from gravel; sandstone, formed from sand; shale, formed from clay; and limestone, formed from soft masses of calcium carbonate. There are many intermediate types. Some wind-deposited sand is consolidated into sandstone.

Seepage . The movement of water through the soil. Seepage adversely affects the specified use.

Sequum. A sequence consisting of an illuvial horizon and the overlying eluvial horizon. (See Eluviation.)

Series, soil. A group of soils that have profiles that are almost alike, except for differences in texture of the surface layer or of the underlying material. All the soils of a series have horizons that are similar in composition, thickness, and arrangement.

Shale. Sedimentary rock formed by the hardening of a clay deposit.

Sheet erosion. The removal of a fairly uniform layer of soil material from the land surface by the action of rainfall and surface runoff.

Shrink-swell. The shrinking of soil when dry and the swelling when wet. Shrinking and swelling can damage roads, dams, building foundations, and other structures. It can also damage plant roots.

Silica. A combination of silicon and oxygen. The mineral form is called quartz.

Silica-sesquioxide ratio. The ratio of the number of molecules of silica to the number of molecules of alumina and iron oxide. The more highly weathered soils or their clay fractions in warm-temperate, humid regions, and especially those in the tropics, generally have a low ratio.

Silt. As a soil separate, individual mineral particles that range in diameter from the upper limit of clay (0.002 millimeter) to the lower limit of very fine sand (0.05 millimeter). As a soil textural class, soil that is 80 percent or more silt and less than 12 percent clay.

Siltstone. Sedimentary rock made up of dominantly silt-sized particles.

Similar soils. Soils that share limits of diagnostic criteria, behave and perform in a similar manner, and have similar conservation needs or management requirements for the major land uses in the survey area.

Sinkhole. A depression in the landscape where limestone has been dissolved.

Site index. A designation of the quality of a forest site based on the height of the dominant stand at an arbitrarily chosen age. For example, if the average height attained by dominant and codominant trees in a fully stocked stand at the age of 50 years is 75 feet, the site index is 75 feet.

Slickensides. Polished and grooved surfaces produced by one mass sliding past another. In soils, slickensides may occur at the bases of slip surfaces on the steeper slopes, on faces of blocks, prisms, and columns; and in swelling clayey soils, where there is marked change in moisture content.

Slick spot. A small area of soil having a puddled, crusted, or smooth surface and an excess of exchangeable sodium. The soil is generally silty or clayey, is slippery when wet, and is low in productivity.

Slippage . Soil mass susceptible to movement downslope when loaded, excavated, or wet.

Slope. The inclination of the land surface from the horizontal. Percentage of slope is the vertical distance divided by horizontal distance, then multiplied by 100. Thus, a slope of 20 percent is a drop of 20 feet in 100 feet of horizontal distance.

Sloughed till. Water-saturated till that has flowed slowly downhill from its original place of deposit by glacial ice. It may rest on other till, on glacial outwash, or on a glaciolacustrine deposit.

Small stones . Rock fragments less than 3 inches (7.6 centimeters) in diameter. Small stones adversely affect the specified use of the soil.

Soil. A natural, three-dimensional body at the earth's surface. It is capable of supporting plants and has properties resulting from the integrated effect of climate and living matter acting on earthy parent material, as conditioned by relief over periods of time.

Soil separates. Mineral particles less than 2 millimeters in equivalent diameter and ranging between specified size limits. The names and sizes of separates recognized in the United States are as follows:

Separate	Millimeters
Very coarse sand	2.0 to 1.0
Coarse sand	1.0 to 0.5
Medium sand	0.5 to 0.25
Fine sand	0.25 to 0.10
Very fine sand	0.10 to 0.05
Silt	0.05 to 0.002
Clay	less than 0.002

Solum. The upper part of a soil profile, above the C horizon, in which the processes of soil formation are active. The solum in soil consists of the A, E, and B horizons. Generally, the characteristics of the material in these horizons are unlike those of the underlying material. The living roots and plant and animal activities are largely confined to the solum.

Stone line. A concentration of coarse fragments in a soil. Generally, it is indicative of an old weathered surface. In a cross section, the line may be one fragment or more thick. It generally overlies material that weathered in place and is overlain by recent sediment of variable thickness.

Stones. Rock fragments 10 to 24 inches (25 to 60 centimeters) in diameter.

Stony. Refers to a soil containing stones in numbers that interfere with or prevent tillage.

Stripcropping. Growing crops in a systematic arrangement of strips or bands which provide vegetative barriers to wind and water erosion.

Structure, soil. The arrangement of primary soil particles into compound particles or aggregates. The principal forms of soil structure are;

platy (laminated), prismatic (vertical axis of aggregates longer than horizontal), columnar (prisms with rounded tops), blocky (angular or subangular), and granular.

Structureless soils are either single grained (each grain by itself, as in dune sand) or massive (the particles adhering without any re-gular cleavage, as in many hardpans).

Subsoil. Technically, the B horizon; roughly, the part of the solum below plow depth.

Subsoiling. Breaking up a compact subsoil by pulling a special chisel through the soil.

Substratum. The part of the soil below the solum.

Subsurface layer. Any surface soil horizon (A, E, AB, or EB) below the surface layer.

Summer fallow. The tillage of uncropped land during the summer to control weeds and allow storage of moisture in the soil for the growth of a later crop. A practice common in semiarid regions, where annual precipitation is not enough to produce a crop every year. Summer fallow is frequently practiced before planting winter grain.

Surface layer. The soil ordinarily moved in tillage, or its equivalent in uncultivated soil, ranging in depth from about 4 to 10 inches (10 to 25 centimeters). Frequently designated as the ""plow layer," or the ""Ap horizon."

Surface soil. The A, E, AB, and EB horizons. It includes all subdivisions of these horizons.

Swamp - An area of low, saturated ground, intermittently or permanently covered with water, and predominantly vegetated by shrubs and trees, with or without the accumulation of peat. Compare - marsh, bog, fen.

Taxadjuncts. Soils that cannot be classified in a series recognized in the classification system. Such soils are named for a series they strongly resemble and are designated as taxadjuncts to that series because they differ in ways too small to be of consequence in interpreting their use and behavior.

Terminal moraine. A belt of thick glacial drift that generally marks the termination of important glacial advances.

Terrace. An embankment, or ridge, constructed across sloping soils on the contour or at a slight angle to the contour. The terrace intercepts surface runoff so that water soaks into the soil or flows slowly to a prepared outlet.

Terrace (geologic). An old alluvial plain, ordinarily flat or undulating, bordering a river, a lake, or the sea.

Texture, soil. The relative proportions of sand, silt, and clay particles in a mass of soil. The basic textural classes, in order of increasing proportion of fine particles, are; *sand, loamy sand, sandy loam, loam, silt loam, silt, sandy clay loam, clay loam, silty clay loam, sandy clay, silty clay, and clay*. The sand, loamy sand, and sandy loam classes may be further divided by specifying ""coarse," ""fine," or ""very fine."

Till plain. An extensive flat to undulating area underlain by glacial till.

Tilth, soil. The physical condition of the soil as related to tillage, seedbed preparation, seedling mergence, and root penetration.

Toe slope. The outermost inclined surface at the base of a hill; part of a foot slope.

Topsoil. The upper part of the soil, which is the most favorable material for plant growth. It is ordinarily rich in organic matter and is used to topdress roadbanks, lawns, and land affected by mining.

Trace elements. Chemical elements, for example, zinc, cobalt, manganese, copper, and iron, are in soils in extremely small amounts. They are essential to plant growth.

Tuff. A compacted deposit that is 50 percent or more volcanic ash and dust.

Upland (geology). Land at a higher elevation, in general, than the alluvial plain or stream terrace; land above the lowlands along streams.

Valley fill. In glaciated regions, material deposited in stream valleys by glacial meltwater. In nonglaciated regions, alluvium deposited by heavily loaded streams.

Variant, soil. A soil having properties sufficiently different from those of other known soils to justify a new series name, but occurring in such a limited geographic area that creation of a new series is not justified.

Variegation. Refers to patterns of contrasting colors assumed to be inherited from the parent material rather than to be the result of poor drainage.

Varve. A sedimentary layer of a lamina or sequence of laminae deposited in a body of still water within a year. Specifically, a thin pair of graded glaciolacustrine layers seasonally deposited, usually by meltwater streams, in a glacial lake or other body of still water in front of a glacier.

Weathering. All physical and chemical changes produced in rocks or other deposits at or near the earth's surface by atmospheric agents. These changes result in disintegration and decomposition of the material.

Well graded. Refers to soil material consisting of coarse grained particles that are well distributed over a wide range in size or diameter. Such soil normally can be easily increased in density and bearing properties by compaction. Contrasts with poorly graded soil.

Wilting point (or permanent wilting point). The moisture content of soil, on an ovendry basis, at which a plant (specifically a sunflower) wilts so much that it does not recover when placed in a humid, dark chamber.

MODEL QUESTIONS

SOIL AND SOIL COMPONENTS

Objective Questions

1. A branch of science that deals with the study of origin of soil, its classification and its description is known as

   a) Edaphology b) Ecology c) Pedology d) Geology

2. What is the percentage (volumetric) of mineral matter in the inorganic (mineral) soil?

   a) 30% b) 35% c) 40% d) 45%

3. Water content in the dry soil is

   a) 25% b) $< 25\%$ c) $> 25\%$ d) $\geq 25\%$

4. A mass of unconsolidated debris varying in composition from place to place and covering bedrock is known is

   a) Regolith b) Soil Profile c) Soil horizon d) Soil debris

5. Soil is the weathered layer of the soil earth's crust

   a) Low weathered layer b) Uppermost weathered layer

   c) Middlemost weather layer d) Lower-middle weathered layer

Answer

1) c; 2) d; 3) b; 4) a; 5) b

Fill in the blanks

1. .... is the study of the soil from the stand point of higher plants, its growth and production
2. The soil consists of 4 major components ....
3. Normal soil contents .... per cent mineral matter
4. Organic matter content is approximately .... % in normal inorganic soil
5. The soil air contains a number of gases viz., ...
6. The layer down below the soil is called ....

7. The soil is completely weathered whereas sub soil is....................
8. Soil is fertile but sub soil is considered as ....

Answer

1. Edaphology; **2.** Mineral, Organic matter, Soil water and Soil air; **3.** 45%; **4.** 5%; **5.** Nitrogen, Oxygen and Carbon dioxide; **6.** Sub soil; **7.** Less weathered; **8.** Less fertile

True or False

1. Soil is a dynamic natural body.
2. Soil is a living body
3. Pedology is considered important for the study of higher plants and its growth and production
4. Soil is the more or less loose and crumby part of the outer earth curst
5. A normal inorganic (mineral) soil contains 60% mineral matter
6. Microorganism and microorganism live in soil in large numbers.
7. Organic matter in the soil is not a source of energy for microorganism.
8. In the pore space, moisture content of the soil increases, the air content decreases and vice versa.
9. Soil water is not medium for supplying nutrients to plants.
10. Soil organic matter improves the physical properties of the soil and organic matter is a major source of plant nutrients.

Answer

1. True; **2.** True; **3.** False; **4.** True; **5.** False; **6.** True; **7.** False; **8.** True; **9.** False; **10.** True

PHYSICAL PROPERTIES OF THE SOILS

Objective Questions

1. Diameter of clay particles is

   a) Less than 0.002 mm　　b) 0.002 – 0.05

   c) 0.001 -0.01 mm　　d) 0.007 -0.09 mm

2. Clay particles are smallest in size and possess

   a) Macro pore spaces　　b) Micro pore spaces

   c) Both types of pore spaces　　d) None

3. Water holding capacity of Sand is

   a) High　　b) Medium　　c) Low　　d) Very high

4. Arrangement of soil particles is referred as

   a) Soil structure　　b) Soil texture　　c) Soil Mass　　d) Soil organism

5. Natural aggregates of soil particles are called

   a) Clod　　b) Peds　　c) Texture　　d) Structure

6. In which soil structure infiltration, permeability and aeration are good
   a) Platy b) Prism like
   c) Block like d) Sphere like (granular)
7. Particle density of a clay soil is
   a) 2.6 b) 2.7 c) 2.4 d) 2.8
8. Red yellow colour of soil is associated with
   a) Organic matter b) Silica c) Lime d) Ferric oxides
9. Which soil warmed up quickly?
   a) Sandy soil b) Clayey soil c) Loam soil d) Silt soil
10. In which soil bluish and greenish colour occurred
   a) Well drained soil b) Ill drained soil c) Intermediate soil

Answer

1) a; 2) b; 3) c; 4) a; 5) b; 6) d; 7) d; 8) d; 9) a; 10) b

Fill in the blank

1. Mechanical analysis helps in deciding the ....
2. Diameter (size) of silt is between .... mm
3. Plasticity, swelling and cohesion is very high in ....
4. The arrangement of soil particles into certain defined pattern is called ....
5. In structure formation organic matter plays an .... part.
6. Puddling of the soil is generally beneficial to the production of .... crop.
7. Soil density is expressed in two well accepted concept-particle density and ....
8. The bulk density of sandy soil is about .... gm /cu cm.
9. In micro pores, movement of air and water is restricted (less) in comparison to .... pores
10. Cohesion is attraction of like materials, adhesion is the attraction of ....
11. Black to dark grey colour of soils are mainly due to ....
12. Sandy soils, in general, warm up quicker than .... soils

Answer

1. Textural classes; **2.** 0.02 -0.002 mm; **3.** Clay soil; **4.** Structure; **5.** Important; **6.** Rice; **7.** Bulk density; **8.** 1.6 gm/cu cm; **9.** Macro; **10.** Unlike materials; **11.** Organic matter; **12.** Clayey

True or False

1. Quartz dominates fine sands in comparison to clay
2. Clay particles are smallest in size and possess fine pore space
3. Silt feels very plastic and stickiness when wet.
4. Blocky structure is generally found in sub soil (B. horizon)

5. Puddling increases the percolation loss in the rice filed.
6. Particle density of silt is more than clay.
7. Bulk density of silt loam is more than clay
8. Sandy soils have more pore spaces than clayey soils
9. Stickiness is the capacity to be mould
10. Bluish and greenish colours of soil are due to the presence of ferrous compounds
11. The main source of the soil heat is the energy of sun's rays (radiant energy)
12. Percentage of pore spaces is highest in clay
13. Dark coloured soils usually warm up more readily
14. In the soil, specific heat of mineral matters is less than that of organic matter

Answer

1. True; 2. True; 3. False; 4. True; 5. False; 6. False; 7. True; 8. False; 9. False; 10.True; 11. True; 12.True; 13. True; 14. True.

SOIL AIR

Objective Questions

1. Decomposition of organic matter produces
   a) O_2 b) CO_2 c) N_2 d) NH_3
2. Plant roots and various micro organisms require
   a) CO_2 b) O_2 c) N_2 d) NH_4
3. Poorly aerated soils are significant sources of
   a) Methane gas b) Ammonia gas c) Nitrogen gas
4. Methane gas, is in part responsible for
   a) Global warming b) Global cooling c) global freezing

Answer

1) b; 2) b; 3) a; 4)a

Fill in the black

1. As the moisture content of the soil increases, the air content ....
2. Soil air and atmospheric air .... in composition
3. The quantity of oxygen in soil air is less than that in ....
4. Mass flow of air is apparently due to .... Between the atmosphere and the soil air
5. The amount of nitrogen in soil air is .... %
6. Soil air contains .... % oxygen
7. The amount of carbon dioxide in soil air is .... %
8. Decomposition of organic matter produces ....
9. Depletion of oxygen in soil air .... nitrification

10. Deficiency of oxygen .... nutrient absorption
11. Poor aeration results in the development of toxin and other injurious substances such as....................
12. In the water logged soil, aeration become ....

Answer

1. Decrease; 2. Differ; 3. Atmospheric air; 4. Pressure difference; 5. 79.8; 6. 18.2; 7. 0.25; 8. CO_2; 9. Reduces; 10. Check; 11. Ferrous oxide, H_2S gas etc; 12. Insufficient

True or False

1. Pore spaces are filled partly with soil water and partly with soil air
2. Soil air and atmospheric air do not differ in composition
3. Soil air contain a much greater proportion of carbon dioxide than atmospheric air
4. Sandy soil contains much lower oxygen percentage than clay soil
5. Oxygen content in upper layer of soil is usually less than that of lower layer
6. The concentration of CO_2 is usually greater in sub soil (lower layer)

Answer

1. True; 2. False; 3. True; 4. False; 5. False; 6. True

SOIL WATER

Objective questions

1. Maximum amount of capillary water remaining in soil after removal of gravitation water is called as
   a) Field capacity b) Saturation point
   c) Holding capacity d) Real moisture
2. Major source of water used by the plants is ...
   a) Capillary water b) Hygroscopic water
   c) Gravitation water d) Inter space water
3. Capillary water will not be available to the plants, if it is held at a tension of .....
   a) 10 atmosphere b) Less than 15 atmospheres
   c) More than 15 atmospheres d) Less than 5 atmospheres
4. Water is available to the plant at atmosphere tension
   a) 31 b) 1000 c) 15-1/3 d) 10000
5. Water is not available to plant but may be available to certain bacteria at atmosphere tension
   a) 31 b) 1000 c) 10000
6. At what PF wilting percentage arrives ......
   a) 4.5 b) 4.2 c) ≥ 4.5 d) 6.0

7. Hygroscopic coefficient correspond to PF

   a) 4.2 b) 4.5 c) 6.0 d) 6.0

8. Generally plant looses water through

   a) Evaporation b) Transpiration c) Absorption d) Percolation

Answer

1) a; 2) a; 3)b; 4)c; 5)a; 6)b; 7)b; 8) b

Fill in the blank

1. Gravitational water occupies the --------------- soil pores
2. Gravitational water moves down readily under the force of ---------------
3. Capillary water is held in the --------------- pores.
4. Infiltration refers to the downward entry or movement of water into the ------------ surface
5. Loss of nutrients along with percolation water is called --------------
6. Permeability is greater in the soil having large number of ----------------- pores
7. Water vapour moves from high vapour pressure to ------------- vapour pressure
8. Evaporation is the change from liquid to the ------------ state
9. Plant looses water through the stomata of leaves is called ------------
10. Maximum water retention is found in ----------- soil

Answer

1. Large (Macro); 2. Gravity; 3. Micro (capillary); 4. Soil; 5. Leaching; 6. Macro; 7. Low; 8. Vapour; 9. Transpiration; 10. Clay

True or False

1. Two basic forces – cohesion and adhesion are responsible for water retention in the soil
2. Gravitational water is available and useful to plants.
3. Capillary water is held in the capillary pores
4. Hygroscopic water is essentially non liquid and moves primarily in the vapour form
5. The available water includes the whole of the hygroscopic water plus a part of the capillary water below the wilting point
6. The water beyond the field capacity stage is said to be superfluous (super available) water
7. The difference between the energy state of soil water and pure water in known as soil water potential
8. In the capillary tube, moisture moves from low potential to high potential
9. In plants, transpiration is not a physiological phenomena
10. Soil water regulates physical, chemical and biological activities in the soil.

Answer

1. True; 2. False; 3. True; 4. True; 5. False; 6. True; 7. True; 8. False; 9. False; 10. True

SOIL COLLOIDS

Objective Questions

1. Illite, montmorillonite and kaolinite are
   a) Kind of soil microbes b) Clay minerals
   c) Kind of protozoa d) Kind of soil
2. Soil colloid is a part of
   a) Sand b) Silt c) Loam d) Clay
3. Inorganic colloid is
   a) Humus b) Silicate clay c) None
4. Organic colloid is
   a) Humus b) Lignin c) Silicate clay d) Cellulose
5. 2:1 type clay is
   a) Montmorillonite b) Kaolinite c) Chlorites d) None
6. 2:1:1 type clay is
   a) Montmorillonite b) Kaolinite c) Chlorites d) None
7. Which clay type has maximum cation exchange capacity
   a) Kaolinite b) Montmorillonite c) Illite
8. Expanding type clay mineral is
   a) Montmorillonite b) Kaolinite c) Illite
9. Swelling and plasticity properties is maximum in
   a) Illite b) Kaolinite c) Montmorillonite

Answer

1) b; 2) d; 3) b; 4) a; 5) a; 6) c; 7) b; 8) a; 9) c

Fill in the blanks

1. Soil particles less than --------------- mm zize possess colloidal properties
2. Clay colloidals are ---------------- charged
3. Illite is a ----------------- type of clay
4. Kaolinite clay is ---------------- expanding type
5. Cation exchange capacity of montmorillonite is ------------- me /100 gm
6. Colloidal particles are found to be in continual motion called ---------- movement
7. Surface area of clay colloid is ------------- than clay
8. Chlorites is a ---------------- type clay
9. Humus is the product is decomposition of ------------ and ------------ residues

Answer

1. 0.001; 2. Negatively; 3. 2:1; 4. Non; 5. 80-100; 6. Brownian; 7. Larger; 8. 2:1:1; 9. Plant, animal

True or False

1. Soil colloids are referred as the dispersed system
2. Clay colloids are positively charged
3. Three layer (2:1) type clay crystal have two outside layers made of silicon and Oxygen (SiO_2) and the middle layer of aluminium and oxygen (Al_2O_3)
4. Illite is in between of kaolinite and montmorillonite type with regards to soil properties
5. Cation exchange capacity of kaolinite and montmorillonite type with regards to soil properties
6. Organic soil colloids have higher cation exchange capacity than inorganic colloids
7. Colloidal particles are not electrically charged
8. On the colloidal surface, divalent ions (Ca^{++}) are held more strongly than monovalent ion (Na^{+})
9. Colloidal clay does not possess the properties of cohesion and adhesion

Answer

1. True; 2. False; 3. True; 4. True; 5. False; 6. True; 7. False; 8. True; 9. False

ION EXCHANGE

Objective Questions

1. Ion exchange takes place in .....

   a) Sand b) Silt c) Loam d) Colloids
2. The cation exchange capacity is expressed in terms of ...

   a) gm b) cm c) milli equivalent / 100 gm d) None
3. In normal fertile soils the total exchangeable bases constitute about

   a) 30 - 40 % b) 80 -90% c) 40 -50 % d) 50 -60
4. There is definite relationship between cation exchange and anion exchange

   a) One increases other decreases b) Both increases

   c) Both decreases d) none

Answer

1) d; 2) c; 3) b; 4) a

Fill in the blanks

1. Phenomenon of the exchange of cations between soil and salt solution is known as ------------------

2. Cation exchange capacity for clay soils usually exceeds 30 me/100 gm while the value ranges from ---------- for sandy soils.
3. When exchangeable sodium from more than 10 -15 % of the total exchangeable cation it gives rise to ----------------
4. Unavailability of phosphate to plants is known as ----------------
5. When the proportion of exchangeable hydrogen (H) is high it gives rise to ------------
6. Montmorillonite has higher cation exchange capacity (CEC) in comparison to -------

Answer

1. Cation exchange; 2. 0-5; 3. Alkaline soils; 4. Phosphate fixation; 5. Acid soil; 6. Kaolinite or Illite

True or False

1. Soil colloids are not the seat of reaction
2. Cation exchange reactions are reversible
3. The active calcium ions replace the hydrogen ions by mass cation
4. Cation exchange does not make available the nutrients, applied in fertilizer form
5. Sandy soils tend to have higher cation exchange capacity than clayey soils
6. Higher organic matter content in a soil have high cation exchange capacity (CEC)
7. Hydrogen and aluminium tend to dominate alkaline soils
8. Cations such as Ca^{++}, Mg^{++} called exchangeable bases, neutralize soil acidity
9. The capacity of soil to exchange cations is the best index of soil fertility
10. At the pH commonly prevailing in cultivated soils – Nitrate (NO_3) Chloride (Cl) nad Sulphate (SO_4) ions are easily lost by leaching.

Answer

1. False; 2. True; 3. True; 4. False; 5. False; 6. True; 7. False; 8. True; 9. True; 10. True

SOIL REACTION

Objective Questions

1. Acidity and alkalinity of soil is indicated by
   a) soil organic matter content b) soluble salt content of the soil
   c) soil reaction d) available soil potassium
2. A pH value of 5.0 indicates that the soil reaction is
   a) acidic b) alkaline c) neutral d) highly alkaline
3. Alkaline soil must have pH value of
   a) 7.5 b) 6.5 c) 5.5 d) above 8.5
4. pH scale is
   a) 1-14 b) 1-7 c) 0-14 d) 7-14

5. Activity of H^+ at 1 x 10^{-8} the pH of solution would be
   a) 6 b) 5 c) 8 d) 10
6. At POH 10, soil would be
   a) acidic b) very acidic c) neutral d) alkaline
7. H^+ concentration in soil solution causes
   a) active acidity b) alkalinity c) salinity d) none

Answer

1) c; 2) a; 3) d; 4) c; 5) c; 6) b; 7) a

Fill in the blanks

1. Soil reaction is measured by .... of a suspension of soil.
2. A neutral reaction is produced by an equal activity of H and .... Ions.
3. The most convenient method of expressing the relationship between H^+ and H^- is..........
4. pH 7.0 shows a neutral reaction, when the pH is less than 7.0, the solution is ....
5. The pH of a water-logged soil will be ....
6. The continual use of ammonium sulphate makes the soil ....
7. The precipitation of .... and .... acids called "acid rain".
8. In highly acidic soils, the availability of some of the nutrients such as .... etc. is increased.
9. Buffering refers to resistance to a change in ....

Answer

1. pH; 2. OH; 3. pH; 4. acidic; 5. high; 6. acidic; 7. sulphuric and nitric; 8. aluminium, iron, manganese; 9. pH

True or False

1. Acidity is due to the excess of OH ions over H ions and alkalinity is due to the excess of H ions over OH ions.
2. At neutrality, the hydrogen ion concentration is 0.0000001 or 1x10^7 grams of hydrogen per litre of solution.
3. Decomposition of organic matter produces carbonic acid.
4. The pH value of acid rain may range form 6.5 to 7.5.
5. Most microorganism function at their best within a pH range 6.0 to 7.5.
6. Plant absorbs most of their nitrogen in the form of ammonia.
7. Fixation of phosphate takes place even when the soil is alkaline (high pH).
8. In acid soils, availability of Ca and Mg nutrients are always high.
9. The availability of sulphur is not affected by soil reaction as sulphur compounds are soluble in the whole pH range.
10. Sandy soils poor in organic matter are more highly buffered than clay soils.

Answer

1. False; 2. True; 3. True; 4.False; 5. True; 6. False; 7. True; 8. False; 9. True; 10. False

ACID SOIL

Objective Questions

1. H^+ remain freely present in soil solution and produces

   a) active acidity b) exchangeable acidity
   c) potential acidity d) none

2. Adsorption of H and Al ions in the soil colloids make soil

   a) neutral b) acidic c) saline d) alkakine

3. Hilly soils without lime are usually

   a) saline b) alkaline c) acidic d) neutral

4. A pH value of 4.5 indicates that the soil reaction is

   a) alkaline b) neutral c) saline d) acidic

5. Availability of Fe, Mn, Zn, Cu in acid soil is

   a) very less b) less c) more d) not aggected

6. Reclamation of acid soil is by

   a) lime b) gypsum c) pyrite d) none

7. If pOH is 11, then soil would be

   a) slightly acidic b) highly acidic c) saline d) alkaline

Answer

1) a; 2) b; 3) c; 4) d; 5) c; 6) a; 7) b

Fill in the blanks

1. Soil acidity is common in all regions where precipitation is ....
2. In very acidic soils, exchangeable .... and .... ions are present in large quantities.
3. In acid sulphate soil pH is less than ....
4. The use of ammonium sulphate and ammonium nitrate fertilizers in the soil, increase ....
5. Availability of phosphorus nutrient .... In acid soils.
6. Application of lime in the acid soils, makes the phosphorous more ....
7. Lime requirement of clay soil is .... in comparison to amendment of loam and sandy soils.
8. The amount of lime required to be added to acidic soil to raise the pH to a desired value is known as................
9. In acid soil use of basic fertilizers like sodium nitrate, basic slag etc. reduces the .... in the soils.

Answer

1. High; 2. Hydrogen and aluminium; 3. 4.0; 4. Acidity; 5. Decreases; 6. Available; 7. High; 8. Lime requirement; 9.Acidity

True or False

1. Acid soil is a base saturated soil.
2. Soil acidity is common in all regions where rainfall is low.
3. Aluminium becomes soluble at very acid soil conditions.
4. In acid sulphate soils, the acidity is due to dissolved or free acidic.
5. The sulphuric acid is produced by oxidation of ferric and aluminium sulphate.
6. The magnitude of leaching is the main factor for the development of acid soil.
7. Some acid soils have developed from parent rock materials which are of basic nature.
8. For reasonable crop production on an acidic soil, pH range from 6.0-7.0 is considered good.
9. The acid soils in India are located in Punjab, Haryana, Rajasthan and Gujarat.

Answer

1. False; 2. False; 3. True; 4. True; 5. False; 6. True; 7. False; 8. True; 9. False

SALINE AND ALKALINE SOIL

Objective Questions

1. A pH value of 6.0-8.5 indicates that the soil is

   a) Acidic b) normal to saline c) neutral d) none
2. Alkaline soil must have pH value of

   a) 7.5 b) 6.5 c) 7.5 d) above 8.5
3. Soluble salt content of soil is determined by measuring its

   a) Electrical conductivity b) pH value
   c) Organic matter content d) sulphar content
4. Degree of soil salinity is indicated by its

   a) Total soluble salt content b) organic matter content
   c) Nitrogen content d) phosphorus content
5. pH of a soil is 8.5, ESP is greater than 15 and EC is greater than 4 mmhos/cm, the soil

   would be

   a) Acidic b) saline c) alkaline d) saline–alkaline soil
6. In saline soil, abundant element is

   a) Ca b) Mg c) Na d) K
7. Reclamation of alkaline soil is done by

   a) Gypsum b) lime c) basic slag d) none

8. Saline soil is reclaimed by

   a) Lime b) gypsum c) basic slag d) none

Answer

1) b; 2) d; 3) a; 4) a; 5) d; 6) c; 7) a; 8) d

Fill in the blanks

1. In saline soil, ESP is less than .... of the total CEC
2. In alkaline soil, electrical conductivity (EC) if solution extract is less than .... mmhos/ cm.
3. In alkali soil, ESP is greater than ....
4. The process of breakdown of H clay under alkaline condition (in degraded alkali soil) is known as ....
5. Dispersion of soil particles (deflocculation) occurred in .... soils.
6. Saline soils are generally formed in the area of .... rain fall
7. In degraded alkali soil, the soil pH of the surface layer s .... Than 7.0.
8. Alkaline soils are generally reclaimed by the use of chemical ....
9. The high pH in alkali soil causes a reduction in availability of plants.

Answer

1. 15%; 2. 4; 3. 15%; 4. Solodization; 5. Alkaline; 6. Low; 7. Less; 8. Gypsum; 9. Nutrients

True or False

1. Saline soils remain in a flocculated condition
2. Saline soils usually have a surface crust of black salts.
3. In saline soil, total soluble salt content is less than 0.1%
4. Alkali soils are black due dispersion of the organic matter.
5. Electrical conductivity (EC) of alkali soil is usually more than 4 mmhos /cm.
6. Excessive salts in the soil solution of saline soil, prevents absorption of nutrients in adequate amount by the roots.
7. High pH in alkali soil increases the availability of plant nutrients
8. Barely, sugar beet and paragrass are highly salt tolerant crops.
9. The addition of organic matter increases acidity, thus helping in lowering the pH in alkaline soil.
10. Use of sulphur in the alkaline soil, tends to increase the alkalinity.

Answer

1. True; 2. False; 3. False; 4.True; 5. False; 6. True;7. False; 8. True; 9.True; 10. False

CALCAREOUS SOIL

Objective Questions

1. Abundant element in calcareous soil is
   a) Ca b) K c) Na d) Zn
2. There is effervescence when calcareous soils come in contact with
   a) Water b) Acid c) Alkali d) none
3. Tue calcareous soils in India are located in
   a) Punjab and Haryana b) Karnataka and Tamil Nadu
   c) Eastern UP and North Bihar d) Maharastra and Gujarat
4. Calcareous soils remain in
   a) Deflocculated condition b) Flocculated condition
   c) Compact condition d) Dispersed condition

Answer

1) a; 2)b ; 3)c; 4)b

Fill in the blanks

1. Calcareous soil contains sufficient ....
2. Calcareous soil usually having about pH 8.0 but pH may range from ....
3. Generally, there is no need of chemical .... for reclamation of calcareous soil
4. Calcareous soil in situ formed by weathering of parent .... material

Answer

1. Calcium carbonate; 2. 7.0 – 8.3; 3. Amendments; 4. Rock

True or False

1. Calcareous soil is formed in arid as well as humid regions.
2. There is formation of potassium carbonate due to weathering of rocks.
3. Calcareous soil of secondary origin is developed in low lying areas or in basin shaped areas.
4. Flocculation in heavy (clay) soil increases pore spaces.
5. In calcareous soil, availability of phosphorus, potassium and zinc increases.
6. Addition of organic matter in calcareous soil increases pH of the soil.

Answer

1. True; 2. False; 3. True; 4. True; 5. False; 6. True

ORGANIC SOILS (HISTOSOLS)

Objective Questions

1. Organic matter content is more than 20%, the soil is classified as;
   a) inorganic soil b) organic soil c) mineral soil d) none

2. Area under organic soils in the world is under
   a) 1% b) 2% c) 3% d) 4%
3. Organic soils are;
   a) saline b) alkaline c) saline –alkaline d) acidic
4. Organic soil is reclaimed by
   a) gypsum b) lime c) sulphur d) none

Answer

1) b; 2) a; 3) d; 4) b

Fill in the blanks

1. In the mineral soils, organic matter content may vary from ....%
2. Organic soils having markedly decomposed original plant materials are termed as................
3. Organic soils are highly acidic with a pH value less than ....
4. Histosols have a higher buffering .... than mineral soil

Answer

1. 0 -20%; 2. Muck soil; 3. 5.5; 4. Capacity

True or False

1. Organic soils, which have markedly decomposed plant materials are called peat soils.
2. In soil classification system, organic soils are identified as the order Histosols
3. The bulk density of organic soils in quite low in comparison to mineral soils. \
4. The colour of organic soil is red.
5. Compared to mineral soils, organic soils have high water holding capacity.
6. The organic soils possess 12:1 C/Nitrogen ratio compared to 20 : 1 for mineral soils.
7. Nitrogen content in organic soils are high in comparison with a mineral soil
8. Both the phosphorus and potassium, content of an organic soil are low as compared to a mineral soil.

Answer

1. False; 2. True; 3. True; 4. False; 5. True; 6. False; 7.True

SOIL MICROORGANISMS

Objective Questions

1. Blue green algae fixes atmospheric nitrogen in
   a) ground nut b) rice c) pea d) wheat
2. Nitrification is performed by bacteria
   a) nitrobacter b) azatobacter c) rhizobium d) bacillus

3. Rhizobium bacteria associated with which part of the plant
   a) leaf b) stem c) root hairs d) all parts
4. Rhizobium japanigam fixes nitrogen in the crop
   a) soyabean b) pea c) clover d) alfalfa
5. In rice field, how much nitrogen is fixed per hectare by blue green algae;
   a) 40-50 kg b) 60-70 kg c) 70-80 kg d) 20-30 kg
6. Oxidation of Sulphur into sulphate form is mineralized by the action of bacteria
   a) azotobacter b) sulphur bacteria c) bacillus d) rhizobium
7. Blight disease of rice, apple and pear is caused by
   a) bacteria b) fungai c) algae d) actinomycetes
8. Smuts and rusts of cereal crops are caused by
   a) bacteria b) algae c) fungi d) none

Answer

1) b; 2)a; 3)c; 4)a; 5)d; 6)b; 7)a; 8)c

Fill in the blanks

1. Algae contain chlorophyll and therefore, they manufacture their ----------------
2. Heterotrophic bacteria depend up ------------- for their food.
3. Symbiotic nitrogen fixing bacteria (Rhizobium) fix nitrogen in association with -------------- plants
4. Most important non symbiotic nitrogen – fixing aerobic bacteria is -----------------
5. The conversion of ammonia to nitrate is known as ----------------------
6. Microorganisms are most active and multiply fast at the optimum temperature which ranges between ------------
7. Symbiotic bacteria, compared to asymbiotic bacteria, fix------------------ nitrogen.
8. Vesicular-arbuscular mycorrihizae (VAM) penetrates -------- cell wall.
9. Ammonia is converted to nitrate by ------------- bacteria.

Answer

1. Own food; 2. Organic matter; 3. leguminous; 4. azatobactor; 5. Nitrification; 6.30-37°C; 7. More; 8. Root; 9. nitrosomonas

True or False

1. Bacteria are the most abundant in soil, next in order are actinomycetes, followed by fungi
2. Autotrophic bacteria depend upon organic matter or living body for their food.
3. Fungi contain chlorophyll thereby, they manufacture their own food.
4. Protozoa microorganisms are of animal nature.
5. The conversion of ammonia to nitrate in known as denitrification
6. Azotobacter bacteria live in the root nodules and fix nitrogen.
7. Azotobacter is an anaerobic nitrogen fixes

8. Clostridium bacteria are anaerobic nitrogen fixer
9. Mutually beneficial association between certain fungi and roots of higher plants is called mycorrhizae.
10. Under anaerobic conditions toxic substances such as methane, hydrogen sulphides are formed due to improper decomposition of organic matter.
11. Microorganisms are most active and multiply fast at the temperature ranges between 5^0C to 55^0C.

Answer

1. True; 2.False; 3. False; 4. True; 5. False; 6. False; 7. False; 8. True; 9. True; 10. True; 11. False

ORGANIC MATTER

Objective Questions

1. Major (original) source of organic matter in the soil is
   a) Animal residues b) plant residues c) compost d) green manure
2. Maximum quantity of compound (organic) available in organic matter is
   a) carbohydrate b) lignin c) protein d) fats
3. Maximum quantity of element found in organic matter is
   a) oxygen b) hydrogen c) carbon d) ash
4. First to be decomposed in the organic matter is
   a) Sugar and water soluble Nitrogen compounds b) protein
   c) Starch d) cellulose
5. Slowest to be decomposed in the organic matter is
   a) Lignin b) starch c) hemicellulose d) cellulose
6. Nitrogen is lost in the form of gas by
   a) Leaching b) percolation c) erosion d) denitrification
7. Which substance is responsible for black colour in humus?
   a) humic b) humic acid c) fulvic acid d) none
8. C: N ratio of humus is
   a) 10:1 b) 15:1 c) 18:1 d) 20:1

Answer

1) b; 2) a; 3) c; 4) a; 5) a; 6) d; 7) b; 8) a

Fill in the blanks

1. Average organic matter content in mineral soils may vary from ------------------%
2. The process of conversion of ammonia to nitrite and then to nitrate in known as ------ ----------------------
3. The process of conversion of proteins to amino acids is known as ------------------
4. When fresh plant residues are added to the soil, they are rich in carbon and poor in -- ----------------------

5. A wide ratio of C: N (35:1) indicates ------------------------- decomposition.
6. Addition of organic matter in the soil reduces losses of ------------- by erosion.
7. The colloidal surface of humus are --------------------- charged
8. The water holding capacity of humus is ----------------- silicate clays
9. Upon decomposition the organic matter of soils changes to -----------------
10. Main component of all organic matter is --------------------

Answer

0-5%; 2. Nitrification; 3. Amminization; 4. Nitrogen; 5. Little or No; 6. Soil; 7. Negatively; 8. More than; 9. Humus; 10. Carbon

True or False

1. Organic matter in the soil comes from the remains of plants and animals.
2. Cellulose decomposes faster than hemicellulose
3. Lignin is deposited on the cell wall to impart strength.
4. Heterotrophic bacteria simplify the complex organic compounds, the autotrophic bacteria (Sulphur bacteria) oxides it into sulphate form.
5. The organic matter is not a source of energy for the growth of soil microorganism.
6. Organic matter serves as a reservoir of chemical elements that are essential for plant growth.
7. Earthworm cannot flourish well in the soil having enough organic matter.
8. Carbon cycle keeps the amount of carbon dioxide constant in the air.
9. Plants absorb nitrogen in the form of amide.
10. Plants take up atmospheric carbon dioxide through photosynthesis.

Answer

1. True; 2. False; 3. True; 4. True; 5. False; 6. True; 7. False; 8. True; 9. False; 10.True.

ORGANIC MANURES

Objective Questions

1. Which of the following manures contain highest proportion of available nitrogen and phosphorus

   a) Farmyard manure b) compost

   c) manure of night soil d)cow dung manure

2. In which of the following cakes nitrogen content is highest

   a) Groundnut b) linseed c) castor d) neem

3. Compost improves the soil condition

   a) Physical b) chemical

   c) biological d) physical, chemical and biological

4. Which crop is best for green manuring
   a) sunhemp b) pea c) gram d) groundnut
5. Which of the following is responsible in preparation of vermicompost
   a) centipedes b) milipedes c) earthworms d) termite
6. Which one is not leguminous green manuring crop?
   a) dhaincha b) mung c) indigo d)bhang

Answer

1) c; 2) a; 3) d; 4) a; 5) c; 6) d

Fill in the blanks

1. Farm yard manure, compost and green manure are ------------------ organic manure
2. Oil seed cakes and bone meal are referred as --------------- organic manure
3. Old animals excrete --------------------- nutrients in comparison to young animals
4. Nitrogen lost in gaseous form is known as ----------------------- loss.
5. Dung, urine nitrogenous fertilizers etc are added in compost pit to hasten decomposition process is known as -----------------------
6. Growing a green manure crop and incorporating in the same field is called ----------- method of manuring.
7. Berseem is grown in --------------------- season
8. Buffering checks rapid changes in ------------------
9. Water holding capacity is --------------------- by addition of organic matter in the soil.
10. Organic manure finds soil particles into structural units called ---------------------

Answer

1. Bulky; 2. Concentrated; 3. More; 4. Volatilization; 5. Starter; 6. Situ; 7. Rabi; 8. pH/ soil reaction; 9. Increased; 10.Aggregate

True or False

1. Urea is a synthetic organic product
2. The loss of urine may be increased by spreading litter under the cattle
3. Bangalore method is a improved method of compost making which overcome many of disadvantages of the Indore method
4. Dhaincha (Sesbania) is not a leguminous green manure crops
5. Indigo (Indigoferea) is a leguminous green manure crop
6. Ipomea is a green leaf manuring crop
7. Green manuring does not check weed growth
8. Surface run off and erosion are reduced by addition of organic matter in the soil
9. Organic manures have very long term effect on the soil because nutrients are released very fastly.
10. Earthworms prefer a well aerated but moist habitat.

Answer

1. True; 2. False; 3. True; 4. False; 5. True; 6. True; 7. False; 8. True; 9. False; 10.True.

REFERENCES

BOOKS

1. Biswas T.D and Mukeherjee S.K. 1987. Text Book of Soil Science – Tata McGraw Publishing Co. Ltd., New Delhi.
2. Brady, N.C. 2002. The Nature and Properties of Soils (13th Ed.) McMillan Co., New York. Indian Publisher –Eurasis Publishing House (P) Ltd., Ramnagar, New Delhi – 55.
3. Buol, S.W., F.B. Hole and R.J. McCracken 1980. Soil Genesis and Classification. Oxford and IBH Publishing Co., New Delhi.
4. Daji.A.J., 1970. A text book of Soil Science –Asia Publishing House, Madras.
5. Jenny, H. 1941. Factors of Soil Formation. A system of Quantitative Pedology. McGraw Hill Book Company, INC, Newyork.
6. Joffe, J.S. 1936. The ABC of Soils, Pedology Publication, New Jersy.
7. Tamhane R.V., Motiramani D.D., Bali. Y.B. and Donahue R.L. 1970. Soils – their chemistry and fertility in tropical Asia – Prentice Hall of India (P) Ltd., New Delhi.
8. White. H. 1989. Introduction to the Principles and Practices of Soil Science, Oxford Publishers, London.
9. Soil Survey Staff 1999. Soil taxonomy: A basic system of soil classification for making and interpreting soil surveys, 2nd edn, and Agriculture handbook number 436. United States Department of Agriculture Natural Resources Conservation Service, Washington, DC.
10. USDA, NRCS 2003. Keys to soil taxonomy, 9th edn. United States Department of Agriculture/Government Printing Office, Washington, DC

Journals

1. Aldrich, D.G. 1987. Some Personal Reflections on Soil Science and Agriculture, 1936 to 1986. Soil Science Society of America Journal 51:1401-1405.
2. Alexander, M. 1986. Highlights of Research in Division S-3—Soil Microbiology and Biochemistry since 1961. Soil Science Society of America Journal 50:839-840.
3. Barnhisel, R.I. 1986. Highlights of Activities in Division S-9—Soil Mineralogy. Soil Science Society of America Journal 50:840-841.

4. Beaton, J.D. 1986. Highlights of Activities in Division S-8—Fertilizer Technology and Use. Soil Science Society of America Journal 50:272-273.
5. Bhattacharya, P. and Chakraborty, G. 2005. Current status of organic farming in India and other countries, *Indian Journal of fertilizers, Vol.1 (9), December 2005, Pp: 111-123*
6. Bicki, T.J. and J.P. Tandarich. 1989. The roots of pedology: A response to "Pedology, a field or laboratory science" by R.B. Daniels. Soil Science Society of American Journal 53(6): 1920-1921.
7. Campbell, C.A. and Zentner, R.P. 1993. Soil organic matter as influenced by crop rotations and fertilization in an Aridic Haploboroll. Soil Sci. Soc. Am. J. 57: 1034-1040.
8. Campbell, C.A., Jame, Y.W., Akinremi, O.O. and Cabrera, M.L. 1995. Adapting the potentially mineralizable N concept for the prediction of fertilizer N requirements. Fert. Res. 42: 61-75.
9. Campbell, C.A., Lafond, G.P., VandenBygaart, A.J., Zentner, R.P., Lemke, R., May, W.E. and Holzapel, C.B. 2011. Effect of crop rotation, fertilizer and tillage management on spring wheat grain yield and N and P content in a thin Black Chernozem: a long term study, Can. J. Plant Sci. 91: 467-483
10. Cassel, D.K. 1986. Highlights of Research in Division S-1, Soil Physics. Soil Science Society of America Journal 50:1093-1094.
11. Curtin, D., Selles, F., Wang, H., Zentner, R.P. and Campbell, C.A. 2000. Restoring organic matter in a cultivated, semiarid soil using crested wheatgrass. Can. J. Soil Sci. 80: 429-435.
12. Ellis, B.G. 1986. Highlights of Research in Division S-2—Soil Chemistry. Soil Science Society of America Journal 50:555-556.
13. Feller, C., L.J.M. Thuriès, R.J. Manlay, P. Robin, and E. Frossard 2003. "The Principles of Rational Agriculture" by Albrecht Daniel Thaer (1752-1828). An Approach to the Sustainability of Cropping Systems at the Beginning of the 19th Century. Journal of Plant Nutrition and Soil Science 166: 687-698.
14. Gardner, W.H. 1986. Early Soil Physics into the Mid-20th Century. Advances in Soil Science 4: 1-101.
15. Kamprath, E.J. 1986. Highlights of Research in Division S-4—Soil Fertility and Plant Nutrition. Soil Science Society of America Journal 50:556-557.
16. Landa, E.R. 2004. Albert H. Munsell: A sense of color at the interface of art and science. Soil Science 169: 83-89.
17. Mermut, A.R. and H. Eswaran 2001. Some major developments in soil science since the mid-1960s. Geoderma 100: 403-426.
18. Monreal, C.M. and Janzen, H.H. 1993. Soil organic-carbon dynamics after 80 years of cropping a dark brown Chernozem. Can. J. Soil Sci. 73:133-136.
19. Ploeg, R.R. van der, W. Böhm and M.B. Kirkham 1999. On the Origin of the Theory of Mineral Nutrition of Plants and the Law of the Minimum. Soil Science Society of America Journal 63: 1055-1062.
20. Schnitzer, M. (1991) Soil organic matter-the next 75 years. Soil Sci:151:41
21. Schoenau, J. and Campbell, C.A. 1996. Impact of crop residues on nutrient availability in conservation tillage systems. Can. J. Plant Sci. 76: 621-626.

22. Simonson, R.W. 1991. The U.S. Soil Survey - Contributions to Soil Science and its Application. Geoderma 48: 1-16.
23. Simonson, R.W. 1997b. Early Teaching in USA of Dokuchaiev Factors of Soil Formation. Soil Science Society of America Journal 61(1): 11-16.
24. Stone, E.L. 1986. Some Highlights of Division S-7, Forest and Range Soils. Soil Science Society of America Journal 50:1094-1095.
25. Tandarich, J.P., R.G. Darmody, L.R. Follmer and D.L. Johnson 2002. Historical Development of Soil and Weathering Profile Concepts from Europe to the United States of America. Soil Science Society of America Journal 66(2): 335-346
26. Van Doren, D.M. 1986. Highlights of Research in Division S-6—Soil and Water Management and Conservation. Soil Science Society of America Journal 50:271-272.
27. Warkentin, B.P. 2008. Soil Structure: A History from Tilth to Habitat. Advances in Agronomy 97: 239-272.
28. Wilding, L.P. 1986. Highlights of Activities in Division S-5—Soil Genesis, Morphology and Classification. Soil Science Society of America Journal 50:1377-1378.
29. Winiwarter, V. and W.H.E. Blum 2008. From marl to rock powder: On the history of soil fertility management by rock materials. Journal of Plant Nutrition and Soil Science 171: 316–324.
30. Yaalon, D.H. 1989. Forerunners and Founders of Pedology as a Science. Soil Science 147: 225-226.

Websites

1. soil.gsfc.nasa.gov/index.htm Soil physical properties: geography.cst.cmich.edu.
2. Soil formation, rocks & minerals, soil profile: www. Seafriends. org. nz/ enviro/ soil/ geosoil .htm //library .thinkquest .org/
3. soil humus –http://users.ids.net/nofari/tnf_humus.htm.
4. APEDA (2010) www.apeda.com accessed on 20th January, 2010
5. Bhattacharya (2006) Organic agriculture (www.danet.nic.in/ncof/).